Degeneration and Regeneration in the Nervous System

Degeneration and Regeneration in the Nervous System

Edited by

N.R. Saunders

and

K.M. Dziegielewska

Anatomy and Physiology
The University of Tasmania
Australia

CRC Press is an imprint of the
Taylor & Francis Group, an **informa** business
A TAYLOR & FRANCIS BOOK

First published 2000 by Taylor & Francis

Published 2019 by CRC Press
Taylor & Francis Group
6000 Broken Sound Parkway NW, Suite 300
Boca Raton, FL 33487-2742

First issued in paperback 2019

ISBN 13: 978-0-367-44748-9 (pbk)
ISBN 13: 978-90-5823-022-5 (hbk)

Visit the Taylor & Francis Web site at
http://www.taylorandfrancis.com

and the CRC Press Web site at
http://www.crcpress.com

Contents

Foreword

The question of why the mammalian central nervous system does not regenerate after injury is of extraordinary interest at many levels. In terms of descriptive biology, it is remarkable how great the discrepancy is between nerve cells that can and cannot repair their connections after their axons have been lesioned. In an invertebrate such as the leech, in fishes and in frogs the central nervous system does show effective regeneration and restoration of function after complete transection. Thus, a leech can swim again after its nervous system has regenerated after being cut in two, and a frog can catch flies with its tongue after its optic nerve has grown back to the tectum. In these "simple" animals the wiring is far more complex than in any man-made circuit, yet somehow fibres grow to find their targets and form effective synapses upon them. In this they resemble their counterparts in the mammalian peripheral nervous system. What makes the mammalian central nervous system so different in this regard? At the cellular and molecular level, differences between non-regenerating and regenerating neurons and the satellite cells that surround them are the focus of intense research. Detailed information is accumulating about molecules that enhance or inhibit growth, as well as their receptors. And at the level of clinical medicine there is the essential question about whether and when treatments can be devised for patients with central nervous system injuries so that functions can be restored.

Recent experiments at all of these levels have provided unexpected new findings and insights. Yet one of the most striking features of the field of regeneration today is how many key questions remain. For example, while we have clues, the mechanisms that prevent regeneration in mammalian CNS are still not fully known. Why is the proportion of axons that actually elongate so small, even when the application of suitable techniques does give rise to successful growth across a lesion? What changes in molecular mechanisms of growth occur in immature mammals during development, that later prevent regeneration in the adult? While it seems reasonable to guess that understanding of growth promoting and inhibiting mechanisms will continue to proceed rapidly, a baffling question remains. It arises from our ignorance about normal development of the nervous system. At present it is not known how specific synapses form, so that one type of cell is selected as a target while another one sitting just next door is ignored. If hope is to be offered to patients with spinal cord lesions, axons must not only grow (obviously a prerequisite for repair) but they must reform useful connections with the appropriate targets. In the best of all possible worlds no errors would be made. One also can imagine a scenario in which incorrect connections are formed and subsequently tuned by use; pain fibres would, one hopes, not re-form connections in patients.

All of us who work on these problems have to face inevitable and quite natural questions about prospects for therapy. A convenient analogy seems to me to be the repair of a watch. A desirable requirement would surely be to have an understanding of how the watch works and what the various components are doing. Without that knowledge one can still hope for some new insight or fluke that will allow the repair to be made. It would, however, be dangerous to promise how soon the watch will work again until the failure has been diagnosed and only one or two parts remain to be replaced. Because we are not even remotely at this stage in our knowledge of the nervous system, predictions about how and when seem unrealistic. (This analogy is of course flawed: the nervous system has to do the job on its own once one has provided the appropriate conditions). Whereas hope represents an essential stimulus for investigators and patients, premature optimism or worse still salesmanship of unsubstantiated ideas constitute dangerous sources of disappointment and discouragement. At present it seems impossible to predict whether realistic, effective treatments and methods of application with tolerable side-effects are just around the corner, probable in the near future or far off. Too many pieces of the puzzle still seem to be missing.

What the chapters in this book do clearly show is how fast the field has moved. One can hope that this readable and comprehensive assessment of contemporary research on regeneration at every level will help to refine important questions and point to new approaches. Best of all perhaps, one can hope that it may soon become out of date. Whether breakthroughs are near or far off, there is a most encouraging conclusion to be drawn from this book: it is now clear what type of research needs to be done to approach problems of regeneration and what strategies can be thought of for treating patients. And this is very different from just 20 years ago.

John Nicholls, FRS
SISSA,
Trieste,
Italy

Contributors

P N Anderson
Department of Anatomy and Developmental Biology
University College London
London
UK

P F Bartlett
Neurobiology Group
The Walter and Eliza Hall Institute of Medical Research
Parkville, Victoria
Australia

M B Bunge
Department of Neurological Surgery and Neurology
The University of Miami School of Medicine
Miami, FL
USA

L D Beazley
Neurobiology Laboratory
Department of Zoology
The University of Western Australia
Nedlands, WA
Australia

M A Bisby
Medical Research Council of Canada
Ottawa, Ontario
Canada

G J F Brooker
Neurobiology Group
The Walter and Eliza Hall Institute of Medical Research
Parkville, Victoria
Australia

D J Brown
Victorian Spinal Cord Service
Austin and Repatriation Medical Centre
Heidelberg, Victoria
Australia

A Conti
Department of Neurology and the David Mahoney Institute of Neurological Sciences
The University of Pennsylvania Medical Center
Philadelphia, PA
USA

S A Dunlop
Neurobiology Laboratory
Department of Zoology
The University of Western Australia
Nedlands, WA
Australia

K M Dziegielewska
Anatomy and Physiology
The University of Tasmania
Hobart, Tasmania
Australia

C H Faux
Neurobiology Group
The Walter and Eliza Hall Institute of Medical Research
Parkville, Victoria
Australia

I A Ferguson
Department of Human Physiology
The Flinders University School of Medicine
Adelaide, SA
Australia

M T Fitch
Department of Neurosciences
Case Western Reserve University School of Medicine
Cleveland, OH
USA

A R Harvey
Department of Anatomy and Human Biology
The University of Western Australia
Nedlands, WA
Australia

P M Hughes
CNS Inflammation Group
School of Biological Sciences
The University of Southampton
Southampton
UK

T J Kilpatrick
Neurobiology Group
The Walter and Eliza Hall Institute of Medical Research
Parkville, Victoria
Australia

P D Kitchener
Department of Anatomy and Cell Biology
The University of Melbourne
Parkville, Victoria
Australia

A R Lieberman
Department of Anatomy and Developmental Biology
University College London
London
UK

J J Lu
Department of Human Physiology
The Flinders University School of Medicine
Adelaide, SA
Australia

V H Perry
CNS Inflammation Group
School of Biological Sciences
The University of Southampton
Southampton
UK

R A Rush
Department of Human Physiology
The Flinders University School of Medicine
Adelaide, SA
Australia

N R Saunders
Anatomy and Physiology
The University of Tasmania
Hobart, Tasmania
Australia

M E Selzer
Department of Neurology and the David Mahoney Institute of Neurological Sciences
The University of Pennsylvania Medical Center
Philadelphia, PA
USA

J Silver
Department of Neurosciences
Case Western Reserve University School of Medicine
Cleveland, OH
USA

J D Steeves
Collaboration on Repair Discoveries
Department of Zoology
The University of British Columbia
Vancouver, BC
Canada

W Tetzlaff
Collaboration On Repair Discoveries
Department of Zoology
The University of British Columbia
Vancouver, BC
Canada

A M Turnley
Neurobiology Group
The Walter and Eliza Hall Institute of Medical Research
Parkville, Victoria
Australia

P Wilson
School of Biological Sciences
The University of New England
Armidale, NSW
Australia

X F Zhou
Department of Human Physiology
The Flinders University School of Medicine
Adelaide, SA
Australia

Introduction

The aim of this book is to bring authoritative, critical accounts of different aspects of the topic of "Degeneration and Regeneration in the Nervous System" to a wide audience, but particularly graduate students and their instructors, as well as those working in the field. The overall aim of scientists working on degeneration and regeneration in the nervous system is to understand why the injured brain and spinal cord are unable to repair themselves and to develop strategies that can eventually be used clinically to promote repair following such injuries. The clinical relevance of such studies is obvious, nevertheless it is important to be aware of the current nature of the problems facing clinicians who look after patients with nervous system injuries.

The authors were asked to limit the number of references they cited and to concentrate on a critical appraisal of a more limited number of key papers; readers will be able to judge for themselves how well this has been achieved. Too many modern day reviews consist of a catalogue of findings culled from the summaries of original articles rounded off with a lengthy bibliography. This has produced reviews that are indigestible to read and often not very informative. What is needed in a review, particularly for those less familiar with the field, is an appraisal of key papers, both good and less so, with enough information given about the individual papers (particularly with respect to the methods used) so that the reader can understand what was done and be guided by the expert reviewer to their own conclusion about the value of the results in the paper and how valid are the authors' own conclusions. Too often it is impossible to tell chalk from cheese (let alone that the cheese is full of holes) in the brief summaries of conclusions that seem to be becoming the standard for modern reviews. Even those readers familiar with a field may be helped by a more critical appraisal of apparently familiar work. One of the more interesting aspects of editing this book has been to see some variability in the acceptance of the results and conclusions from the same papers by different authors. This has been left in, although it may seem like unnecessary overlap, as it illustrates very clearly the problem of persuading reviewers of the need to make critical (which does not always mean negative) appraisals of the published data.

It is hoped that these articles will also be of value to those working in the field. The articles represent an up to date statement of where most of the current approaches to degeneration and regeneration in the nervous system stand at the end of the 20th Century. There is much that is new and the pace over the last 10 years has quickened considerably. Yet, as Doug Brown shows in the first article, there is still a long way to go before our increased knowledge of why the central nervous system has such limited capacity to regenerate and of what can be done experimentally to overcome this, can be applied in patients. The remaining articles review most of the major aspects of experimental approaches to promoting regenerative repair in the nervous system. The emphasis is on the central rather than the peripheral nervous system, because it is here that the problem of repair has been so intractable. But the articles by Bisby and by Hughes & Perry remind us that although the situation for repair in the peripheral nervous system is more favourable, there is still a lot that is poorly understood. The articles dealing with the central nervous system show that there are grounds for optimism that some means of promoting repair of the brain and spinal cord will be found in the foreseeable future. However a note of caution is appropriate. There have been many false dawns and prophets who have disappeared almost without trace during this century. It is not appropriate for scientists to make over optimistic claims on the basis of any of our present experimentally based knowledge. In this respect things do not seem to have changed very much with time. Thus Ramon y Cajal (1928) in his monograph *Degeneration and Regeneration of the Nervous System* commented on "......the feverish thirst for novelty, the suggestion of fashionable theories, and other causes entirely foreign to a calm and disinterested pursuit of the truth." We can all dream and hope for a cure, for example for spinal injury; but it is unfair on patients, their families and friends, for scientists to be making firm predictions of a timetable for establishing a cure. Nor should scientists make claims which go beyond that which can be supported by their data. This consideration was a main reason for producing this book.

Some emphasis in this volume has been given to spinal cord injury. This is partly because there has been a large amount of recent work and interest in the problem, much of it stimulated by the valiant work of Christopher Reeve in promoting interest in and raising funds for research. What happened to Christopher Reeve (Reeve, 1998 *Still Me*, Random House, New York) also illustrates why, although all central nervous system injuries are a terrible misfortune, spinal injuries are particularly so. With brain injuries, providing essential brainstem function is spared, the plasticity of the brain (discussed by Kitchener & Wilson) is such that some remarkable recoveries have occurred. However, there is still not nearly enough known to prevent the death or severe disablement which follows many head injuries and most of the treatments used are not supported by evidence from "gold standard" clinical trials (Bullock and Povlishock, 1996, *Journal of Neurotrauma* **13**, 641–734). In contrast to the plasticity of the brain, in the spinal cord the degree of plasticity is much less, although perhaps currently underestimated. Once long tracts are severed or compressed to the point of axotomy, they will not recover and there is usually insufficient overlap in function in the spinal cord for the missing functions to be taken over by surviving tracts, should there indeed be any. The spinal cord is such a narrow structure, normally well protected by the bone forming the spinal canal, that any injury sufficient to damage it in part, may well be severe enough to damage it completely.

In the general strategy of devising spinal repair procedures that could eventually be applied in patients, there are at least four problems to be overcome:

(a) Central nervous system neurons show a variable response in their ability to produce neurites in response to injury, in contrast to peripheral nervous system neurons, which show a consistent ability to do this.

(b) Following damage to the CNS, as for example in a spinal cord injury, any neurites that do appear at the site of injury are unable to cross it, which in patients may involve a substantial length of spinal cord.

(c) Once methods for promotion of growth of axons across the site of injury are available in a clinically applicable form, the axons may have to grow considerable distances to reach appropriate targets and may require specific guidance cues to direct them to functionally appropriate targets.

(d) Having reached appropriate targets, effective functional re-innervation of the targets should occur.

It is not entirely clear how separable are these four components of successful repair. However several chapters in this volume suggest that there is increasing evidence that they may indeed be substantially separate processes and that achievement of one will not automatically lead to success with the others. Thus the work described by Beazley & Dunlop on regeneration in the lizard shows clearly that while axotomised fibres can regrow to the appropriate targets in the visual system in this species, no functionally effective innervation occurs. Beazley & Dunlop describe different features of a wide range of species from cold-blooded vertebrates to mammals. Particularly with respect to effectiveness of target re-innervation, there appears to be a spectrum of regeneration. This goes from amphibia and lampreys (also discussed by Conti & Selzer, with particular respect to the underlying subcellular structures that may be responsible for regenerative outgrowth of injured axons) which can regenerate not only new axons but also functional connections with appropriate targets, through lizards that show excellent axonal growth, but inappropriate target innervation, to mammals in which neither regenerative axon growth nor appropriate target innervation normally occur. The evolutionary significance of this progressive loss of regenerative ability (an ability which is even more marked in invertebrates) through the animal kingdom is unclear.

Anderson & Lieberman deal with the problem of the variable response of different neuronal populations to injury and the different strategies (implanted cells or different growth factors) that are effective in experimentally promoting fibre growth in different neuronal populations. Ferguson and his co-workers describe some possible contributions of neurotrophic factors in regeneration and propose a role for the low affinity NGF receptor p.75.

Bartlett and his colleagues give an important account of the new field of stem cells, which, as they point out, have long been considered important in the haemopoietic system, but have been largely neglected, until recently, in the nervous system.

Saunders & Dziegielewska deal with a more extreme form of variable response to central nervous system injury that is developmentally regulated: providing the mammalian spinal cord is sufficiently immature it will respond to a complete transection by a substantial degree of repair and recovery, but later in development it will not.

In separate chapters Bunge and Harvey deal with different aspects of bridging the gap at the site of a brain or spinal cord injury.

In recent years there has been much interest in the possibility that the initial inflammatory response that is part of the response to a brain or spinal cord injury may contribute to the inability of the injured central nervous system to regenerate. Some aspects of this problem are considered by Fitch & Silver in the central nervous system, and by Hughes & Perry in peripheral nerve. The rather longer standing studies of various mechanisms thought to have an inhibitory effect on neurite outgrowth following injury are covered by Tetzlaff & Steeves, together with a wide-ranging review of other mechanisms that will need to be undersood if methods of repairing brain and spinal cord injuries that can be applied to patients are to be developed. These include the nature of the neuronal response to injury and how it might be modified to make it more effective. In this respect they provide a valuable comparison between the responses to injury of central neurons and those with peripheral processes.

One difficulty about interpreting much of the voluminous literature in this field is that many studies deal with only one feature of the problem. Thus for example in the case of spinal injury, there are many studies of behaviour following an experimental injury, although only in recent years has an attempt been made to standardise some of the tests proceedures, so that different studies can be adequately compared (American Paralysis Association guidelines, *Experimental Neurology* **107**, 113–117). There are many other studies that deal with the morphology of the site of injury and in more recent years there has been the advent of pathway tracing techniques; these have sometimes been included in morphological studies of the consequences of spinal injury. Relatively few papers have studied the recovery of impulse conduction across the site of injury. Even fewer studies have incorporated all of the three elements of assessment of the response of the immature spinal cord to injury. Some laboratories have published data on more than one approach, but in separate papers, not always in the same journal, although sometimes published simultaneously. This may be a reflection of pressures on editorial space and on scientists to publish papers that can be counted, rather than to present a comprehensive analysis of the problem in a single paper.

An interpretational difficulty needs to be borne in mind when considering the many studies that have used immature animals: different species have been used for spinal injury studies, but there is a tendency to refer to newborn or neonates as though they were the same, irrespective of species. A moments reflection on the different state of maturity of a newborn cat or dog and a newborn marsupial such as an opossum, should be enough to emphasize the importance of not unthinkingly transferring information from one species to another. A further point for consideration is the nature of the lesions used in spinal injury studies. It has been repeatedly stressed by many authors that spinal injuries in humans are very variable and usually involve contusion/ compression rather than section of the cord either in part or whole. Standardised contusion injuries have been devised for studies in adult animals, but many of the experimental injury studies in animals usually appear to have involved either local crush (forceps) or cut injuries and are generally referred to as "transections". It is important to realise that an important feature of a good experimental study lies in the reproducibility of the methods used. The very heterogeneity of human spinal injuries makes it extremely difficult to adequately evaluate different treatments let alone understand the nature of the response to injury in the spinal cord (cf Brown, Chapter 1, this volume).

We should like to thank the authors for their enthusiastic support of this project and particularly for their willingness to consider modifications to their manuscripts. We should particularly like to thank Tracey Walls for her editorial assistance in preparing the manuscripts for the press.

Norman R Saunders
Katarzyna M Dziegielewska
11 June 1999

Abbreviations

ALS, amyotrophic lateral sclerosis
AP1, activator protein-1
ASIA score, American Spinal Injury Association motor score
ATP, adenosine triphosphate

Bcl-2, B-cell lymphoma (protein)
BDNF, brain-derived neurotrophic factor
BMP-2, bone morphogenic protein
BrDU, bromo-deoxy uridine

C, cervical (spinal cord level)
CGRP, calcitonin gene related peptide
ChAT, choline acetyl transferase
CNS, central nervous system
CNTF, ciliary neurotrophic factor
CR-3, complement type-3 receptor
CREB, cyclic AMP response element binding protein

DCN, dorsal column nuclei
DiI, 1,1′-dioctadecyl3,3,3′3′-tetramethylindocarbocyanine perchlorate (carbocyanine dye)
DRG, dorsal root ganglia

E, embryonic day
EGF, epidermal growth factor

f-actin, filamentous-actin
FEV_1, forced expiratory volume in one second
FGF-1, acidic fibroblast growth factor-1
FGF-2, basic fibroblast growth factor-2
FGFR-1, fibroblast growth factor receptor-1
FIM - functional independence measure
FVC - forced vital capacity

GABA, gamma amino butyric acid
g-actin, globular actin
Gal-C, galactocerebrocide
GAP-43, growth associated protein with apparent molecular weight of 43 kilodaltons
GDNF, glial derived neurotrophic factor
GFAP, glial fibrillary acidic protein
GGFs, glial growth factors

GM-1 ganglioside, monosialotetrahexosylganglioside
GMA - glyceryl methacrylate
GM-CSF, granulocyte-macrophage colony stimulating factor

HEMA, 2-hydroxyethyl methacrylate
HPMA, N-(2-hydroxypropyl) methacrylamide
HRP, horseradish peroxidase
HSPG, heparan sulphate proteoglycan
5HT, 5-hydroxytryptamine

ICAM-1, intercellular adhesion molecule-1
ICU, intensive care unit
IGF-1, insulin-like growth factor-1
IL-Iβ, interleukin-Iβ
IL-6, interleukin-6
IL-10, interleukin-10
IN-1, IN-2, neurite inhibitory protein antibodies (monoclonal)
IV, intravenous

LFA-1, leukocyte functional antigen-1
LIF, leukemia inhibitory factor
LPS - lipopolysaccharide

MAC-2/galectin-3, cell surface receptor on macrophages
MAG, myelin associated glycoprotein
MAP2, myelin associated protein 2
MBP, myelin basic protein
MCP-1, macrophage chemotactic protein-1
MEP, maximum expiratory pressure
MHC, major histocompatibility (antigen)
MIP, maximum inspiratory pressure
MIP-1α and MIP-1β, macrophage inhibitory protein-1α and -1β
MMPs, matrix metalloproteinases
MND, motor neuron disease

NACSIS, National Acute Spinal Cord Injury Study
N-CAM, neuronal-cell adhesion molecule
NF-H, neurofilament, high molecular weight sub unit
NF-L, neurofilament, low molecular weight sub unit
NF-M, neurofilament, middle molecular weight sub unit
NGF, nerve growth factor
NGFIA, nerve growth factor (transcription factor) IA
NMDA, N-methyl-D aspartate
NT-3, neurotrophin-3
NT-4, neurotrophin-4
NT-4/5, neurotrophin-4/5

P, postnatal day
PAN/PVC, acrylonitrile:vinylchloride copolymer
PDGF, platelet derived growth factor
PNS, peripheral nervous system
PP, polypyrrole
PrV, principle trigeminal nucleus

SI, primary somatosensory cortex
SOCS, suppression of cytokine signalling
SOD 1, super oxide dismutase
SP, substance P
SpVc, spinal trigeminal nucleus caudalis
SpVi, spinal trigeminal nucleus interpolaris
SpVo, spinal trigeminal nucleus oralis

TGF, transforming growth factor
TIMP-1, tissue inhibitor of metalloproteinases-1
TNF, tumour necrosis factor
TNFR, tumour necrosis factor receptor
TOAD 64, turned on after division protein with apparent molecular, weight of 64 kilodaltons
TRH, thyrotropin releasing hormone
Trk A, tyrosine kinase-containing receptor A
Trk B, tyrosine kinase-containing receptor B
TRN, thalamic reticular nucleus

VCAM-1, vascular cell adhesion molecule^{-1}
V gang, trigeminal ganglion
VLA-4, very late antigen-4
VP, ventroposterior Thalamus
VPl, lateral ventroposterior thalamus
VPm, medial ventroposterior thalamus

1

Repair after spinal cord injury: A clinical perspective

Douglas J Brown

Introduction

The management of Spinal Cord Injuries has improved dramatically in the last 50 years, but it is only over the last 10 years or so that there has come a real belief that a cure for the paralysis is possible. Patients with spinal cord injury and their families have been very excited by the prospect of cure and some have been frustrated by the delays in achieving this goal. Clinicians have found themselves in an awkward position sandwiched between the expectations of their patients and the promises of researchers. They have seen the promise of a cure within ten years come and go, and more and more research being published with the bright hope that it will lead to a cure for spinal cord injury. In this uncertain and difficult time it is helpful to look at the clinical impact of laboratory research to try to obtain some perspective on the current situation in order that patients may be appropriately informed and encouraged.

The range of laboratory research has been very broad and covered not only surgical techniques such as peripheral nerve grafts and fetal implants, but also discoveries of neurotransmitters and local hormonal factors which seem to be important. Drugs have been developed which aim to influence some of the possible theoretical causes of acute spinal cord neuronal damage and some have been trialed both in the laboratory and in clinical practice. Much of the work involves acute spinal cord injury, though the majority of patients have long standing injury. The impact of laboratory studies for the latter group, in achieving a cure, has not been well thought out. The consequences of a partial cure may be both good and bad. Bad consequences may include increased spasticity and neurogenic pain. This potential dark side of recovery has not been highlighted and has certainly been overshadowed by the bright prospect of complete cure. It is not unreasonable scepticism to believe that complete cure is unlikely to come in one magic treatment, but rather that there will be some improvement in nerve function and this will be, hopefully, helpful.

The complexity of causes of spinal cord damage indicates the need for a variety of treatments. The aetiology may be classified clinically as traumatic and non traumatic, but pathologically the situation is more complex. Acute damage can be due to accidents which cause direct mechanical damage to the cord by compression or by the cutting of the neural tissue by the broken bones of a fractured spine. Other non progressive causes are commonly acute ischaemia, haemorrhage and abscess. Progressive damage occurs with tumours, particularly metastatic ones, and recurrent inflammatory processes which cause demyelination, such as multiple sclerosis. A third group is the congenital and developmental causes such as spina bifida. The prospect of repair and cure is not equal in all these groups. For example the treatment of the carcinoma is the primary concern, rather than the repair of the secondary spinal cord damage caused by expanding secondaries. With the focus on saving the patient's life, little thought has been given to cure of the paralysis and rightly so. Perhaps the cure of paralysis of other causes will also ultimately benefit this group of patients.

Similarly the focus in multiple sclerosis has been treatment and prevention, but now that treatment is more successful, healing the damage has become more important.

The congenital and hereditary spinal paralyses are a diverse group of pathologies, as yet little understood. Study of these patients may ultimately shed light on neuronal organisation, a knowledge of which may be essential to effect repair of spinal nerves damaged by trauma and acquired diseases. However, at present, research into progressive and congenital causes of spinal cord paralysis has led to little information of clinical value for cases of non progressive spinal cord injury.

In considering the clinical impact of basic research into spinal cord repair in non progressive cases, it is useful to consider treatments aimed at acute spinal cord injury and those that may improve function in chronic spinal cord patients. There may be potential overlap, but currently therapies are not studied as if they have broad applicability.

Methylprednisolone

Corticosteroids have been used in the treatment of central nervous system diseases because of their anti-inflammatory properties which may decrease oedema. They may thereby ameliorate neurological dysfunction caused by pressure from neighbouring swollen tissue as well as from disruption of neuronal connexions within the oedematous tissue. However, in 1980 Demopoulos and co-workers suggested that the principal molecular basis for post traumatic neuronal degeneration is oxygen free radical-induced lipid peroxidation. They examined the effect of very high doses of methylprednisolone in experimental studies to support their theory. Hall *et al.* (1992) subsequently published studies on dose-response and time-action characteristics of this putative action of methylprednisolone. Thus, in acute spinal cord injury, methylprednisolone is the drug which has undergone the most thorough clinical studies and has gone on to enter clinical practice, though its use is not accepted by all.

The National Acute Spinal Cord Injury Study (NASCIS, Bracken *et al.*, 1984) was established in the United States in 1975 to investigate drugs and drug regimens in the treatment of acute spinal cord injuries, typically within the first hours after injury. To date three large studies have been undertaken and all have involved methylprednisolone. Methylprednisolone was chosen because a number of spinal cord injury studies using animals have shown benefit and there were theoretical advantages over dexamethasone, which was already in use in some central nervous system disease states.

In the first such study, NASCIS I (Bracken *et al.*, 1984), the effect of methylprednisolone in two dosage regimens was compared —1g bolus intravenously (IV) followed by 250mg IV six hourly for 10 days versus 100mg bolus IV followed by 25mg IV six hourly for 10 days. No placebo was included. Nine centres admitted 330 patients over a 33 month period. Of the nine centres involved, six were specialised spinal cord units. The protocol for this, as for subsequent NASCIS trials, was strict and every effort was made to ensure the reproducibility and comparability of neurological assessment between centres and investigators. Patients were entered within 48 hours of injury, most within 12 hours. Detailed neurological examinations were made on admission to the study, at six weeks and six months. Extensive statistical comparisons did not demonstrate any difference in neurological outcome between the two groups, including the time of institution of treatment after injury. Of complications studied, only the rate of trauma and wound infections was significantly higher in the high dose methylprednisolone group. Fatalities, pulmonary embolus, sepsis and decubitus ulcers were more common in the high dose group, but did not reach statistical significance. Although the case fatality rate did not reach statistically significant levels in the high dose group, in conjunction with the lack of any benefit, it was decided to terminate the trial earlier than planned. The scientific validity of this decision can be questioned, but not the lack of benefit. In the absence of a control group in NACSIS I it cannot be stated that treatment was better than no treatment. It is not clear why a placebo group was not used.

Based on animal data which became available towards the end of the trial (Hall & Braugler 1982) the authors speculated that the dose of methylprednisolone used may not have been therapeutic. The doses used were 1.3 and 14.3 mg/kg in contrast to animal studies which indicated a need for doses of 15–30mg/kg. Thus the authors foreshadowed their next studies.

The results of NASCIS II were published in the popular press in 1990 and this caused considerable outrage. Clinicians were in no position to make informed decisions and were caught in a difficult situation with their patients until a peer reviewed paper was published in the New England Journal of Medicine approximately one month later in May, 1990 (Bracken *et al.*, 1990). The controversy may have contributed to the scepticism which has led to a lack of enthusiasm for the advocated treatment. The way the paper was written did nothing to encourage clinicians, as key clinical questions were not addressed in the context of careful statistical analysis of the data.

Essentially the study was of similar design to that of NASCIS I and was conducted with the same strictness of protocol and neurologic evaluation. Ten centres enrolled 487 patients who were randomised to three study groups: (i) very high methylprednisolone (30mg/kg bolus IV followed at one hour by 5.4mg/kg/hr for 23 hours), (ii) naloxone (5.4mg/kg bolus IV followed at one hour by 4mg/kg/hr for 23 hours) and (iii) placebo. The opiate receptor antagonist naloxone hydrochloride was included because of animal studies of experimental spinal cord injury which had shown neurologic benefit in those animals treated with the drug (Faden *et al.*, 1981a,b; Young *et al.*, 1985). The combination of high dose methylprednisolone and naloxone was not used because of excess mortality in cats treated with this combination. Patients were entered within 12 hours of injury and neurological evaluations were made on admission, at six weeks, six months and (published later) at one year (Bracken *et al.*, 1992). Once entered into the trial, patients were treated according to their treating physicians' usual protocols and these were not considered or assessed in the study.

In contrast to NASCIS I, the extensive statistical analysis of NASCIS II showed a significant improvement in neurologic scores of motor function, pinprick and light touch sensation in the group given methylprednisolone when compared with naloxone and placebo groups. The naloxone regimen used in this study had no consistent neurological efficacy. Complications were not statistically significantly different between the groups, although the rates of wound infection and gastrointestinal bleeding were slightly increased (but the differences were not statistically significant) in the methylprednisolone group. Mortality was the same in all three groups. In the follow up report of one year evaluation of patients entered into NASCIS II (Bracken *et al.*, 1994), the investigators showed that the results reported as six weeks and six months were sustained at 12 months; however, it was noted that the neurologic outcomes of patients commencing treatment with either methylprednisolone or naloxone more than eight hours after injury were worse than those taking placebo, indicating a clinical treatment window. A point perhaps not always emphasized enough.

Following the report of NASCIS II, very high dose methylprednisolone was accepted as a standard of care by many, as potentially beneficial by others and of unproven value by some. Subsequent correspondence clarified a number of aspects of the study such as the equal distribution of surgery in all groups (Lyons *et al.*, 1990; Yarkony & Roth, 1990; Senegor, 1990; Stifel & Brown, 1990; Bracken *et al.*, 1990) but left many issues unclear – the timing of surgery in relation to injury, spinal level of injury, etc. Nevertheless the methylprednisolone regimen was approved for the treatment of acute spinal cord injury by the regulating bodies of many countries, including the UK and Canada.

It was initially used by all spinal cord units in Australia and a system of early intervention of treatment by hospitals of first referral was established. However, at the end of 1992 the Therapeutic Goods Administration of the Australian Federal Government refused to approve an application for spinal cord injury to be included in the list of conditions for which methylprednisolone was indicated. The reason given was that the single reported study (NASCIS II) "did not demonstrate significant improvement in neurologic function six months after the treatment" (Hill, 1992). An appeal against this decision was not upheld. Consequently the clinical situation with respect to the role of methylprednisolone became unclear and its use is not based on consensus amongst clinicians in that country.

Three retrospective studies have been published following NASCIS II. In 1995 George *et al.*, published a retrospective analysis of patients admitted to two trauma centres between 1989 and 1992 (George *et al.*, 1995). They compared patients treated with the very high dose methylprednisolone regimen described in NASCIS II (n=80) with an historical group of patients treated without methylprednisolone (n=65). They found no differences in admission trauma score between the two groups. Hospital and Intensive Care Unit (ICU) stays were not different. They were able to compare functional independence measure (FIM) scores obtained during rehabilitation for 45 patients treated with methylprednisolone and 25 patients who were not. Admission FIM scores for mobility were not different but at discharge the methylprednisolone group fared significantly worse (5.16 versus 4.57, $P<0.05$).

This study lacks the tight control of clinical evaluations needed to eliminate random bias which can occur with retrospective analyses. The validity of FIM scores is particularly subject to random bias unless clinicians are carefully instructed in the use of this measurement scale.

A similar study by Gerhart *et al.*, (1995) showed no difference in neurological outcomes between those treated with methylprednisolone and historical controls.

A more recent retrospective study attempted to quantify adverse events in patients treated by the NASCIS II methylprednisolone dosage regimen compared to an historical control group which had not received methylprednisolone (Gerndt *et al.*, 1997). In contrast to the study by George *et al.* (1995), intensive care days were increased in the steroid group as was the incidence of pneumonia, which occurred 2.6 times as frequently in the treated group. The increase in ventilation and days in ICU is probably a reflection of the increased incidence of pneumonia. There was no difference in other adverse parameters including mortality. On the positive side there was a decrease in rehabilitation days in the methylprednisolone group.

These three studies suffer a number of methodological problems, including retrospectivity and the use of historical controls. The differences in outcomes, such as length of ICU stay, may reflect a variety of factors which have changed with time, such as resource changes which have led to different solutions, eg., transfer from ICU to high dependency status with decreased ICU length of stay may have occurred in one case and in another case a policy of early admission to ICU may have led to an increase in length of stay there. Without good analysis of these other variables, it is impossible to do more than consider these studies as cautionary or potentially encouraging views of an aspect of practice. The place of methylprednisolone cannot readily be understood unless all the important variables are accounted for.

Thus the studies by George *et al.* (1995), Gerhart *et al.* (1995) and Gerndt *et al.* (1997) do not help the clinician decide whether or not to use methylprednisolone, nor do they help them explain the risks and benefits to patients and relatives.

A study from Japan published in 1994 (Otani *et al.*, 1994, quoted by Bracken *et al.*, 1997) was reported to have replicated the findings of NASCIS II, but as the paper, was in Japanese and published in a local journal, this trial has not been evaluated by a process of international peer review nor has it been made available in English, the international scientific language. It cannot therefore, at this stage, be considered to support the use of methylprednisolone in acute spinal cord injury. There is a need for another study which fulfils the criteria of peer review and international availability to confirm or contradict NASCIS II.

A recently published study, NASCIS III, compares the very high dose methylprednisolone regimen of NASCIS II with the same regimen extended to 48 hours and with a group given the very same high dose methylprednisolone initial bolus, but followed by six hourly boluses IV of tirilazad 2.5 mg/kg (Bracken *et al.*, 1997). Both drugs are thought to act by inhibition of lipid peroxidation and thus protect neuronal membranes from the secondary metabolic effects of injury (Demopoulous *et al.*, 1980; Hall, 1992; Demopoulous *et al.*, 1982) As in previous studies, all other treatments were left to the treating physicians and no account of these was analysed. However the authors did attempt to assess function by performing the FIM at six weeks and six months. The scientific validity of this in the absence of base line assessment is unclear.

NASCIS III showed significantly improved motor function at 6 weeks and 6 months after injury in those patients treated with the very high dose methylprednisolone regimen to 48 hours when therapy was commenced 3–8 hours after injury. For those who started methylprednisolone within three hours of injury, no benefit was gained beyond 24 hours. Results for the group given methylprednisolone initially followed by tirilazad (a 21-aminosteroid analogue) were equivalent to the 24 hour high dose methylprednisolone regimen. Complications of severe sepsis and severe pneumonia were significantly higher in the 48 hour methylprednisolone group, but fatalities were equivalent in all groups.

Because of its cost and lack of additional benefit tirilazad cannot be recommended in the dosage regimen used in this study. The authors suggest that tirilazad may prove beneficial if used in different dosage regimen and the reason for pursuing this is its decreased serious complication rate when compared to the 48 hour methylprednisolone dosage regimen.

The lack of placebo group and the presumption that very high dose methylprednisolone for 24 hours serves as a treatment standard complicates the evaluation of this study (*vide supra*) and its impact on current practice is as yet unclear.

Key Points

- **High dose Methylprednisolone effective in improving neurological scores if started eight hours after injury and continued for 24 hours (NASCIS II).**
- **High dose Methylprednisolone should be continued to 48 hours if started between 3–8 hours of injury (NASCIS III).**
- **Results of both studies not yet confirmed by independent research published in peer reviewed international journals**

GM-1 ganglioside

Gangliosides have been studied extensively in animal models and clinical trials have been undertaken in stroke patients (Battistin *et al.*, 1985; Argentino *et al.*, 1989). In 1991 results from a pilot study of monosialotetrahexosylganglioside (GM-1 ganglioside) in the treatment of acute spinal cord injury were published (Geisler *et al.*, 1991). The claimed action of GM-1 ganglioside is that it enhances regeneration and sprouting of neurons and restores neuronal function after injury (Gorio, 1988) unlike methylprednisolone and tirilazad which are thought to act to protect neuronal membranes after injury. Therefore the study protocol differed considerably from that used in the NASCIS evaluation of methylprednisolone. In particular, dosage was started within 72 hours of injury (mean 48 hours) and was administered for a much longer period (mean 26 days). The drug was given intravenously daily in 100mg doses. The study group was small (n=37:18 drug, 16 placebo), but was followed for one year after spinal cord injury. The study showed significant improvement in American Spinal Injury Association (ASIA) motor score for the GM-1 ganglioside patients compared to the placebo group. Individual muscle recoveries were analysed to show that the improvement in the GM-1 ganglioside group was related to improvement in the motor strength of muscles that were initially paralysed rather than in those that were only partially paralysed. No serious side effects of treatment were reported. There was no difference in complication rate noted between the two groups.

The study has been criticised on a number of grounds, but principally the small number mitigates against a conclusive statement regarding the efficacy of GM-1 ganglioside. The principal problem with the small numbers is that randomisation did not eliminate bias between the two groups (Walker, 1991; Landi & Ciccone, 1992; Geisler *et al.*, 1992). In particular the initial ASIA scores for the GM-1 ganglioside group revealed a greater deficit compared to the placebo group, but both groups were functionally equivalent at one year. Thus the question is whether or not this would have been so without active drug treatment, ie., would the group starting off with poorer motor function have improved more because they could have achieved more, or not?

Despite the late start in treatment with GM-1 ganglioside, the end result seems to have been equal to the methylprednisolone regimen of NASCIS II. Could this be the treatment of choice for those who reach an acute centre more than 8 hours after injury?

A larger scale trial is underway and hopefully this will provide more evidence for or against the value of this drug in regeneration and recovery of damaged neurons following acute spinal cord injury and help define its place in the clinical management of this group of patients.

4-Aminopyridine

Most research work has been devoted to treatment and prevention of metabolic deterioration after acute spinal cord injury (sometimes referred to as secondary injury). It is widely believed that much of the damage occurs at the time of injury and this is supported by patient reports of instantaneous paralysis. It is likely that quite different strategies will need to be developed for patients who have suffered extensive neuronal loss at the time of injury or have long standing paraplegia or quadriplegia. The needs of the latter group have not been a focus of research, however some work continues to be done to address this issue.

One such approach has been to explore the place of the potassium channel blocking agent, 4-aminopyridine, in chronic spinal cord injury. The action of this drug is completely different to the other drugs listed in Table 1. It is a potassium channel blocking agent with the capacity to restore conduction in demyelinated neurons within the spinal cord (Zeidman *et al.*, 1996). A study of six patients with chronic spinal cord injury was reported in 1993 by Hayes *et al.* (1993) A small study, reported by Hansebout (1993), showed temporary neurological improvement in five of six patients with incomplete spinal cord injury in a placebo-controlled, double-blind, crossover trial of eight patients. A further study by Hayes *et al.* (1994) involved six subjects and provided evidence of increased nerve conduction in chronic spinal cord injury patients. The effect appears to be temporary. In an uncontrolled study of 19 patients with stable chronic spinal cord injury an electrophysiological effect (increased amplitude of motor evoked potentials) of 4-aminopyridine was demonstrated (Qiao *et al.*, 1997). The drug was not continued for a long period in any of these studies to see whether the benefits in muscle power were able to be sustained.

A clinical application for the drug is not clear. However a recent study by Potter *et al.* (1998) showed that twice daily oral administration of 4-aminopyridine led to clinical improvement for the duration of the study in 3 incomplete quadriplegic patients. These benefits included reduced pain, increased muscle power, voluntary control of bladder and bowel and improved male genital function. Another recent paper (Segal & Brunnemann, 1997) may indicate a place for 4-aminopyridine in patient treatment. The authors showed improvement in respiratory function in quadriplegic patients given an oral, immediate-release preparation of 4-aminopyridine in a dose of 10mg. Improvement began six hours after administration and persisted for at least 12 hours. No improvement was reported in the paraplegic patients. The study is important as it may indicate a clinically useful application for 4-aminopyridine in quadriplegic and high paraplegic patients with bronchitis or pneumonia, especially those with incomplete spinal cord lesions. The increases in mean values of forced expiratory volume in one second (FEV_1), forced vital capacity (FVC), maximum inspiratory pressure (MIP) and maximum expiratory pressure (MEP) may be of great help in the management of acute respiratory problems in chronic spinal cord patients. There may be a place for the drug in the management of acutely injured patients as well, but there are no studies in this situation yet.

Clearly much more work needs to be done to define the clinical applications of 4-aminopyridine and controlled trials with greater numbers need to be undertaken. However the drug does have promise for the management of respiratory conditions in spinal cord injury and may also have some role in bowel and bladder management and other areas of neurologic deficit, not yet explored in human studies.

Other drugs

As many as there have been theories propounded to explain neurologic deterioration after acute spinal cord injury, so there have been treatments proposed. A list of drugs and drug groups which have been studied is given in Table 1 and those which have been trialed are summarised in Table 2 (Faden, 1997; Zeidman *et al.*, 1996). Most have been shown to have some benefit in some trials. Most have not been studied enough to warrant large clinical trials. Some have been used in small studies eg., GM-1 ganglioside (Geisler *et al.*, 1991) and thyrotropin-releasing hormone (TRH) (Pitts *et al.*, 1995).

Table 1 Drugs and drug groups which are being studied in experimental spinal cord injury

Corticosteroids
Antioxidants & free radical scavengers
Gangliosides
N-methyl-D-aspartate (NMDA) receptor antagonists
Taxol
Opioid receptor antagonists
TRH and analogues
Melanocortins
Calcium channel antagonists
Platelet activating factor antagonists
Monoamine modulators
Arachidonate metabolism modulators

Table 2 Treatments which have undergone prospective clinical trials.

	Number of patients	Benefit	Reference
Methylprednisolone			
NASCIS I	330	–	Bracken *et al.*, 1984
NASCIS II	487	+	Bracken *et al.*, 1990
NASCIS III	499	+	Bracken *et al.*, 1992
Naloxone NASCIS II	487	–	Bracken *et al.*, 1990
Tirilazad NASCIS III	499	+	Bracken *et al.*, 1990
GM-1 ganglioside	37	+	Geisler *et al.*, 1991
TRH	20	+	Pitts *et al.*, 1995
4-aminopyridine	6	+	Hayes *et al.*, 1993
	8	+	Hansebout *et al.*, 1993
	6	+	Hayes *et al.*, 1994
	19	+	Quaio *et al.*, 1997
	17	+	Segal and Brunnemann 1997
Omental Transposition	23	–	Clifton *et al.*, 1996

+ Benefit; – No benefit

TRH and TRH analogues are thought to act in a variety of potentially beneficial ways (Faden, 1997) to antagonise a variety of autodestructive factors, including endogenous opioids, to improve spinal cord blood flow and restore ionic homeostasis. From a clinical point of view the situation regarding TRH is confusing. There appears to be no clinical follow-up of the small study reported in 1995 of 20 acute spinal cord injury patients studied between 1986 and 1988 (Pitts *et al.*, 1995). There were 11 drug treated and nine placebo treated patients. Seven patients did not complete the study timetable of follow up for 12 months. Interpretation of the results is made difficult by the small numbers and loss of patients from the trial. Nevertheless statistical analysis suggested that TRH was significantly beneficial as assessed by the Sunnybrook scale of functional severity of spinal cord injury (Tator *et al.*, 1982) at four months for patients with incomplete spinal cord lesions on admission when compared with the placebo group. Clearly a larger scale trial is necessary to determine whether or not there is a place for TRH in acute spinal cord injury management.

Progression to large scale clinical trials is inhibited by drug costs, availability in a suitable treatment form, by side effects, by difficulties in developing appropriate treatment protocols for humans and by the cost and clinical interest required to conduct such trials. There is a need for funding mechanisms to bring to clinical trial promising drugs or other forms of treatment. The number of drugs being evaluated in experimental situations which may lead to repair of damaged spinal cords is such as to indicate that current *ad hoc* funding mechanisms for such trials will be inadequate for proper evaluation of these treatments.

X-Irradiation

Experimental work in rats subject to x-irradiation in the third week after transection of the spinal cord is reported to result in electrophysiological recovery of neuronal function following structural regeneration (Kalderon & Fuks, 1996a,b). The clinician's interest is aroused by the time frame for treatment, which is much later than the drug treatments which have undergone clinical trial and by the apparent success in allowing neuronal re-growth with the establishment of functional connections by axotomized corticospinal neurons. This work needs to be replicated by other workers and treatment protocols developed for human trials. The complication rate, particularly in the longer term and for larger animals, needs to be established before a clinical trial can be undertaken. However, the clinical tool, x-ray therapy, is already available. If x-irradiation inhibits cellular events which prevent recovery, it may allow natural recovery processes to take place. Perhaps clinical treatment in the future will commence with methylprednisolone, continue with GM-1 ganglioside and conclude with x-irradiation. Such treatment protocols have been developed and refined in oncology, but we are still a long way from such a sophisticated approach to the treatment of acute spinal cord injury.

Surgical options

Surgery of the traumatised spinal column has not been shown to improve neurologic outcome, though it has value in improving vertebral alignment and allowing earlier mobilisation. While there is no evidence for improved spinal cord function as a result of surgery, operation is considered indicated by many if there is further neurological deterioration at a later date, if root lesions fail to recover and for cauda equina injuries where the nerve roots may be trapped in a traumatic split in the dura. There is still no consensus of opinion as to whether there is a place for surgery for the cord in acute spinal trauma. A scientific controlled trial is the only way to settle the question. While a trial of non-operative compared with operative treatment is still awaited, a large multicentre trial is underway in North America. This trial is comparing outcomes after early surgery (within hours of injury) against later operation (>1 day after injury). The results should be available in the next 1–2 years.

The focus of much basic research has been on surgical techniques which may be useful in promoting re-growth of axons across the damage cord by means of nerve grafts or other channels, or by filling the gap with embryonic or fetal tissue which may circumvent host rejection and establish functional connexions between the proximal and distal segments of the divided cord (Table 3). These approaches are discussed in many of the chapters in this book.

At this stage, studies are confined to animal work and often combine surgery with growth factors and other enhancing techniques (Houle & Ziegler, 1994; Bregman *et al.*, 1997; Chen *et al.*, 1996; Liu *et al.*, 1995). Work has been done in both brain and spinal cord, although there appear to be only two brief reports of this technique

Table 3 Surgical techniques undergoing laboratory investigations

- Nerve graft and cellular tube grafts
- Embryonic and fetal transplants
- Omental transposition
- Artificial bridges with or without growth factors or neural cells

being applied in a human study (Wirth III *et al.*, 1998; Thompson *et al.*, 1998). A number of review articles are available including several in this volume. They provide an overview of the current state of research in this area and thus indicate the difficulties of bringing any of these techniques to human trial (Tessler, 1991; Anderson & Howland, 1995).

Omental transposition

Surgical techniques have generally been aimed at assisting recovery from acute spinal cord trauma as noted above. One surgical technique which has been applied to humans has been omental transposition. Following experimental work in animals (Goldsmith *et al.*, 1985) omental transposition has been advocated as a treatment of acute and chronic spinal cord injuries in order to improve neurologic outcome. A number of human cases have been operated upon and case reports and small uncontrolled series have been reported. Unfortunately most of these do not provide sufficient data to enable scientific conclusions to be drawn as to the efficacy of this surgery (Goldsmith *et al.*, 1986; Abraham *et al.*, 1987; Raphael *et al.*, 1992; Goldsmith, 1994). Other reports are not available in English and therefore readership is restricted. However a number of cases are reported to have improved after this surgery (Goldsmith, 1994).

A careful trial of 15 patients with stable, chronic spinal cord injury (>1.5 years from injury), who underwent a omental transposition, showed no significant improvement from pre-surgery status when followed for one year after operation (Clifton *et al.*, 1996). Of four patients in a preliminary study, one died from a pulmonary embolus, one was lost to follow up and two were studied at one and four and half years post operatively. In the second series of 11 patients, there were no fatalities, but significant complications occurred in five patients. No improvement was noted in the 13 patients from the two series or when the 11 from the second series was compared to eight non operated match controls followed for a similar period. Indeed there was a trend, not significant, for the operated group to suffer some sensory decline and one patient had marked loss of power in both biceps muscles beginning four months after surgery.

The lack of benefit and the significant complication rate do not indicate a place for this treatment as currently advocated by some.

Summary

A comparison of Table 1 with Tables 2 & 3 shows how little of basic laboratory research has reached clinical trial in humans. Of the new treatments for spinal cord injury, methylprednisolone is most advanced in finding a place in the treatment armamentarium. GM-1 ganglioside and TRH show promise and the former is undergoing a large clinical trial. The potassium channel blocking agent, 4-aminopyridine, may have promise in the management of incomplete spinal cord injury patients and this requires further study.

The most important task facing basic researchers is to bring their work to the situation where it can be translated into clinical trials in patients with spinal cord injuries. The problem that will arise at that stage is that our current funding arrangements for clinical trials are *ad hoc*. There needs to be a concerted push for funding of large clinical trials. To date the big trials have been undertaken in the United States (NASCIS trials), but it may be important to widen the base of Spinal Cord Injury Units that are able to take part in such trials and it may need to become multi-national as well as multi-centred. This issue has been identified by the International Spinal Research Trust, which has included clinical trials as an important aspect of its strategic plan. The next decade should be a decade of clinical trials and all funding bodies should be gearing to emphasise treatment of CNS diseases, including spinal cord injury, as a major thrust of their work.

Key Points

- The results of large scale clinical trials of GM-1 ganglioside awaited.
- 4-Aminopyridine may have a place in the management of patients with incomplete chronic spinal cord injury. More studies are needed.
- TRH and TRH analogues may have a place in clinical management of acute spinal cord injury, but large scale clinical trials are needed to confirm this.
- Evaluation of the effective early surgical treatment of the disrupted vertebral column on spinal cord outcome is awaited.
- No surgical technique has been shown to be helpful in the repair of chronic spinal cord injury.
- Issues of funding and subject numbers need to be addressed so that large scale clinical trials can be undertaken of new, laboratory studied treatments of promise.

References

Abraham J Paterson A Bothra M Mofti A B and Taylor G W (1987) Omental-myelo-synangiosis in the management of chronic traumatic paraplegia: a case report. Paraplegia 25, 44–49

Anderson D K Howland D R and Reier P J (1995) Fetal neuronal grafts and repair of the injured spinal cord. Brain Pathology 5, 451–457

Argentino C Sacchetti M Toni D Savoini G D'Arcangelo E Erminio F Federico F Ferro Milone F Gallai V Gambi D Mamoli A Ottonello G A Ponari O Rebucci G Senin V and Fieschi C (1989) GM-1 ganglioside therapy in acute ischemic stroke. Stroke 20, 1143–1149

Battistin L Cesari A Galligioni F Marin G Massarotti M Paccagnella D Pellegrini A Testa G and Tonin P (1985) Effects of GM-1 ganglioside in cerebrovascular diseases: a double-blind trial in 40 cases. European Neurology 24, 343–351

Bracken M B Collins W F Freeman D F Shepard M J Wagner F W Silten R M Hellenbrand K G Ransohoff J Hunt W E Perot P L Grossman R G Green B A Eisenberg H M Rifkinson M Goodman J H Meagher J W Fischer B Clifton G L Flamm E S and Rawe S E (1984) Efficacy of methylprednisolone in acute spinal cord injury. Journal of the American Medical Association 251, 45–52

Bracken M B Shepard M J Collins W F Holford T R Young W Baskin D S Eisenberg H M Flamm E Leo-Summers L Maroon J C Marshall L F Perot P L Piepmeier J Sonntag V K H Wagner F C Wilberger J E and Winn H R (1990) A randomized controlled trial of methylprednisolone or naloxone in the treatment of acute spinal cord injury: Results of the Second National Acute Spinal Cord Injury Study. New England Journal of Medicine 322, 1405–1411

Bracken M B Shepard M J Collins W F Hofland T R Young W Piepmeier J Leo-Summers L Baskin D S Eisenberg H M Flamm E Marshall L F Maroon J Wilberger J Perot P L and Sonntag V K H (1990) A radomized controlled trial of methylprednisolone or naloxone in the treatment of acute spinal cord injury letter. New England Journal of Medicine 323, 1209

Bracken M B Shepard M J Collins W F Holford T H Baskin D S Eisenberg H M Flamm E Leo-Summers L Maroon J C Marshall L F Perot P L Piepmeier J Sonntag V K H Wagner F C Wilberger J L Winn H R and Young W (1992) Methylprednisolone or naloxone treatment after acute spinal cord injury: 1-year follow-up data: Results of the Second National Acute Spinal Cord Injury Study. Journal of Neurosurgery 76, 23–31

Bracken M B Shepard M J Holford T R Leo-Summers L Aldrich E F Fazl M Fehlings M Herr D I Hitchon P W Marshall L F Nockels R P Pascale V Perot P L Piepmeier J Sonntag V K H Wagner F Wilberger J E Winn R H and Young W (1997) Administration of methylprednisolone for 24 or 48 hours or tirilazad mesylate for 48 hours in the treatment of acute spinal cord injury: Results of the Third National Acute Spinal Cord Injury Randomized Controlled Trial. Journal of the American Medical Association 277, 1597–1604

Bregman B S McAtee M Dai H N and Kuhn P C (1997) Neurotrophic factors increase axonal growth after spinal cord injury and transplantation in the adult rat. Experimental Neurology 148, 475–494

Chen A Xu X M Kleitman N and Bunge M B (1996) Methylprednisolone administration improves axonal regeneration into Schwann cell grafts in transected adult rat thoracic spinal cord. Experimental Neurology 138, 261–276

Clifton G L Donovan W H Dimitrijevic M M Allen S J Ku A Potts III JR Moody F G Boake C Sherwood A M and Edwards J V (1996) Omental transposition in chronic spinal cord injury. Spinal Cord 34, 193–203

Demopoulous H B Flamm E S Pietronigro D D et al. (1980) The free radical pathology and the micro circulation in the major central nervous system disorders. Acta Physiologica Scandinavica (suppl. III) 492, 91–119

Demopoulous H B Flamm E S Seligman M L et al. (1982) Further studies in free- radical pathology in the major central nervous system disorders: Effects of very high doses of MP on the functional outcome Morphology and chemistry of experimental spinal cord impact injury. Canadian Journal of Physiology and Pharmacology 60, 1415–1424

Faden A I Jacobs T P and Holaday J W (1981a) Opiate antagnonists improves neurological recovery after spinal cord injury. Science 211, 493–494

Faden A I Jacobs T P Mougeye E and Holaday J W (1981b) Endorphines in experimental spinal cord injury: therapeutic effects of naloxone. Annala of Neurology 10, 326–332

Faden A I (1997) Therapeutic approaches to spinal cord injury. Advances in Neurology 72, 377–386

Geisler F H Dorsey F C and Coleman W P (1991) Recovery of motor function after spinal cord injury — a randomised placebo-controlled trial with GM-1 ganglioside. New England Journal of Medicine 324, 1829–1838

Geisler F H Dorsey F C and Coleman W P (1992) GM-1 ganglioside for spinal cord injury. New England Journal of Medicine 326, 494

George E R Scholten D J Buechler C M Jordan-Tibbs J Mattice C and Albrecht R M (1995) Failure of methylprednisolone to improve the outcome of spinal cord injury. American Surgery 61, 659–663

Gerhart K A Johnson R C Menconi J Hoffman R E and Lammertse D P (1995) Utilisation and effectiveness of methylprednisolone in a population – based sample of spinal cord injured persons. Paraplegia 33, 316–321

Gerndt S J Rodriguez J L Pawlik J W Taheri P A Wahl W L Michaels A J and Papadopoulos S M (1997) Consequences of high-dose steroid therapy for acute spinal cord injury. Journal of Trauma 42, 279–284

Goldsmith H S Steward E and Duckett S (1985) Early application of pedicled omentum to the acutely traumatised spinal cord. Paraplegia 23, 100–112

Goldsmith H S Neil-Dwyer M S and Barsoum L (1986) Omental transposition to the chronically injured human spinal cord. Paraplegia 24, 173–174

Goldsmith H S (1994) Brain & Spinal Cord re-vascularisation by omental transposition. Neurological Research 16, 159–162

Gorio A (1988) Gangliosides as a possible treatment affecting neuronal repair process. Advances in Neurology 1, 29–37

Hall E D and Braugler J M (1982) Glucocorticoid mechanisims in acute spinal cord injury: a review and therapeutic rationale. Surgical Neurology 18, 320–327

Hall E D (1992) The neuroprotective pharmacology of methylprednisolone. Journal of Neurosurgery 76, 13–22

Hansebout R R Blight A R Fawcett T S and Reddy K (1993) 4-aminopyridine in chronic spinal cord injury: a controlled double-blind crossover study in eight patients. Journal of Neurotrauma 10, 1–18

Hayes K C Blight A R Potter P J Allatt R D Hsieh J T C Wolfe D C Lam S and Hamilton J T (1993) Pre-clinical trial of 4-aminopyridine in patients with chronic spinal cord injury. Paraplegia 31, 216–224

Hayes K C Potter P J Wolfe D L Hsieh J T C Delaney G A and Blight A R (1994) 4-aminopyridine sensitive neurological deficits in patients with spinal cord injury. Journal of Neurotrauma 11, 433–446

Houle J D and Ziegler M K (1994) Bridging a complete transection lesion of adult rat spinal cord with growth factor – treated nitrocellulose implants. Journal of Neurology Transplantation & Plasticity 5, 115–124

Hill S (1992) Quoted from a letter to the Managing Director Upjohn Pty. Ltd. From Drug Evacuation Branch Therapeutic Goods Administration Commonwealth Dept. of Heath Housing and Community Services.

International Spinal Research Trust Annual Review 1996/97. P.3

Kalderon N and Fuks Z (1996a) Structural recovery in lesioned adult mammalian spinal cord by X-irradiation after the lesion site. Proceedings of the National Academy of Sciences USA 93, 1179–1184

Published erratum appears in Proceedings of the National Academy of Sciences USA 1996. 93, 14992

Kalderon N and Fuks Z (1996b) Severed corticospinal axons recover electrophysiologic control of muscle activity after x-ray therapy in lesioned adult spinal cord. Proceedings of the National Academy of Sciences USA 93, 11185–1190

Published erratum appears in Proceedings of the National Academy of Sciences USA 1996. 93, 14993

Landi G and Ciccone A (1992) GM-1 ganglioside for spinal cord injury. New England Journal of Medicine 326, 493

Lyons M K Partington M D and Meyer F B (1990) A randomized controlled trial of methylprednisolone or naloxone in the treatment of acute spinal cord injury letter New England Journal of Medicine 323, 1207–1208

Liu L S Khan T Sayers S T Dauzvardis M F and Trausch C L (1995) Electrophysiological improvement after co-implantation of carbon filaments and fetal tissue in the contused rat spinal cord. Neuroscience Letters 200, 199–202

Pitts L H Ross A Chase G A and Faden A I (1995) Treatment with thyrotropin-releasing hormone (TRH) in patients with traumatic spinal cord injuries. Journal of Neurotrauma 12, 235–243

Pottor P J Hayes K C Hsieh J T C Delaney G A and Segal J L (1998) Sustained improvements in neurological function in spinal cord injured patients treated with oral 4-aminopyridine: three cases. Spinal Cord 36, 147–155

Qiao J Hayes K C Hsieh J T Potter P J and Delaney G A (1997) Effects of 4-aminopyridine on motor evoked potentials in patients with spinal cord injury. Journal of Neurotrauma 14, 135–149

Raphael H Malpica A Espinoza M and Moromizato P (1992) Omental transplantation in the management of chronic traumatic paraplegia: case report. Acta Neurochirugica. 114, 145–146

Segal J R and Brunnemann S R (1997) 4-aminopyridine improves pulmonary function in quadriplegic humans with long standing spinal cord injury. Pharmacotherapy 17, 415–423

Senegor M (1990) A randomized controlled trial of methylprednisolone or naloxone in the treatment of acute spinal cord injury. Letter, New England Journal of Medicine 323, 1208

Stifel H G and Brown M (1990) A radomized controlled trial of methylprednisolone or naloxone in the treatment of acute spinal cord injury. Letter, New England Journal of Medicine 323, 1208–1209

Tator C H Rowed D W and Schwartz M L (1982) Sunnybrook Spinal Injury Scales for assessing neurological injury and neurological recovery In: Early Management of Acute Spinal Cord Injury. C.H. Tator (ed.) Raven Press New York Chap. 2 pp 7–24

Tessler A (1991) Intraspinal transplants. Annals of Neurology 29, 115–123

Thompson F J Uthman B Mott S Remson E J Wirth III E D Fessler R G Reier P J Behrman A Trimble M and Anderson D K (1998) Neurophysiological assessment of neural tissue transplantation in syringomyelia patients. Society for Neuroscience Abstracts 24, 33.16

Wirth III E D Fessler R G Reier P J Thompson F J Uthman B Behrman A Remson E J and Anderson D K (1998) Feasibility and safety of neural tissue transplantation in patients with syringomyelia. Society for Neuroscience Abstracts 24, 33.17

Walker M D (1991) Acute spinal cord injury. New England Journal of Medicine editorial 324, 1885–1887

Yarkony G M and Roth E J (1990) A radomized controlled trial of methylprednisolone or naloxone in the treatment of acute spinal cord injury. Letter, New England Journal of Medicine 323, 1208

Young W Flamm E S Demopoulous H B Tomasula J J and De Crescito V (1985) Effect of naloxone on post traumatic ischemia in experimental spinal contusion. Journal of Neurosurgery 63, 319–326

Zeidman S M Ling G S Ducker T B and Ellenbogen R G (1996) Clinical applications of pharmacologic therapies for spinal cord injury. Spinal Disorders 9, 367–380

2

Recovery from injury in the immature mammalian spinal cord

Norman R Saunders and Katarzyna M Dziegielewska

Introduction

The notion that the immature mammalian central nervous system (CNS) may repair and regenerate itself more effectively than that of the adult has been current for all of this century. There are numerous references to sprouting of injured axons in brain and spinal cord of immature animals in Ramón y Cajal's 1928 monograph which summarised data from experimental observations made on injured spinal cords in young cats and dogs over the three previous decades. All of Ramón y Cajal's work and most of the subsequent studies dealt with postnatal common laboratory animals, particularly rats and cats. There are few references in the literature to spinal injuries in fetuses, but most of the results of attempts at such experiments seem to have been negative (eg Hooker & Nicholas, 1930). However, there is a small number of reports of successful "regeneration" of injured axons by the outgrowth of neurites and some recovery of function in fetuses (eg Migliavacca, 1928). This perhaps accounts for the fact that there seems to have been a long standing belief that the immature CNS has some capability for regenerative repair that is not present in the adult; the wealth of negative experiments seem to have been largely ignored in favour of the few experiments with more positive outcomes. This is well illustrated by the articles collected together by WF Windle in a monograph entitled "Regeneration in the Central Nervous System" (Windle, 1955) and by Lloyd Guth's articles (eg 1975). However there were dissenting voices, apparently based on an inability to reproduce the few positive results published (eg Hess, 1955; p176 in Windle, 1955). Also Clark's extensive papers (eg 1942) seem to have been influential in establishing a belief amongst many neuroscientists and neurologists that there is nothing special about the ability of the the developing CNS to repair itself or to sustain repair in the adult, if immature CNS tissue is implanted into injured adult CNS.

In several key areas of spinal injury research, conclusive positive results have been late coming for types of experiments that were attempted many years ago and yielded predominantly negative results. The later success does not seem to be particularly due to developments in technology, but more to the persistence of those involved. For example using implants of fetal CNS in adult CNS injury sites, Clark (1940, cited in

Clark 1942) was unsuccessful whereas Björklund (1991) has demonstrated growth of host axons into such implants. There is a similar contrast with results with lesions of embryonic mammalian spinal cord (cf Hooker & Nicholas, 1930 and Saunders *et al.*, 1992, 1998).

The purpose of this review is to make a reasonably detailed analysis of key papers that define what is established (or not yet clear) about essential features of the response of the immature central nervous system to injury. This survey is largely confined to studies in mammals (marsupial and eutherian, the latter sometimes being inappropriately referred to as placentals). Some reference is also made to experiments in chick embryos, as these represent the first well documented successful repair of a completely transected spinal cord in a warm-blooded species (Clearwaters, 1954; Hasan *et al.*, 1991, 1993; Shimidzu *et al.*, 1990). Cold-blooded animals are dealt with in the chapters by Beazley & Dunlop and by Conti & Selzer in this volume.

Key Points

- Early attempts to transect spinal cord *in utero* mostly resulted in failure to recover, although there were a few positive results reported.
- Notwithstanding the preponderance of negative evidence, there has been a longstanding belief that the immature CNS is better able to repair itself, following injury.
- Numerous experiments involving implantation of immature CNS into injured adult, but especially newborn (eutherian) CNS indicate that CNS neurons have an ability to sprout in response to injury, although there is a widespread belief that CNS neurons are generally unable to respond to injury by regenerating new axons.

The main points to be covered in this review are:

(1) Is it indeed correct, as asserted by many scientists throughout this century but denied by others, that the immature spinal cord (and CNS in general) of mammals has a greater ability to repair itself than when adult?
(2) If so, to what extent is this greater ability for repair due to regeneration of neurites from the axons damaged by experimental injury, as opposed to growth of new axons through (or around) the site of the lesion that were not present at the time of injury, ie what are the relative contributions to repair from regeneration and from normal growth and development of the spinal cord?

There is an important general point that is worth noting here. Much of the work on spinal injury has been understandably preoccupied with finding ways of getting neurite growth through the site of a lesion. There has been much less work on the problem of obtaining fibre growth from the lesion to appropriate and functionally effective targets in the spinal cord and brain. The achievement of fibre growth across a lesion will be of no practical value if it is not followed by functionally useful innervation. Thus important additional questions are:

(3) What evidence is there that if nerve fibres grow across the site of the injury they continue to grow along the spinal cord and not only reach appropriate targets, but these targets can be innervated in a functionally effective manner?
(4) Can any functional recovery that occurs be attributed to this growth of fibres and succesful innervation or could it be due to reorganisation of undamaged spinal cord, the capacity for which might be greater in the case of immature animals than in the adult?

There have been numerous studies involving partial lesions of the spinal cord. Presumably one reason for their use is that young animals with such lesions do show substantial degrees of functional recovery and development, although the very fact that the animals are still developing, often makes it difficult to sort out what is recovery and what is some feature of normal development, perhaps modified by the injury. These partial lesion experiments may also be difficult to evaluate, partly because of the inevitable variability in the extent of the injury and also because of the difficulty of distinguishing between a regenerative response from the injured axons as against collateral sprouting from uninjured axons at or near the site of injury. There is also the possibility that the isolated (caudal) parts of the spinal cord may reorganise local circuits, particularly in the immature animal. Thus it may be difficult to establish unequivocally that an improvement in function was due to regeneration from the damaged axons, rather than local "plasticity". This is a general problem with all spinal injury studies but is probably more so in the immature nervous system. A point worth noting is that the partial lesions are often referred to as "hemisections" or "over hemisections" although this does not always mean that the axis of the lesion is the same. Thus in some cases one half of the cord has been transected as in the classical Brown-Sequard lesion (ie either the left or right side of the cord has been injured). In other cases it is more the dorsal or ventral half of the cord that has been lesioned. This of course has profound implications for the type and degree of recovery that might be expected, given the way in which fibre tracts are organised in the spinal cord. The partial lesion experiments have however, revealed one important distinction between different features of the response to spinal cord injury. This is that there are developmentally regulated processes which independently determine whether axons in an injured spinal cord can grow directly across a lesion as distinct from being able to grow through the uninjured parts of the spinal cord due to a period in the immature spinal cord when there is a locally "permissive" environment for fibre growth.

It is not the intention of this review to determine priority for the various points that do appear to be established concerning the repair capabilities of the immature spinal cord. Rather it is the aim to select papers where the experimental approach has been sufficiently comprehensive yet detailed, that it is possible to make a reasonable satisfactory appraisal of the conclusions, without the reader having to wade through large numbers of related papers.

Key Points

Main questions to be reviewed

- Is the immature spinal cord (CNS) better able to repair itself than when adult?
- If neurites do grow across or around the site of injury will they continue to grow beyond the site of injury and do they make connections with appropriate targets?
- To what extent is any recovery, if it occurs, due to regeneration from damaged axons, to growth of new axons as part of normal development or to "plastic" rearrangement of the surviving neuronal circuits?

Spinal injuries in immature Eutherian mammals

Complete transection of the spinal cord

Chambers (1955) and Clemente (1955) reviewed briefly many of the experiments carried out in the first half of the century on spinal injury in immature animals including fetuses. They cite both negative and positive outcomes, but give the impression that they both support the belief in the greater potential for regenerative recovery from injury in immature animals. Since then a huge number of studies has been carried out in the newborn of cats and rats in particular. Many of these concerned only one feature of recovery, often behaviour. Thus it may be necessary to examine several papers from the same laboratory in order to have the fullest available picture of what has been established. Stelzner and colleagues have, over many years, studied effects of complete spinal transection in postnatal rats of various ages. Key early papers are: Stelzner *et al.* (1975); Weber & Stelzner (1977); Bernstein *et al.* (1981); Bernstein & Stelzner (1983).

Stelzner *et al.* (1975) review briefly some of the earlier papers that contain evidence for recovery/fibre regeneration in immature animals. In their own study they compared the functional effects of mid-thoracic spinal transection in neonatal (postnatal day, P0) to postnatal day 5 (P5) and weanling rats (P21-26). They also described some limited morphological studies. Lesions were made by crushing with watchmakers forceps or cutting with iridectomy scissors using "clean surgical technique". The completeness of the transection was checked by eye and in the case of cut cords, by passing an angled needle through the site of injury. Histological examination of the spinal cords following behavioural testing showed that all lesions were complete. In neither the group lesioned as neonates nor those lesioned as weanlings, was there any evidence of fibre growth across the site of injury. In animals lesioned as neonates, the lesions extended over a longer length of cord and consisted of dural sacs of fluid without structural contents. The cord on either side of the lesion appeared immature in the animals when adult and there was little gliosis. In contrast, in the animals lesioned as weanlings, the lesions were more compact and there was substantial gliosis at both ends of the lesion; but the decrease in spinal cord cross-section was less than in the cords of animals lesioned in the neonatal period. Another age-related difference was that in the neonatal animals there was no sign of demyelination (presumably because there was little or no myelin present at this stage of development). In contrast, in the animals lesioned as weanlings there was a rim of demyelinating fibres as well as a greater degree of gliosis in the weanlings when adult.

Behavioural tests consisted of (a) observation of posture and locomotion, (b) observation of hindquarter responses to being placed on different surfaces including one with 1 cm^2 openings or on an elevated grid with variable sized openings, (c) placing, hopping and stepping responses and (d) responses to various sensory stimuli. However only brief descriptions were given for each of these tests without any attempt to quantitate them, although Weber & Stelzner (1977) devised a scoring system for various qualitative indices of behaviour. It was only in later studies that detailed attempts were devised to develop comprehensive quantitative methods for evaluation of recovery and development following spinal injury in immature animals (for examples and references to numerous studies see Diener & Bregman, 1998a; Saunders *et al.*, 1998; Wang *et al.*, 1998).

In summary, the main descriptive findings of Stelzner *et al.* (1975) were as follows: A small number of responses that were considered to involve supraspinal activity was observed in the neonatal animals before operation. Postoperatively these disappeared. Immediately postoperatively there was depression of spinal reflexes caudal to the lesion. These recovered by two days after operation. By 4–5 days postoperatively the hind limbs supported the hindquarters and the feet rested on their plantar surfaces. Up to about two weeks of age the transected animals dragged their hindquarters, but then showed evidence of stepping, whereas in normal littermates diagonal limb progression (an indicator of forelimb-hindlimb coordination) was apparent from six days of age. The presence and rate of hindlimb stepping increased in the operated animals at 15 to 21 days but unlike normal controls these steps were rapid and were not coordinated with the forelimbs. A summary of this data can be found in Table 1 of Stelzner *et al.* (1975). More detailed behavioural analysis with quantitation of some aspects was published by Weber & Stelzner (1977). From these data and those on morphology it is clear that complete spinal transection in these animals resulted in permanent severance of the cord and although the neonatally lesioned animals showed a wider behavioural repertoire than the lesioned weanlings when adult, this behaviour was attributable solely to local segmental spinal responses. Both Stelzner *at al.* (1975) and Weber & Stelzner (1977) concentrated in their behavioural analysis on locomotion. Both studies mention tests of vestibular function in the Methods section of these papers, but neither study reported any results. This is important for comparison of later papers describing the effects of implanting fetal spinal cord into cords of neonates with partial or complete spinal transection. The behavioural results of such experiments are often difficult to evaluate because of the concentration on locomotor performance, which may be driven largely if not entirely by local spinal rhythm generators (Rossignol, 1996). There seems to have been no attempt in many of the earlier complete spinal transection experiments to oppose the severed ends of the spinal cord at the time of operation. So in the light of later evidence that there is a developmentally regulated period before which spinal repair in a mammal is possible (see section (c) "Structure and function in adult opossums following complete neonatal spinal transection") it may be that fibre growth across the lesion before this critical time did not occur because the gap to be bridged was too great. In a related paper (Bernstein *et al.*, 1981) it was shown in transected cords of both neonatal and weanling rats that numerous neurites did indeed grow out from both sides of the lesion. But they did not make any progress across the lesion, as also shown much earlier by Ramón y Cajal (1928). From currently available knowledge it is likely that at least part of the explanation for this lies in the unfavourable environment that the CNS may provide for sustaining neurite outgrowth.

Key Points

Spinal repair in immature (neonatal) eutherians.

- Repair, in terms of axon growth across the lesion, does not occur if the spinal cord is completely transected, even in a species such as rat which are very immature at birth.
- Neonatal, but not adult, animals may show some degree of functional recovery following complete transection of the cord eg hindlimb weight support and rhythmic movements of the hind limbs.
- Neurites will grow into and possibly through the site of the lesion, if the transected spinal cord (complete or partial lesion) is implanted with fetal spinal cord. Different experimenters report different degrees of neurite outgrowth; opinions differ on their likely contribution to functional recovery. See also Box on p. 27.

Partial transection of the spinal cord

Presumably because of the finding that even in newborn mammals such as cats and rats, neurites do not grow across the site of a complete spinal transection, several groups began in the late 1970s, early 1980s, to experiment with partial lesions with or without implantation of fetal CNS tissue, in an attempt to provide the neurites growing out from damaged spinal axons with a substrate through which they could grow. This approach and other forms of bridging have been extensively studied by a number of laboratories in the last 20 years, including investigations in adult spinal or brain lesioned animals that will not be dealt with here as they are reviewed in chapters in this volume by Bunge and by Harvey.

If a partial spinal cord or unilateral pyramidal tract lesion is made in neonatal rats (Stelzner *et al.*, 1979), cats (Bregman & Goldberger, 1983a,b,c), hamsters (Kalil & Reh, 1982) or postnatal opossums after the age of about P20 (see Xu & Martin, 1991 for references), without implantation of fetal spinal cord, no axons grow through the lesion; rather they grow around it via the undamaged spinal cord tissue. This has been shown in extensive studies of lesions in the brainstem or spinal cord in the young of a number of different mammalian species. For example Kalil & Reh (1982) in the neonatal hamster (chosen because it is even less mature than rats when born) lesioned the pyramidal tract on one side in the medulla, rostral to the decussation of the pyramids. They showed that fibres labelled by injection of ^{14}C proline into the developing motor cortex did not cross the site of the lesion; instead they grew prolifically in aberrant tracts but crossed the midline to do so, although at a level that is rostral to the lesion and therefore also rostral to the normal site of decussation. Interestingly the axons, although growing via an aberrent route, appeared to find their way to normal target zones in the spinal cord.

Behavioural studies on these animals were reported separately (Reh & Kalil, 1982b). Hamsters with pyramidal lesions made at different postnatal ages were evaluated behaviourally using grid crossing and seed shelling tests. Only the seed shelling test (average time to shell and eat 10 sunflower seeds) results were presented in any detail; but descriptive comments about the gait analysis were also included. Animals with lesions made at P4–8 when tested at P60 or later appeared behaviourally similar to unoperated adults. Whereas adults with pyramidal tract lesions or adults that had been lesioned at P21 were severly impaired with respect to their gait for about 2 weeks post lesion, but were by then indistinguishable from unoperated controls. However the seed shelling test proved more sensitive and quantiative deficits (ie increased time to perform the task) persisted for at leat 80 days beyond the time at which the lesions had been made. In the case of the neonatal animals their normal performance on the seed shelling test when adult deteriorated on re-lesioning. Reh & Kalil (1982b) correlated this with destruction of the aberrant pathway described in Kalil & Reh (1982). The results also show that fibre growth can occur after injury to the spinal cord, as shown by others, particularly in neonatal cats and rats. However, these experiments did not directly deal with the problem of whether the growing axons were regenerating from those damaged at the time of injury or were arriving later as part of normal cord development. Reh & Kalil (1982b) concluded that the growing axons were unlikely to be late arriving because their earlier studies had been interpreted as showing that the corticospinal tract axons grow as a single wave through the medulla (Reh & Kalil, 1982a). This was based on earlier counts of axon profiles in electron micrographs. The counts showed quite a lot of variation and did not take acount of the possibility that axotomy might induce cell death or that some cell death may occur following successful innervation of targets by some neurons and not others.

The possibility that cell death and consequent loss of axons might mask late arrival of new axons was not considered by Kalil & Reh (1982). They suggested that their morphological studies (Reh & Kalil, 1982a) which included axon counts at different neonatal ages, could be interpreted as indicating that the pyramidal fibres grew as a single front and that they were all present at the pyramid at the age when the lesion was made. However, it was subsequently demonstrated by Merline & Kalil (1990) that massive cell death occurs in the cortical neurons from which the axons of the pyramidal tracts arise if a pyramid is lesioned before P14 in the hamster; the amount of cell death was greater the younger the animal was at the time the lesions were made. It thus seems likely that the aberrant fibre growth described by Kalil & Reh (1982) was due to later arriving new axons rather than regeneration from injured axons. The difficulty of distinguishing between regenerating and late arriving axons will be dealt with in more detail below.

The results of Kalil & Reh (1982) are important because they show that growing axons can apparently find their normal targets (as implied by the normal functional development described) in the spinal cord even when the route is disrupted by a lesion. Thus local effects at the site of the injury can be separated from later developing inhibitory properties of the more mature spinal cord (see chapter 4 by Fitch & Silver).

Martin and his colleagues (eg Xu & Martin, 1991) also studied growth of axons around a tract lesion; in their case they used lesions of the rubrospinal tract in North American opossums (*Didelphis virginiana*, referered to here as *Didelphis* to distinguish

Key points

Partial spinal lesions (hemisections)

- Neurites grow around the lesion, but not through it unless implanted with appropriate fetal CNS.
- Some recovery of function may occur following a partial lesion of the spinal cord. Not always clear to what extent the neurites which grow around the lesion contribute to recovery.
- Not usually clear whether the neurites are regenerating from damaged axons, are newly growing axons (see Box on p35) or are collateral sprouts from undamaged axons.
- Extent of lesion often variable.

it from the South American opossum *Monodelphis domestica*). Earlier studies by Martin's group (papers cited in Xu & Martin, 1991) had shown that rubrospinal axons can grow around a lesion of the tract made at around P20 and that there is a critical period of development after which such growth ceased. Xu & Martin (1991) used a double labelling method designed to distinguish between regenerating axons from injured fibres and late arriving new fibres growing as a result of normal develoment. This is such an important point that several of the key papers that have used a double labelling approach will be reviewed in a later section. Xu & Martin (1991) showed that, as in the case of the corticospinal tract, many of the neurons died following axotomy and most of the axons growing around an injury to the rubrospinal tract were late arriving new fibres. However the finding of a few rubrospinal neurons that were labelled with both dyes is important evidence, now supported by other similar experiments in various species, that at least some of the axons injured in an immature spinal cord are capable of regeneration.

Numerous functional studies have been carried out in immature animals following partial spinal cord lesions with and without implants of for example fetal CNS tissue. Amongst more recent studies are those of Bregman & Goldberger (1993a) who described experiments in neonatal and adult cats subjected to the partial lesions (described in their paper as "hemisections"). Drawings of the extent of the lesions show that these were quite variable, in some cases involving slightly over half of the spinal cord and in others only about 1/3rd of the cross sectional area. Behavioural tests included a detailed analysis of gait. Both neonatal and adult cats recovered well from the operation, but there were no overall consistent differences in performance. For some tests the neonatally operated animals did better than the adult operated animals and for other tests the reverse was true or there was no difference. The authors concluded that true "sparing of function" appeared to be restricted to motor patterns that are not developed at the time the lesion was made. Bregman and colleagues have carried out a large number of sudies on the morphological and behavioural recovery of neonatal rats with partial (overhemisection) spinal lesions. The most recent are those of Diener & Bregman (1998a,b) involving cervical lesions and complex behavioural analysis. Animals with fetal spinal cord implants were found to have near normal features of upper limb control, in contrast to severe impairment in animals with lesions but no implants. One problem with all such partial lesion experiments, as mentioned above, is that the extent of the lesions was quite variable and some animals were excluded because their lesions were judged not to meet criteria for inclusion in the study. However it is not clear whether these exclusions were made "blind" with respect to the behavioural analysis, which leaves open the possibility of unconscious bias in selection of the animals.

Structure and function of the mature spinal cord following complete transection and implantation of fetal CNS in the neonate

In spite of a very large number of reported experimental attempts to demonstrate structural and functional recovery during the developmental period following complete transection of the immature spinal cord, in some cases with implantation of fetal CNS, there are in fact very few reports that this has been done successfully. As indicated in the Introduction, although there is a widespread belief that the immature mammalian nervous system is better able to respond and repair itself following injury than is the case in the adult CNS, almost all of the published evidence reviewed above shows that a complete spinal cord transection, even in an animal as immature as a hamster or a rat at birth, is not followed by any significant structural or functional recovery unless the lesion is only partial and/or implanted with fetal CNS tissue. There are, however, two recently published studies in which a segment of spinal cord (ie a complete transection) in neonatal rats has been removed and replaced with a similar segment of fetal spinal cord (Iwashita *et al.*, 1994; Miya *et al.*, 1997) and both structural and behavioural evaluations were carried out when the animals reached adulthood. There is also a similar study that has been reported for neonatal kittens (Howland *et al.*, 1995b).

Both of the papers by Iwashita *et al.* (1994) and Miya *et al.* (1997) represent a considerable technical achievement, although it is not clear why this seems not to have been possible at any time within the last twenty years, since similar approaches have been used in the same experimental animals but only involving partial lesions (see above). Iwashita *et al.* (1994) point to the lack of success in this field, citing in their short paper as many as ten previous attempts which had failed.

In Iwashita *et al* 's (1994) experiments a lower thoracic segment (T10 to T11) of spinal cord was resected in neonatal (P1 to P2) rats. In experimental animals this lesion was filled with embryonic (E14 to E17) rat spinal cord from the same segment. One puzzling feature of their description of the operation is that they indicate that 1.5 to 2mm of spinal cord was removed in the P1 to P2 animals and an equal length was resected in the embryonic animals notwithstanding the fact that the spinal cord of the embryos would have been much shorter. In about half of the operated animals the implanted embryonic spinal cord was said to have a normal dorso-ventral orientation, whereas in some of the controls this was inverted or the rostro-caudal orientation was deliberately reversed. In a few animals the lesion was filled with grafted sciatic nerve and there was also a number of ungrafted controls. Approximately two thirds of the animals in each group survived and of the 22 of 32 "normal orientation" grafts, 14 showed histological integration of the graft into the host spinal cord when examined once the animals had reached early adulthood. Six of these 14 animals were found, on histological examination, to have dorsal and ventral roots with a reasonable degree of development of both grey and white matter within the transplanted region. The other eight animals in this group had only white matter within the graft. In addition to a morphological description of the lesion area in each group of animals these authors also carried out retrograde labelling studies by injecting "Fast Blue" into the cord caudal to the site of the lesion and anterograde transport by injecting wheat germ agglutinin into the cerebral cortex. Iwashita *et al.* (1994) also report on behavioural studies and some electrophysiology. However the space constraints of a "Letter to Nature" are such that there is not always sufficient experimental detail to be able to adequately interpret the results reported. Thus the evidence from behavioural studies in Iwashita *et al.* (1994) is limited. It consists of a small number of video frames illustrating locomotor function for one experimental animal (correctly orientated graft) and one ungrafted control. In no case can all four limbs be seen simultaneously. It is therefore difficult to judge in the case of the grafted animal, whether there was true hindlimb/forelimb coordination. In the text there is a general statement "The animals with successful replacement exhibited good motor performance with hind/forelimb coordination". However no information is given about variations in performance. It is not clear how many animals were tested and no attempt has been made to quantitate the tests. All animals in the control groups and in what is described as the "unsuccessful" experimental group showed functional deficits of various grades. These were subdivided into paralytic cases (flaccid paralysis of hind limbs with loss of bladder and bowel function) and paretic cases in which walking was abnormal, but presumably no flaccid paralysis of hind limbs or loss of bladder and bowel control was present. Interpretation of rhythmic walking movements is complicated, because the local spinal rhythm generators are capable of producing a spinal cord response, elicited by tactile and proprioceptive sensory input from the limbs. This consists of rhythmic walking movements of both the fore-and the hindlimbs (in an animal with a complete spinal lesion between the fore-and hindlimb spinal innervation) that are coordinated between the two hind limbs and betwen the two forelimbs. But there is no coordination between fore-and hindlimbs in lesioned animals (see Rossignol, 1996). Weight bearing in lesioned animals with no implants was only limited. But it requires very careful, quantitative, behavioural analysis if gait measurements are used, to determine whether animals show evidence of forelimb-hindlimb coordination. On the other hand the fact that in the experiments of Iwashita *et al.* (1994) the animals with "successful" fetal grafts could right themseleves when dropped suggests the development of intact supraspinal innervation of the hindlimbs.

There is a major problem in the interpretation of recovery from lesions in spinal cords that are still developing, which has been mentioned above. In brief, the problem is that it is difficult to distinguish between growth of fibres through the lesion that is occurring as part of normal development, ie the fibres that were not present at the site of the lesion to be injured, as against fibres that were present at the site of injury, and were damaged and subsequently regenerated. Iwashita *et al.* (1994) used retrograde and anterograde labelling in order to demonstrate that there was morphological continuity of axons across the site of the grafted lesion. The authors' interpretation (legend to their Figure 2) was that the labelled fibres seen were formed by developmental growth of uncut fibres. However such a conclusion does not support their final conclusion that "reconstruction of the human spinal cord may not be impossible" since there would be no new undamaged fibres to grow across the lesion in a fully developed adult human patient. The same group (Asada *et al.*, 1997) has recently published a further study comparing structural repair and development in neonatal rats with complete transections of the spinal cord and implants of either fetal spinal cord or peripheral nerve. As in the earlier study (Iwashita, *et al.*, 1994) the morphological repair was substantially better in the case of fetal spinal cord implants, but no additional information on the behaviour of these animals was provided.

Key points

Complete transection and implantation of fetal spinal cord Iwashita *et al.* (1994)

- Neonatal rats; segment of thoracic spinal cord removed; lesion implanted with E15 spinal cord.
- Evaluated as adults.
- Morphological repair and development occured, particularly if dorso-ventral orientation of implant was correct.
- White and grey matter differentiated.
- Behavioural studies claimed to demonstrate forelimb-hindlimb coordination, but evidence limited. Righting when dropped suggests supraspinal innervation.
- Retrograde labelling of brain stem neurons from Fast Blue injection caudal to lesion. Authors interpreted this as due to labelling of new, late arriving axons (see Box on p. 35); this does not support their contention that "reconstruction of the human spinal cord may not be impossible", because such newly growing fibres would not be available in the adult spinal cord.

In the paper by Miya *et al.* (1997) the spinal cord of P1 to P2 neonatal rats was transected at the T8 to T9 level and approximately two segments of spinal cord were aspirated. The lesion was implanted with E14 rat spinal cord which was obtained from both cervical and thoracic regions of cord. Attempts were made to maintain rostrocaudal and dorso-ventral orientation but the authors stated that this could not be assured. These animals were compared with animals that had similar lesions but no fetal transplants and with unoperated controls. The authors carried out substantial and wide ranging behavioural tests and a more limited series of morphological observations on the cords from the animals prepared once the behavioural testing was finished. Completeness of the transections was verified by lack of continuity of neural structures studied in serial sagittal sections. The authors comment on the variability of the performance of the animals, both in the spinal transection group and in the spinal transplantation group. However the overall results showed clearly that the transplanted rats did much better than the transected rats on a variety of tests particularly those requiring forelimb-hindlimb coordination and those requiring descending control. Survival of transplant tissue was identified in all of the animals examined and was taken to be present if neurons stained by MAP2 or could be morphologically identified. The laminar structure of normal spinal grey matter was never seen in transplant tissue, in contrast to the results of Iwashita *et al.* (1994). One unexpected feature of the results of Miya *et al.* (1997) was that the transplant animals showed an improvement in intra limb/inter limb coordination, foot placement and balance on the more complex task of crossing a narrow beam and stairway descent than during the less demanding overground locomotion test on a wide runway.

For some behavioural tests the spinal transection animals were more or less unable to perform, eg wide runway and ladder crossing. In the case of the transplant animals there was a wide range of performance on all tests but only rarely did an animal with a transplant approach the ability shown by the control animals. This is in contrast to the results of the neonatal *Monodelphis* experiments (Saunders *et al.*, 1998, that will be described in a separate section below). There are occasional comments on the relation between the morphological findings and behaviour of individual animals. Thus neither density nor distribution of 5HT positive fibres seemed to correlate with motor performance. There does not seem to have been any systematic correlation between the degree of structural repair and development and the functional status of individual animals.

Two animals that had fetal implants and two that had only transections were subject to retransection of the cord when adult. This led to a considerable loss of function when examined at the end of the first week; however some hindlimb function had recovered and all four re-transected rats were able to make some unassisted weight supporting steps, which two trained rats transected only as adults could not do. Thus after the second transection neonatally operated animals, suprisingly, performed better than animals transected for the first time as adults. Wang *et al.* (1998) also report much better locomotor performance of opossums (*Didelphis*) whose spinal cords had been transected in the first week of life and subsequently retransected when adult, as compared with adults whose cords were transected for the first time.

In Miya *et al.* (1997) all histological preparations were examined by two or more investigators who did not know the motor performance of the animal. The completeness of the transection was assessed by absence of continuity between rostral and caudal stumps in serial sections through the lesion site. In contrast to our opossum (*Monodelphis*) experiments (see below) the performance of the transplanted animals on the ladder (grid) test was rather poor. The rats with implants showed a poor level of motor precision; impaired ability to coordinate sensory information between fore- and hindlimbs was thought to account for the poor performance of these animals on this test, although the animals with transplants did better than the spinal rats in terms of weight support and controlled use of the hindlimbs to cross the grid. The performance of our opossums was distinctly better in showing a high degree of coordination of forelimb and hindlimb locomotion and no significant difference in the number of errors made crossing the grid by control animals and animals whose spinal cord had been crushed in the first week of life.

In the morphological section of Miya *et al.* (1995) there is some detailed consideration of the staining pattern for 5HT fibres. The authors point out that in three of the transplanted animals 5HT positive fibres were found within the transplant and this could account for the 5HT staining found caudal to the transplant in these three animals. 5HT staining in the other animals in the absence of staining within the transplant, was clearly likely to have originated from the host animal. However the published evidence on whether or not there may be local spinal 5HT neurons in the spinal cord or whether the finding of staining below the lesion is evidence of supraspinal innervation does not seem clear cut. If the staining found caudal to the lesion originated from the brain stem it ought to have been possible to find fibre staining within the transplant, even if there were no cell bodies. This is not commented upon.

The authors discuss the mechanisms by which transplants might be producing the greater degree of motor function identified in this group of animals. The most obvious one is of course that the presence of the transplant allowed descending systems to grow fibres across the lesion and similarly ascending fibres may have made connections that provide sensory feedback for coordinated movements. However they also considered the possibility that this minimal amount of functional and anatomical recovery could have been enhanced by compensatory mechanisms within the lumbar spinal cord after injury. They also suggested that a major effect of the transplanted tissue may have been via an indirect or trophic function exerted on the spinal cord acting to keep alive axotomised host neurons.

Miya *et al.* (1997) do not appear to discuss the origin of the fibres that grow through the transplants, ie they do not distinguish between regeneration and developmental growth of new fibres.

In the Howland *et al.* (1995b) experiments, kittens received complete low thoracic spinal cord transections and embryonic transplants within 48 hours of birth. Transplant tissue was obtained from E19 to E45 fetal cats. However only transplant tissue from E20 to E26 survived in the host and enhanced locomotion. The authors were careful to analyse results only from animals in which it was shown that the initial lesion was a complete transection as verified by serial sections, that there was surviving transplant and that there was no gross pathology in the host spinal cord. Eleven of the thirteen animals that fell into these categories were subjected to systematic testing of bipedal treadmill locomotion, quadrupedal treadmill locomotion and overground locomotion. Routine histology of serial sections was carried out using cresyl violet and myelin staining. In addition, immunocytochemistry was used to identify calcitonin gene related peptide (CGRP) and substance P (SP). As would be expected for a purely spinal lesion, bipedal treadmill locomotion was shown both by the spinal kittens and by spinal kittens with transplants. By the 4th to 5th postnatal weeks the animals with transplants tended to show some weight-bearing ability in the hindlimbs which was incorporated into their overground locomotion pattern. Some aspects of the performance of the five animals with the best qualitative overground locomotion were quantitated and compared with animals with spinal transection or with unoperated controls. The five spinal transplant animals, particularly two of them, showed evidence of forelimb-hindlimb coordination that would be likely to have required supraspinal input.

Quadrupedal treadmill locomotion was assessed at fifteen weeks of age, at which time all kittens with transplants were capable of full weight support. Most of the kittens with transplants showed forelimb/hindlimb coordination when examined at twenty weeks of age. But details of only five of these are shown in Table 3 of Howland *et al.* (1995b). The greater variety of patterns used by the spinal kittens in overground locomotion was interpreted as suggesting that there was no forelimb/ hindlimb coordination. Analysis of this paper is made more difficult because

Key points
Complete transection and implantation of fetal spinal cord Miya *et al.* (1997)

- Neonatal rats; removal of segment of thoracic spinal cord; lesion implanted with E14 spinal cord.
- Evaluated as adults.
- Grafts incorporated into spinal cord structure, but morphological repair and development very limited.
- White and grey matter not differentiated.
- More comprehensive and detailed behavioural studies than Iwashita *et al.* (1994). Evidence of forelimb-hindlimb coordination and supraspinal control.
- Wide range of behavioural performance in animals with lesion plus transplant, but generally much poorer than unoperated controls.
- Animals retransected as adults performed better than those transected for the first time as adults.
- 5 HT stained fibres below the lesion interpreted as evidence of supraspinal innervation.
- Factors considered as possible contributions to better performance of animals with fetal implants: descending axons with functionally effective connections; local compenstory mechanisms; trophic effects of factors released by implanted fetal tissue.
- Origin of fibres not discussed.

the details of the behavioural studies of the spinal animals used for comparison with the transplant animals have been published in a separate paper (Howland *et al.*, 1995a).

Morphological analysis of the sites of the lesion containing the transplant showed that the transplants lacked general structure of normal spinal cord. There was no evidence of continuity of white matter tracts into or across the transplant, although some regions of the transplant resembled normal dorsal horn. In addition to these areas of characteristic grey matter the transplants contained some lightly myelinated fibres and multiple central canal-like formations lined by cuboidal ependymal cells. Spinal segmental systems were identified by CGRP and SP immunoreactivity. CGRP immunoreactive cell bodies were present in the neuron dense transplants. In normal spinal cord these were either motor neurons within the spinal cord or dorsal horn neurons at a cervical level. The authors reported that the majority of CGRP immunoreactive fibres appeared to be of host dorsal root origin because they could be followed as they crossed the host transplant interfaces but it is not clear whether they were able to trace these fibres from the remote cervical levels down to T12 where the transplant was placed. Supraspinal 5HT and noradrenergic axons were identified within the transplants, although the number of preparations in which they were identified is not indicated. The results section also states that the descending 5HT immunoreactive fibres not only grew extensively into the transplant, but also into the spinal cord caudal to the transplant as far as the L5 segment. However from the summary of the paper it is apparent that this occurred in only one case. Thus the interpretation that the greater locomotor behaviour described in the transplanted animals compared with the spinal animals was due to this fibre growth seems doubtful. The relation between the degree of morphological repair and growth and the degree of behavioural function is not clear. Some of the transplants were described as neuron-sparse but contained immunocytochemically labelled fibres; however the behavioural evaluations indicated that the neuron-rich transplants and the neuron-sparse transplants affected locomotion equally well (see below).

The authors (Howland *et al.*, 1995b) discussed three mechanisms which may have accounted for the recovery or enhanced function mediated by the implanted embryonic spinal cord: (1) the transplant may serve as a source of trophic support to rescue injured neurons that would otherwise die following axotomy; (2) the transplant may function as a relay with descending host axons forming synapses within the transplant and thus connecting with neurons that send axons into the host spinal cord; (3) the transplant may function as a bridge which is crossed by ascending or descending host axons. The authors suggested the development of locomotion mediated by transplants may have been due to a combination of these mechanisms. They suggested that behavioural results indicated that functional connections were established across the transplant lesion site. However, as indicated above, the morphological evidence to support this is limited. It seems that in only one case were (5 HT and noradrenergic) fibres demonstrated to have penetrated into the spinal cord caudal to the lesion. If the authors had also used silver staining methods to pick up all the axons present within their transplants, there might have been a better opportunity to identify additional populations of fibres which had crossed into the spinal cord caudal to the site of the transplant. Additionally, retransection of the spinal cord at the site of the transplant would have given a clear indication of the extent to which the demonstrable locomotor behaviour depended upon fibres crossing the transplant. The proposal that the transplants may have acted primarily as a bridge rather than a relay is based on the finding that behaviour was similar in the neuron-dense and neuron-sparse transplants. However the lack of evidence that fibres actually crossed the spinal transplants, except in one case, leaves this interpretation in doubt.

The degree of both structural repair and functional development reported for these three experiments (Iwashita *et al.*, 1994; Miya *et al.*, 1997; Howland *et al.*, 1995) was quite different. This may have been due to differences in surgical technique, possibly resulting in less inflammatory response in the operations of Iwashita *et al.* (1994) who report the most favourable repair and behaviour. Fitch & Silver (chapter 4 this volume) stress the importance of the local inflammatory response in inhibiting neurite outgrowth from damaged axons. It may also be that the marked differences in repair and behaviour for fetal spinal cord and peripheral nerve implants reported by Asada *et al.* (1998) may be explained by a difference in the inflammatory reponse to these different grafts,

Key points

Complete transection and implantation of fetal spinal cord Howland *et al.* (1995)

- Neonatal cats, low thoracic spinal cord lesion implanted with E20 -E26 spinal cord.
- Evaluated as adults.
- Grafts incorporated into spinal cord structure, but morphological repair and development very limited.
- White and grey matter not differentiated.
- More comprehensive and detailed behavioural studies than Iwashita *et al.* 1994. Evidence of forelimb-hindlimb coordination and supraspinal control. Quantitative behavioural results suggested differences between animals with spinal transection and implant and transection alone; however, behavioural results from only 5 of 13 transplant animals that met morphological criteria for inclusion in study were used.
- Wide range of behavioural performance in animals with lesion plus transplant, but generally much poorer than unoperated controls.
- 5 HT, CGRP and SP stained fibres identified in transplant; considered to be of host origin. In one animal 5 HT fibres below the lesion interpreted as evidence of supraspinal innervation.
- Factors considered as possible contributions to better performance of animals with fetal implants: transplant may act as a bridge for ascending and descending axons; transplant may act as relay; trophic support of axotomised neurons by factors released from implanted fetal tissue.
- Origin of fibres not clear.

particularly since the peripheral nerve grafts were from 2 week old animals, whereas the spinal cord grafts were from fetuses that might be less likely to provoke a local inflammatory response.

Regenerating or newly growing, late arriving axons?

Determining the origin of axons growing around a lesion (or through it if the spinal cord is sufficiently immature) is extremely important. If all such fibre growth were due to new growth of axons not present at the site of injury at the time when it was made, but which subsequently grew as part of normal development, then this would represent an interesting degree of plasticity in the developing nervous system. However, it would not be a useful system in which to study regeneration. Nevertheless even if all of the axon growth were due to newly developing axons this would have other implications for the prospect of succesful spinal repair. One problem frequently overlooked in the literature, is that even if there is success in getting axon growth across a spinal lesion in the injured adult spinal cord (see chapters by Bunge and by Harvey) those axons still need to reach appropriate targets if they are to make a useful contribution to functional recovery following a spinal injury. The finding that this can occur in the injured immature spinal cord provides a valuable system for studying this successful innervation in a preparation in which it happens independently from the local inhibitory effects at the site of the injury (cf chapter 4 by Fitch & Silver).

So far, the main approach to distinguishing between regenerating and late arriving axons has involved the use of injection of two distinguishable dyes at different times with respect to the time of spinal cord lesion. The first dye is injected a few days before or at the time of lesioning the cord at a site caudal to the lesion. Sometime later (usually 30 days or more) a second dye is injected into the cord, caudal to the lesion. The dyes are taken up by the axons in the cord at the level of the dye injection and transported by retrograde axonal transport to the cell bodies of the parent neurons. Only neurons with axons present at the site of injury when the lesion is made will be labelled by the first dye. Late arriving axons will only label their parent neurons with the second dye. Neurons whose axons were present at the time of injury and subsequently regenerated new axons that reached the site of injection of the second dye will be labelled by both dyes. However, it is not straightforward to carry out these experiments and to interpret the results, as has been outlined by Bates & Stelzner (1993) and Hassan *et al.* (1993). A major problem is ensuring that the first dye is only present in the spinal cord for a limited period, otherwise the later arriving fibres will also take it up and give rise to double labelled neurons that would be interpreted as having regenerated. The dye most commonly used for this experiment (Fast Blue) is known to remain in tissue for at least several weeks. Therefore it is important that effective steps are taken to remove it soon after the injured axons have been labelled. Some studies (eg Bernstein-Goral & Bregman, 1993) have involved simultaneous injury and dye injection with removal of the dye by gentle suction of the part of the cord containing the dye. However, a problem that cannot be dealt with entirely satisfactorily is whether or not the removal of dye was fully effective. Bernstein-Goral & Bregman (1993) examined serial sections for the presence of Fast Blue at the time (usually 30 days after spinal cord lesion) when structural studies of the spinal cord were carried out. Their experiments involved partial ("overhemisection") lesions of the spinal cord in rats; the site of the lesion was implanted with fetal spinal cord. Preparations with evidence of Fast Blue in the spinal cord were excluded from the analysis. However this does not deal with the possibility that lesser amount of Fast Blue might have been present for a shorter period (perhaps being gradualy reduced to undetectable levels by macrophage activity).

Bates & Stelzner (1993) carried out a similar study in postnatal (P0-11) and adult rats in which Fast Blue was injected into one side of the cervical spinal cord and which was aspirated two days later to remove the dye and create a partial spinal cord lesion. Bates & Stelzner (1993) applied rigorous criteria for inclusion of preparations in the final analysis of their results: (1) the dorsal funiculus had to be completely severed (2) the Fast Blue had to have been completely removed from the spinal cord (3) the second tracer (diamadino yellow) had to be confined to a site caudal to the lesion. As pointed out with respect to the paper of Bernstein-Goral & Bregman (1993), it was not possible to be sure from examination of sections at 30 or more days after Fast Blue injection that some Fast Blue had not been removed by suction. Bates & Stelzner (1993) dealt with this problem by checking that their suction techniqe was effective. This was done by examining some preparations two days after removal of the dye to ensure that removal was indeed complete. They did the further control of injecting the cord with Fast Blue and then at two days post injection the cord was completely severed thereby preventing the growth of any fibres across the lesion and picking up any residual Fast Blue. This control showed that the Fast Blue found in brainstem neurons could only have originated from the original injection and further showed that these neurons did not die as a consequence of axotomy. One consequence of the rigid criteria applied by Bates & Stelzner (1993) was that only 13 of the original 45 experimental animals could be included in the final analysis. Of these only four were animals that had double-labelled (ie regenerating) neurons. Three of these were injected on P2 and lesioned on P4. The fourth was injected on P4 and lesioned on P6. The results are presented and discussed in terms of the percentage of cells double labelled when compared with those labelled with diamidino yellow (late arriving). However this is a bit misleading because Table I in their paper shows the actual numbers of cells that had been double labelled to have been very small: only 2 cells in two of the preparations. Also it is not clear why the results are presented as a percentage of the neurons with late arriving fibres, since these were not available to be labelled by the first dye. It would be more relevant to calculate how many of the Fast Blue neurons were double-labelled since these were the neurons present at the time of the lesion with fibres that could have regenerated. This can be calculated from Table 1 in Bates & Stelzner (1993) as 0.07, 1.1, 0.4 and 4.3%. However, as pointed out by Bates & Stelzner (1993), the proportion of double labelled cells is likely to have been an underestimate because diamadino yellow is a nuclear stain and therefore not all of the Fast Blue stained cells (in which the cytoplasm is stained) would have contained a nucleus in the plane of section when counted. This could have been avoided by only counting Fast Blue cells in which a nucleus could be seen, but this in turn would have resulted in an underestimate of Fast Blue cells. There does not seem to have been an attempt to quantitate the error involved in using labels that stain different parts of the cell. Nevertheless, the evidence presented by Bates & Stelzner (1993) appears to be some of the best available in support of regeneration *in vivo* following injury to spinal cord axons. It is supported by the studies of Bernstein-Goral & Bregman (1993) and Xu & Martin (1991) and also by a similar double labelling approach in the chick embryo (Hasan *et al.*, 1993). These last authors found that up to about 30% of brainstem neurons were double labelled. Like Bates & Stelzner (1993), Hasan *et al.* (1993) also used the number of cells labelled by the second marker as the baseline for estimating the proportion of double labelled (regenerating) neurons. They did this because the number of neurons labelled with the first dye increased with age between E10 and E13. However, unlike the results of Bates & Stelzner (1993), Hasan *et al.* (1993) found less difference in the numbers of cells labelled singly by each dye; thus the estimated proportion of double labelled cells was less influenced by which comparator was chosen. Hasan et *al.* (1993) used an additional control to show that the first dye was not present in the cord for sufficient time to be taken up by later arriving fibres. This control involved complete section of the cord at the time of injecting the first dye, so that it could not be retrogradely transported to brainstem neurons. Following the injection of the second dye in the older animals after repair and recovery had occurred, none of the first dye was found in brainstem neurons, presumably because it had all disappeared from the cord by the time neurites had crossed the site of the spinal lesion.

The only direct evidence for outgowth of regenerating spinal cord neurites comes from the *in vitro* studies of Nicholls and his collegues who injured axons directly and followed their susequent growth using timelapse video microscopy (Varga *et al.*, 1996 and see below).

Key points

Regenerating or newly growing, late arriving axons?

- At time of injury in immature spinal cord some tracts will not have developed and thus are not present at site of injury (newly growing, late arriving, axons).
- Axons growing across complete lesion of spinal cord in neonatal marsupial or eutherian implanted with fetal spinal cord could be regenerating or newly growing, late arriving axons.
- These two origins of neurites crossing an injury site are distinguished by a double labelling technique.
- One retrogradely transported dye is injected just before or at the time of spinal cord transection. A second dye is injected some weeks later. The first dye marks neurons with axons present at the site of injury when made. The second dye marks neurons that only sent axons through the site of injury some time after it was made. Double labelled neurons had axons at the site when injury was made and subsequently regenerated axons that picked up the second dye as well as the first.
- Results from several labs suggest that regenerating fibres do contribute to those crossing the site of injury; the numbers of regenerating axons appear to depend on the tract and may be quite small.

How immature is immature?

One of the problems with experiments in which the spinal cord of neonatal eutherians (eg rats, cats, hamsters) has been transected is that however early it has been done in the newborn period no fibre growth through the lesion has been demonstrated. This appears to be a function of the stage of development of the spinal cord, since such fibre growth across the lesion has been demonstrated in chick embryo spinal cord (Hasan *et al.*, 1991, 1993; Shimzu *et al.*, 1990) and in embryonic rat spinal cord in culture (Saunders *et al.*, 1992). In these preparations such fibre growth does occur straight through the lesion. In the chick embryo experiments, complete spinal transections were made *in ovo*. Providing the lesions were made early enough (E10–E13) when the chicks hatched, detailed morphological and behavioural studies showed that they were indistinguishable from sham operated controls. Hasan *et al.* (1993) used the double labelling approach described in the previous section and concluded that approximately 20–30% of the fibres that crossed the lesion had regenerated following transection. The so far unanswered, and probably currently technically unanswerable question, is to what extent did the regenerating fibres in these and the mammalian experiments discussed in the previous section, contribute to the functional recovery and development observed?

Recovery from complete spinal cord transection in neonatal marsupials

Growth of axons directly across the lesion

It is clear from the papers discussed above that in eutherian mammals, although neurons that project to and from the spinal cord are capable of growing neurites (some of which are regenerating from damaged axons) growth occurs round the lesion and never across it, unless implanted with a favourable bridge (fetal CNS or peripheral nerve). However, the chick embryo experiments of Shimidzu *et al.* (1990) and Hasan *et al.* (1991, 1993) suggested that if it were possible to injure the spinal cord at ages earlier than the neonate in cats and rats then successful fibre growth across the lesion might be achieved. Early attempts to do this *in utero* were largely a failure (see above and reviews by Windle, 1965; Guth, 1975) although a few brief reports of success were published (eg Migliavacca, 1928). The first unequivocal demonstrations that fibres will grow across a lesion in an immature mammalian spinal cord were those of Treherne *et al.* (1992) in neonatal *Monodelphis* and Saunders *et al.* (1992) in fetal rat. Both of these studies were done in isolated CNS preparations as originally developed by Nicholls *et al.* (1990) for studies of the development of the *Monodelphis* CNS under well controlled conditions.

Growth of neurites across a lesion in immature spinal cord *in vitro*

In addition to demonstrating for the first time that immature spinal cords are capable, when injured, of growing neurites across the site of a complete spinal cord transection (Saunders *et al.*, 1992; Treherne *et al.*, 1992) it was shown that fibre growth occurred across the lesion within four to five days. These preparations provide an ideal situation in which to study and manipulate (for example with growth factors) the early stages of the process of recovery from injury. Opossum CNS preparations have been maintained for periods in excess of 2 weeks (Fernandez & Nicholls, 1998). That this was not some peculiarity of marsupials was shown by Saunders *et al.* (1992) who carried out similar experiments in fetal (E15–16) rat spinal cords *in vitro,* with the important difference that the rat preparations require a higher culture temperature (a reflection of the different body temperatures of these two species). Later stages of recovery and behavioural development and recovery cannot of course be studied. However that has now been shown to be possible *in vivo* in neonatal opossums (see next section). Subsequent studies *in vitro* have investigated the growth of individual neurons following a lesion, the recovery of impulse conduction across the lesion and the formation of new synapses (Varga *et al.*, 1996). These preparations are ideal for such studies and could also be used for investigating the effects of neurotrophins, adhesion and guidance molecules under well controlled conditions that would provide direct access to the CNS rather than having to devise means of circumventing the blood-brain barrier in *in vivo* experiments.

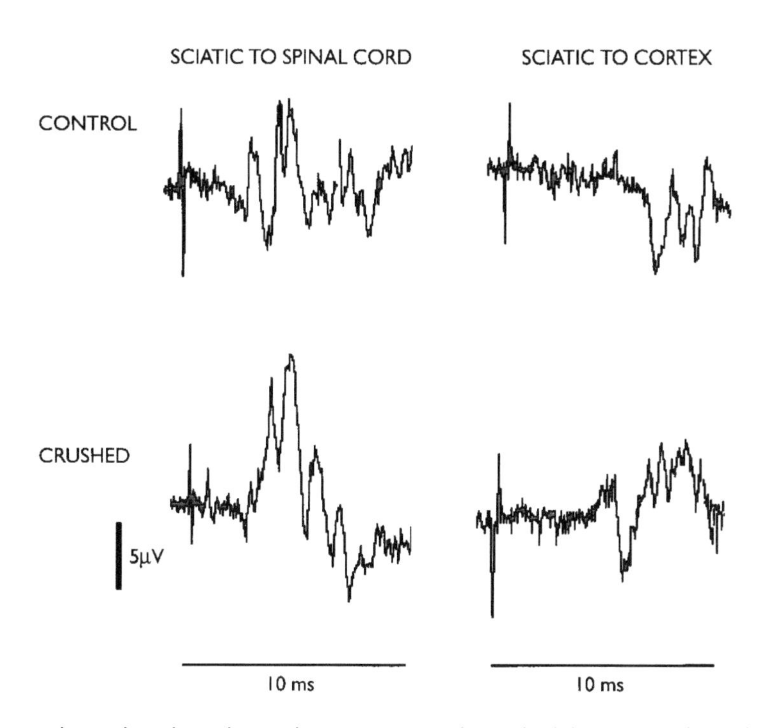

Figure 1 Electrophysiological recordings in an anaesthetized adult *Monodelphis* with spinal cord crush made at P7 (c) and (d) compared with an anaesthetized control adult (a) and (b). In (a) and (c) the stimulating electrodes were placed on the sciatic nerve with recording electrodes on the spinal cord. In (b) and (d) the recording electrodes were moved to the sensory cortex. Similar records were obtained in all 5 operated and 3 control (unoperated) animals that were tested for impulse conduction across the site of the crush lesion. From Saunders *et al.* (1998) in which experimental details can be found.

Structure and function in adult opossums following complete neonatal spinal transection

Saunders *et al.* (1995, 1998) and Fry & Saunders (1999) have carried out comprehensive morphological, electrophysiological and behavioural studies in neonatal *Monodelphis* of their recovery and development following a complete spinal transection. The spinal cords were cut or crushed under sodium pentobarbitone or halothane anaesthesia of both mother and young, under sterile condidtions. At different periods after operation the behaviour of the animals compared with unoperated controls of the same age, was observed and videod. At 3 months, 6 months and 2–3 years of age the animals were studied using comprehensive quantitative methods of behavioural analysis, including footprint analysis, grid crossing, climbing and swimming tests. Some animals at each age were killed and prepared for morphological examination. At 2–3 years of age some of the animals were anaesthetised and conduction of impulses was studied by conventional electrophysiological techniques. These showed that conduction from periphery to cortex and in the reverse direction was present (Figure 1). Morphological examination showed that particularly after a crush lesion (Figure 2), substantially normal structure

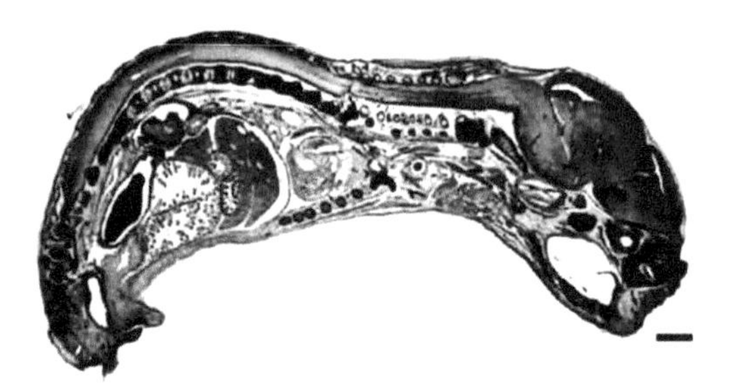

Figure 2 This figure shows a typical example of a spinal cord transection made in neonatal *Monodelphis* This section was prepared from a P7 animal that was terminally anaesthetized within one hour of operation and the spinal cords removed for morphological examination after fixation in Bouin's solution and paraffin embedding. Bar is 1 mm. Modified from Saunders *et al.* (1998).

Figure 3 Longitudinal (horizontal) (a-c) silver stained sections of spinal cords of 3 month old *Monodelphis* at the level of a complete crush lesion made at P7-8. (b) is from a control and (a) and (c) are from operated (crush) animals. The rostral end of the spinal cord is uppermost. All 3 sections are at same magnification. Bar = 1mm. Lesions were made at T1-2. Modified from Saunders *et al.* (1998).

was present at the site of the lesion when the animal had become adult (Figure 3). There was less structural repair following section of the cord by cutting. This was probably because, although the cord was completely severed by both techniques, the degree of separation of the cut ends was likely to have been greater and more variable than after crushing. Also, no pia, archnoid or dural membrane was left intact in the case of cutting, whereas following crushing, even although those layers are thin and immature in the neonatal opossum, there may have been enough present to assist the process of neurite growth across the lesion. Nicholls and his colleagues (Varga *et al.*, 1996) have shown that *in vitro* fibre growth occurs along the pial layer on the outer part of the cord, rather than being guided by radial glial processes that are present at the site of the lesion. In the cut *in vivo* preparations, any delay in the first fibres crossing the lesion would mean that fibres that did eventually cross would encounter a less favourable environment for growth than the earlier entering fibres in the crushed preparations.

The behavioural abilities of these animals that had received complete spinal cord lesions in the first week of life were remarkable. They could walk (Figure 4), cross grids (Figure 5), swim (Figure 6) and climb (Figure 7). Most of the formal behavioural tests did not show any significant differences between controls and animals with crushes (Figure 8). The behavioural analysis of animals with cut spinal cords in this series is not yet complete. Initial findings suggest that the behavioural tests show much more variation and that this may correlate with the degree of structural repair (Fry & Saunders, 1999). Neverthless an impresive degree of approximately normal behaviour was present in most animals that had received cut lesions, notwithstanding that the morphological studies suggest that the structural repair was much less than in animals with crush lesions (Fry *et al.*, 1998 and Fry & Saunders, 1999).

Substantial studies by George Martin and his group in the North American opossum *Didelphis virginiana* also showed that lesions made in the neonatal period were followed by fibre growth directly across the lesion. This group had earlier shown that, following partial lesions made after about one month of age, fibres grew around the lesion through the undamaged spinal cord but not directly across it (Xu & Martin, 1991). Unlike most of the studies in neonatal cats and rats, Martin's group has made extensive use of retrograde and anterograde labelling methods as a means of not only tracing fibres across the site of the lesion but also determining their cellular origin. One minor technical point which may be worth noting when comparisons are being made between the work of this group and those of others, particularly for animals in the first week of life, is that the Martin group designates the day of birth as P1 whereas most other groups, including ourselves, designate the day of birth as P0. All ages in this chapter are designated with respect to the latter convention. In one of the first documented studies of fibre growth across the lesion in neonatal opossum spinal cord *in vivo*, Wang *et al.* (1996) transected the thoracic spinal cord at about T7 on a postnatal days 4, 11, 19, 25 and 32. Thirty to 40 days later the mother and young were reanaesthetised, the spinal cord of the young was exposed and injected bilaterally with the dye Fast Blue approximately four segments caudal to the site of transection. Fast Blue injections were also made at the same spinal cord level in five age matched controls. The brains and spinal cords were processed for morphological examination and the brain stem sites and numbers of labelled neurons were identified and counted. For animals whose spinal cords were transected at P4 the lesion site was barely

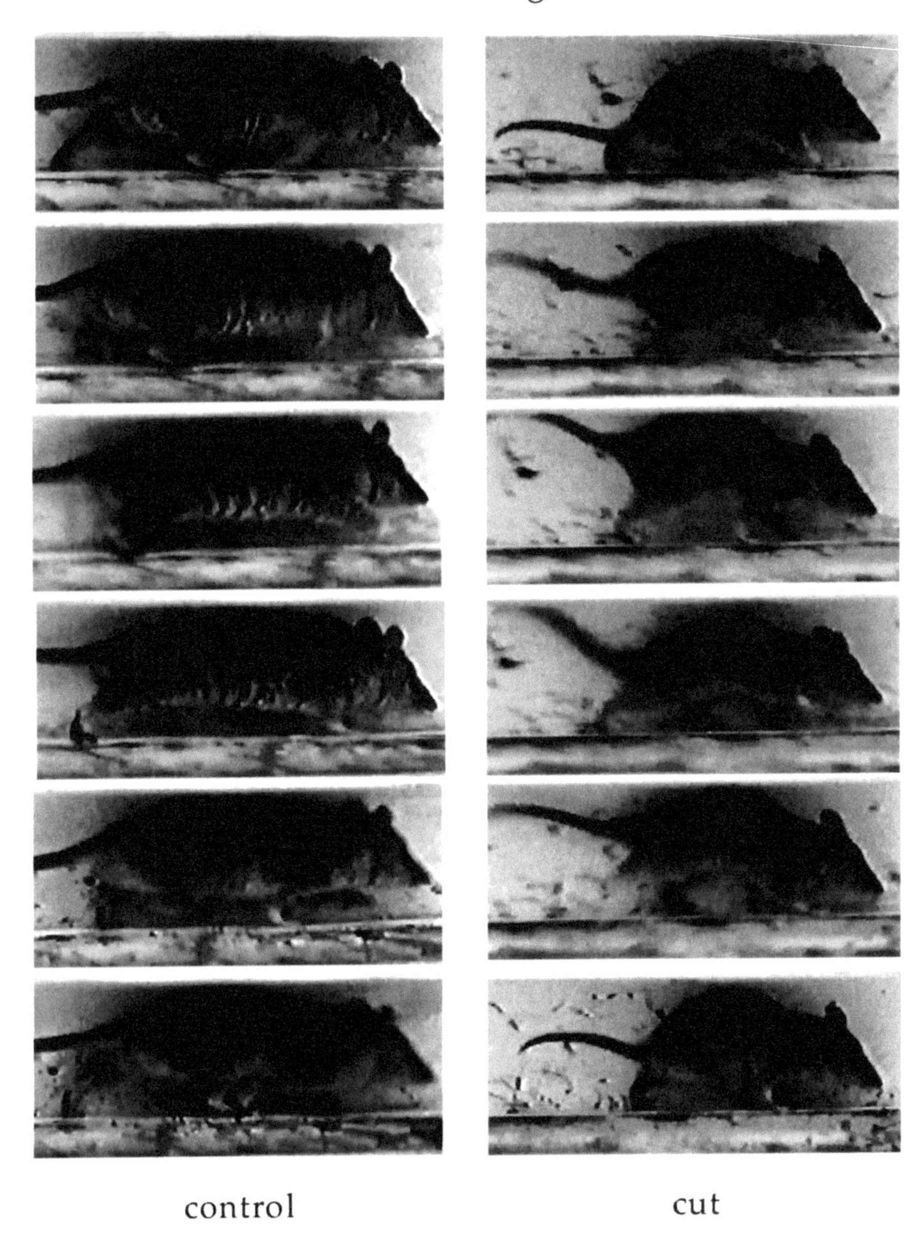

Figure 4 Shows control and operated (cut) animals (young adults) walking in a runway. Cut animals had its spinal cord sectioned at T2-3 at age P7. Walking patterns were quantitatively assessed by the use of footprint analysis. Data and Figures kindly supplied by Mr. M. Lane.

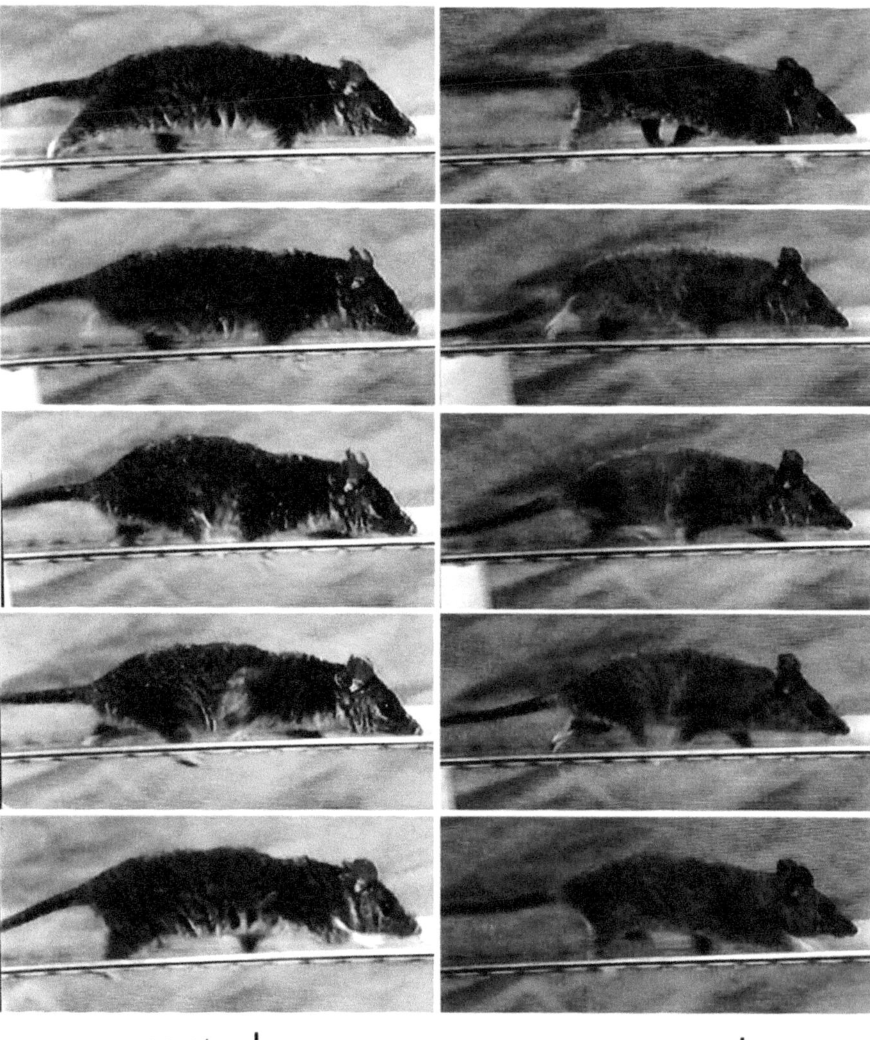

Figure 5 Shows consecutive frames taken from video recordings of crossing a 1.5 cm grid in an adult *Monodelphis* whose spinal cord had been completely crushed at P7 (operated) compared with an unoperated control adult. The animals crossed a grid with bars 1.5 cm apart. The time taken to cross and the number of errors made were recorded (see Figure 8). Modified from Saunders *et al.* (1998) with additional frames.

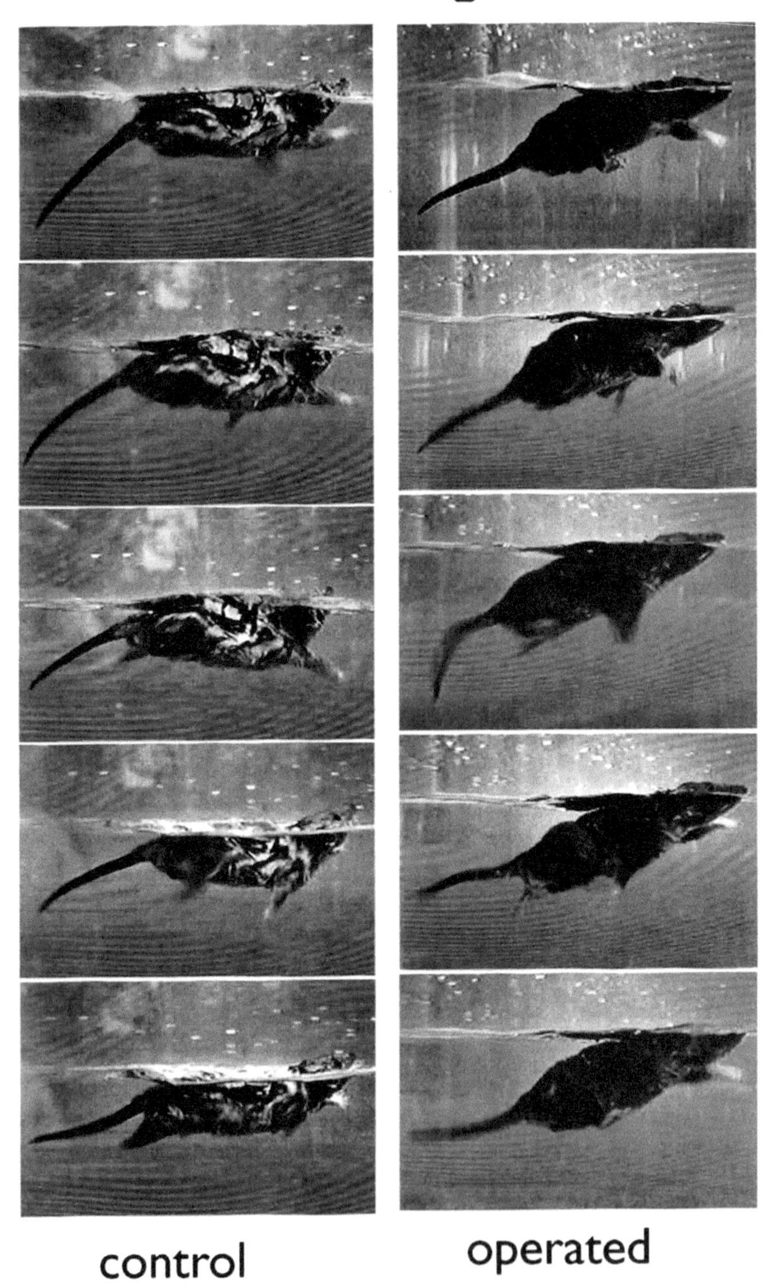

Figure 6 Shows consecutive frames taken from video recordings of swimming in an adult *Monodelphis* whose spinal cord had been completely crushed at P7 (operated) compared with an unoperated (control) adult. Animals made 7 timed swims in a 1.2 m tank (see also Figure 8). Modified from Saunders *et al.* (1998) with additional frames.

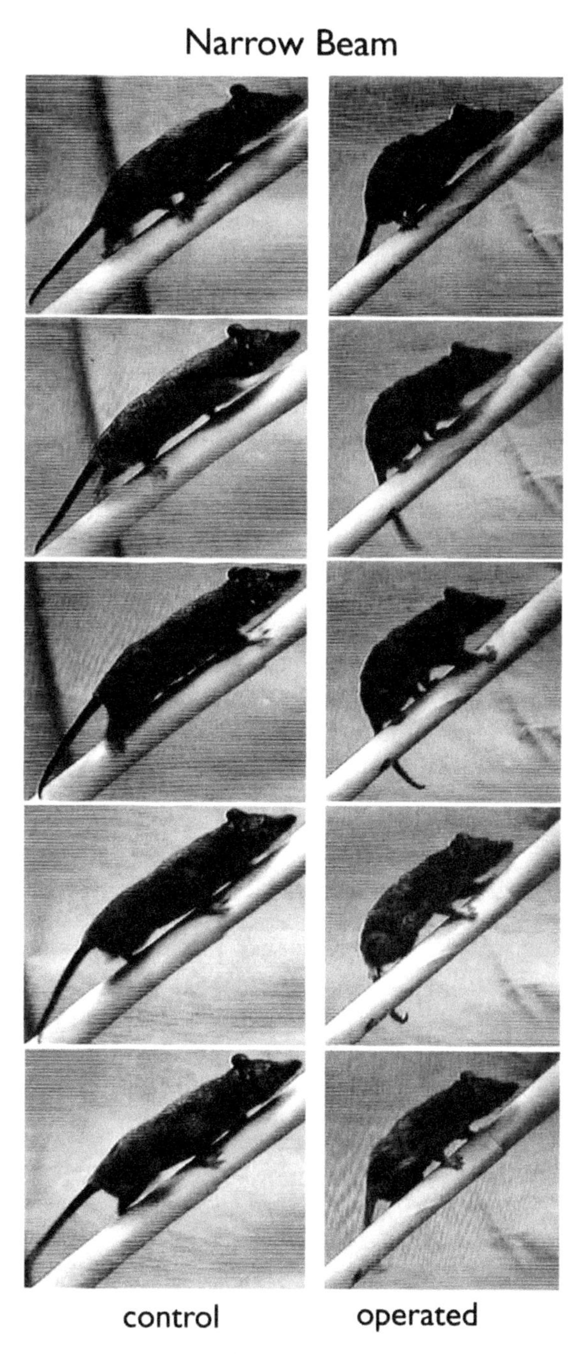

Figure 7 Shows consecutive frames taken from video recordings of climbing in an adult *Monodelphis* whose spinal cord had been completely crushed at P7 (operated) compared with an unoperated (control) adult. The animals climbed a narrow beam inclined at about 45 degrees to the horizontal. The time taken to climb and the number of errors made were recorded (see also Figure 8). Modified from Saunders *et al.* (1998) with additional frames.

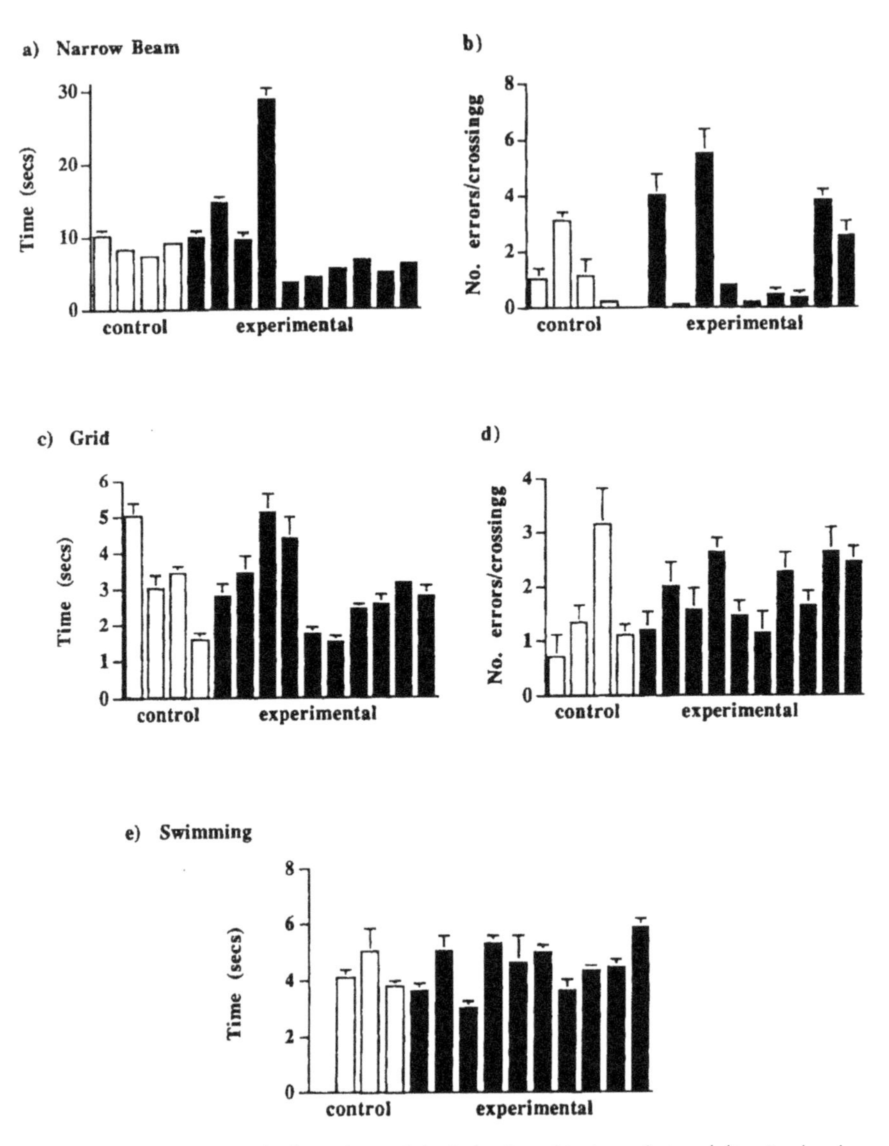

Figure 8 Quantitative results from three of the behavioural tests made in adult animals whose spinal cords had been crushed at P7.
(a) and (b) Climbing a narrow beam (illustrated in Figure 7). The time taken (a) and number of errors (b) are shown for individual operated (filled bars) and control (open bars) animals. Means from 10 climbs, vertical bars are one SEM; where no bar is shown it was too small to be visible. (c) and (d) show results from individual animals crossing a 1.5 cm grid (illustrated in Figure 5). Mean times taken (c) and number of errors (d) in 10 trials are shown; vertical bars are as for (a) and (b).
(e) shows the time taken to swim a standard distance (illustrated in Figure 6). Means of 7 trials; vertical bars are as in (a) and (b).
Note that for these tests 4 controls were used except in the swimming test in which n = 3, because one of the controls was a non-swimmer. Data from Saunders *et al.* (1998).

discernible grossly. On histological examination it was found to contain spinal cord tissue albeit rather disorganised. One of the more obvious abnormalities was fusion of the dorsal horns. Numerous Fast Blue-containing neurons, which were identified at supraspinal levels in nuclei of animals with spinal lesions were similar in number and distribution to those labelled by comparable injections in the age matched non-lesioned controls.

For animals transected at P11 and examined at thirty days or more of age, continuity was present at the site of the lesion, but the site of transection could be easily identified. Recognisable spinal cord was not present histologically. Fast Blue injections into the spinal cord caudal to the lesion, did give rise to labelling in brainstem nuclei but the number of neurons labelled was much less than in the experiments in which the spinal cord transection was made at P4. For spinal cord transections made at P19, a thin area of continuity was present at the lesion site, but no recognisable spinal cord was present histologically. Fast Blue injections labelled fewer cells than at P4 or P11. For a spinal transection at P25 only a thin region of continuity was present at the lesion site 30 days later and again no recognisable spinal cord was seen at the site of the lesion. Fast Blue injection labelled only an occasional brain stem neuron. Transections made at P32 followed by spinal cord injections of Fast Blue at 32 to 35 days later there was no evidence for supraspinal labelling. Thus in the very early neonatal period (P4) there was a substantial growth of fibres across the lesion from supraspinal neurons. The amount of this growth decreased substantially in the subsequent month and was not apparent for lesions made 32 days after birth. Wang *et al.* (1998) discussed whether or not the fibre growth across the lesion could have contributed to hind limb function in similarly lesioned South American opossums reported by Saunders *et al.* (1994, 1995) and also observed in North American opossums in Martin's laboratory. Wang *et al.* (1998) suggested that the published studies do not establish whether or not hindlimb function could depend on growth of supraspinal axons across the lesion and suggested the alternative that it could be due to a compensatory mechanism in the isolated lumbar sacro-spinal cord. However this does not take account of the fact that Saunders *et al.* (1995) reported not only near normal walking in the neonatally transected opossums but that they were also able to climb, a function that would not be possible without vestibular input. The more detailed studies of Saunders *et al.* (1998) confirmed the climing ability of adult *Monodelphis* whose cords had been transected at one week of age and also showed that they could cross grids. These two functions would require effective vestibulospinal conections. The fact that the animals could swim demonstrates that the rhythmic walking movements which these animals showed, were not due to local spinal rhythm generators within the cord (cf Rossignol, 1996). The importance of this test is illustrated by a recent finding in one of our animals whose spinal cord had been cut at one week of age. When adult this animal had some ability to walk, but when swimming used only its forelimbs; in contrast when climbing out of the swimming tank, as soon as its hindlimbs touched the vertical wall of the platform, the animal was able to use its hind limbs. Presumably the caudal rhythm generator was stimulated by the sensory input consequent upon the hind feet touching the side of the platform.

Wang *et al.* (1996) also discussed the problem of the extent to which the fibre growth they described could have been due to regeneration of cut axons as against the normal growth of new fibres in the developing spinal cord that would not have been present at the time when the transection was made. They cited evidence from the chick embryo experiments (Hasan *et al.*, 1993) and dealt with this problem directly in the opossum in a separate paper (Wang *et al.* 1998b). Wang *et al.* (1997) have also studied the growth of axons of the fasciculus gracilis tract across a lesion made at different ages in *Didelphis*. These neurons are also able to grow neurites directly across a lesion but the process ceases earlier in development than for the descending supraspinal tracts. In these experiments the problem of the origin of the fibres (regenerating or late ariving) was not studied. From the developmental timetable of growth of the fasciculus gracilis Wang *et al.* (1997) suggested that regeneration was likely to contribute to the fibre growth, but could not exclude the possibility that it was due to late arriving fibres.

Wang *et al.* (1998a; this paper can be consulted for references to other morphological studies of spinal repair that are not dealt with here) have carried out extensive studies of locomotion in adult *Didelphis* given complete cut transections of the spinal cord at P4. They used the comprehensive BBB (Basso *et al.*, 1995) scale and showed that the operated animals had a substantial degree of normal locomotion when compared to unoperated controls. A few animals were re-transected when adult and compared to adults transected for the first time. There was a substantial loss of locomotor function in the retransected animals when their pre and post re-transection locomotion was compared. This indicated that growth of supraspinal axons contributed to the adult behaviour prior to retransection. However, remarkably, the performance of the retransected animals was subsantially and significantly better than the animals transected for the first time as adults. It is a pity that the behavioural tests were not more comprehensive, including for example swimming and climbing tests. A swimming test would have allowed the possibility to distinguish the importance of local sensory input to the spinal rhythm generators as opposed to other interconnections by fibres that had crossed the lesion and which would have been interrupted by the second lesion. Climbing would have directly tested supraspinal control which is required for balance. However the finding that retransected animals had better function than adults transected for the first time as adults suggests, as discussed by Wang *et al.* (1998a) that substantial reorganisation of local spinal circuits takes place following transection in the newborn period.

Determinants of axon growth following injury in the immature spinal cord

Most of what has been dealt with above has been concerned with a review of key experimental evidence that dealt with the problem of fibre growth and functional development and recovery following injury in the immature spinal cord. The factors which determine the extent of axonal growth across a complete lesion and subsequent functional recovery and development can be summarised as:

(1) Intrinsic growth potential of axotomised and late arriving fibres.
(2) Appropriateness of the pathway through which the fibres should grow, both through the lesion itself and in uninjured cord away from the site of the lesion.
(3) Achievement of functionally effective innervation of targets in the spinal cord (descending fibres) and brain (ascending fibres).

Key points

Spinal repair in the completely transected neonatal marsupial spinal cord

- Complete crush of the spinal cord of the neonatal opossum (*Monodelphis domestica*) *in vitro* was followed by axon growth and action potential conduction across the lesion, within a few days of injury.
- Structural and functional evidence that this early repair included synapse formation has been obtained.
- A proportion of the axons crossing the lesion was shown to be regenerating from damaged axons.
- The axons grew directly through the lesion without support of fetal implants.
- This is not a purely marsupial phenomenon since simialr results have been obtained with fetal (E15) rat spinal cord *in vitro*. Thus it is immaturity that is important.
- *In vivo* in the same species (*Monodelphis*) a complete crush of the thoracicic spinal cord was followed by substantial morphological repair and development. When the animals reached adulthood, their behaviour was barely detectably different from unoperated controls, even with detailed quantitative testing.
- Following a complete transection by cutting, the degree of morphological repair was substantially less than following crushing, probably because of delay in the time for axons to cross the lesion.
- Behaviour in animals following cutting was remarkably normal, but less so than in animals with crush lesions.
- Similar experiments have been carried out in *Didelphis viginiana* using cut lesions. More comprehensive morphological studies defined the developmental period in which growth of axons across the lesion and subsequent recovery was possible. This was different for different tracts.
- Behavioural analysis suggested forelimb-hindlimb coordination.
- In both species retrograde labelling demonstrated the neuronal populations in the midbrain and brainstem that contributed to the axons crossing the lesion.
- Double labelling experiments in *Didelphis* indicated that a significant proportion of the axons crossing the lesion was regenerating.

In a general sense the studies using neonatal opossums by the groups of Saunders (Saunders *et al.*, 1995, 1998; Fry & Saunders, 1999) and Martin (eg Wang *et al.*, 1997, 1998) provide some answers to these points. The finding that these immature marsupials recovered both structurally and functionally to a remarkable extent following a complete transection (particularly when by crushing) of the spinal cord in the first week of life, indicates but does not prove that immature neurons probably have a greater ability to send out long processes than later in life. The near normal function exhibited by these animals when they became adult suggests that the outgrowing neurites were able to follow appropriate paths or at least found a route to an appropriate target. The fact that the growing neurites, whether originating from damaged axons by regeneration, or as new, later arriving fibres, found functionally effective targets, gives great hope for eventually finding ways of stimulating substantial growth of fibres not only across a lesion, but also for those fibres to reach appropriate targets in the adult spinal cord. The finding that function was essentially normal suggests that further study of recovery in postnatal opossums may uncover ways of promoting functionally effective fibre growth in the injured adult spinal cord. The anatomical repair that followed cutting was substantialy less than following crushing, which also suggests that recovery may be achievable if only a fraction of the normal complement of fibres grow across the lesion. This adds to the earlier finding of Blight (1983) who used contusion injuries in adult rats and found that a substantial degree of normal function was retained, even when only about 5–10% of axons remained uninjured. In the case of the opossum experiments, the fibres were ones that had grown across the lesion, rather than ones spared by the lesion.

Key points

Determinants of axon growth and recovery following transection in immature spinal cord.

- **Intrinsic growth potential of axotomised and late arriving axons.**
- **Appropriateness of pathway in lesion and beyond for axon growth.**
- **Achievement of functionally effective targets.**
- **Results of experiments with marsupial species suggest that the last two are achieveable, but the mechanisms remain to be elucidated.**

In order to know whether the recent information from experiments with spinal injuries in newborn opossums will contribute to developing ways of promoting recovery in adult spinal injured patients, clearly much more detailed knowledge of the developmentally regulated phenomena that contribute to repair in the immature spinal cord will be required. The long term hope is that such knowledge could eventually be used to transform the mature spinal cord back to a more immature stage with substantially greater abilities to repair itself.

Acknowledgements

The authors' studies of spinal cord injury are supported by the Motor Accident Insurance Board Foundation (Tasmania), the Motor Accident Insurance Commission (Queensland) and the Australian Research Council Small Grants Scheme.

References

Asada Y Kawaguchi S Hayashi H and Nakamura T (1998) Neural repair of the injured spinal cord by grafting: comparison between peripheral nerve segments and embryonic homologous structures as a conduit of CNS axons. Neuroscience Research 31, 241–249

Basso D M Beattie M S and Bresnahan J C (1995) A sensitive and reliable locomotor rating scale for open field testing in rats. Journal of Neurotrauma 12, 1–21

Bates C A and Stelzner D J (1993) Extension and regeneration of corticospinal axons after early spinal injury and the maintenance of corticospinal topography. Experimental Neurology 123, 106–117

Bernstein D R Bechard D E and Stelzner D J (1981) Neuritic growth maintained near the lesion site long after spinal cord transection in the newborn rat. Neuroscience Letters 26, 55–60

Bernstein D R and Stelzner D J (1983) Plasticity of the corticospinal tract following midthoracic spinal injury in the postnatal rat. Journal of Comparative Neurology 221, 382–400

Bernstein-Goral H and Bregman B S (1993) Spinal cord transplants support the regeneration of axotomized neurons after spinal cord lesions at birth: a quantitative double labelling study. Experimental Neurology 123, 118–132

Björklund A (1991) Neural Transplantation -an experimental tool with clinical possibilities. Trends in Neuroscience 14, 319–322

Blight A R (1983) Cellular morphology of chronic spinal cord injury in the cat: analysis of myelinated axons by line-sampling. Neuroscience 10, 521–543

Bregman B S and Goldberger M E (1983a) Infant lesion effect: I. Development of motor behaviour following neonatal spinal cord damage in cats. Developmental Brain Research 9, 103–117

Bregman B S and Goldberger M E (1983b) Infant Lesion effect: II. Sparing and recovery of function after spinal cord damage in newborn and adult cats. Developmental Brain Research 9, 119–135

Bregman B S and Goldberger M E (1983c) Infant lesion effect: III. Anatomical correlates of sparing and recovery of function after spinal cord damage in newborn and adult cats. Developmental Brain Research 9, 137–154

Chambers W W (1955) Structural regeneration in the mammalian central nervous system in relation to age. In Regeneration in the Central Nervous System, edited by W F Windle, pp. 135–146. Springfield: Charles C Thomas Publisher

Clark W E Le Gros (1942) The problem of neuronal regeneration in the central nervous system. I. The influence of spinal ganglia and nerve fragments grafted in the brain. Journal of Anatomy 77, 20–48

Clearwaters K P (1954) Regeneration of the spinal cord of the chick. Journal of Comparative Neurology 101, 317–326

Clemente C D (1955) Structural regeneration in the mammalian central nervous system and the role of neuroglia and connective tissue. In Regeneration in the Central Nervous System, edited by W F Windle, pp. 135–146. Springfield: Charles C Thomas Publisher

Diener P S and Bregman B S (1998a) Fetal spinal cord transplants support the development of target reaching and coordinated postural adjustments after neonatal cervical spinal cord injury. Journal of Neuroscience 18, 763–778

Diener P S and Bregman B S (1998b) Fetal spinal cord transplants support growth of supraspinal and segmental projections after cervical spinal cord hemisection in the neonatal rat. Journal of Neuroscience 18, 779–793

Fernandez J and Nicholls J G (1998) Fine structure and development of dorsal root ganglion neurons and Schwann cells in the newborn opossum *Mondelphis domestica*. Journal of Comparative Neurology 396, 338–350

Fry E J Knott G W and Saunders N R (1998) Early repair after spinal cord injury in neonatal opossums, *Monodelphis domestica*. Proceedings of the Australian Neuroscience Society 9, 112

Fry E and Saunders N R (1999) Spinal repair in immature animals: a novel approach using the South American opossum, *Monodelphis domestica*. Clinical and Experimental Pharmacology and Physiology. Invited review, in press

Guth L (1975) History of Central Nervous System Regeneration Research. Experimental Neurology 48, 3–15

Hasan S J Nelson B H Valenzuela J I Keirstead H S Shull S E Ethell D W and Steeves J D (1991) Functional repair of transected spinal cord in embryonic chick. Restorative Neurology and Neuroscience 2, 137–154

Hasan S J Keirstead H S Muir G D and Steeves J D (1993) Axonal regeneration contributes to repair of injured brainstem-spinal neurons in embryonic chick. Journal of Neuroscience 13, 492–507

Hess A (1955) p 176 in Windle W F (1955) Regeneration in the Central Nervous System. Springfield: Charles C Thomas Publisher

Hooker D and Nicholas J S (1930) Spinal cord section in rat fetuses. Journal of Comparative Neurology 50, 413–467

Howland D R Bregman B S Tessler A and Goldberger M E (1995a) Development of locomotor behaviour in the spinal kitten. Experimental Neurology 135, 108–122

Howland D R Bregman B S Tessler A and Goldberger M E (1995b) Transplants enhance locomotion in neonatal kittens whose spinal cords are transected: a behavioural and anatomical study. Experimental Neurology 135, 123–145

Iwashita Y Kawaguchi S and Murata M (1994) Restoration of function by replacement of spinal cord segments in the rat. Nature 367, 167–170

Kalil K and Reh T (1982) A light and electron microscopic study of regrowing pyramidal tract fibres. Journal of Comparative Neurology 211, 265–275

Merline M and Kalil K (1990) Cell death of corticospinal neurons is induced by axotomy before but not after innervation of spinal targets. Journal of Comparative Neurology 296, 506–516

Migliavacca A (1928) La rigenerazione del midollo spinale nei feti e nei neonati. Bollettino della Societa Medico-Chirurgica-Pavia 1147–1151

Miya D Giszter S Mori F Adipudi V Tessler A and Murray M (1997) Fetal transplants alter the development of function after spinal cord transection in newborn rats. Journal of Neuroscience 17, 4856–4872

Nicholls J G Stewart R R Erulkar S D and Saunders N R (1990) Reflexes, fictive respiration and cell division in the brain and spinal cord of the newborn opossum, *Monodelphis domestica*, isolated and maintained *in vitro*. Journal of Experperimental Biology 152, 1–15

Ramón y Cajal (1928) Degeneration and Regeneration of the Nervous System. Trans. R M May, Ed. J De Felipe and E G Jones, 1991 History of Neuroscience No.5 Fidia/OUP, Oxford

Reh T and Kalil K (1982a) Development of the pyramid tract in the hamster. II. An electron microscopic study. Journal of Comparative Neurology 205, 77–88

Reh T and Kalil K (1982b) Functional role of regrowing pyramidal tract fibres. Journal of Comparative Neurology 211, 276–283

Robinson G A and Goldberger M E (1986) The development and recovery of motor function in spinal cats. Experimental Brain Research 62, 373–386

Rossignol S (1996) Neural control of stereotypic limb movements. In Handbook of Physiology, Section 12, Exercise: Regulation and Integration of Multiple Systems, Ed L B Rowell and J T Sheperd, American Physiological Society, pp 173–216

Saunders N R Balkwill P Knott G Habgood M D Møllgård K Treherne J M and Nicholls J G (1992) Growth of axons through a lesion in the intact CNS of fetal rat maintained in long-term culture. Proceedings of the Royal Society, London B 250, 171–180

Saunders N R Deal A and Knott G W (1994) Recovery from complete spinal cord crush and subsequent development of locomotor function in neonatal opossum (*Monodelphis domestica*). European Journal of Neuroscience Suppl. 7, 46

Saunders N R Deal A Knott G W Varga Z M and Nicholls J G (1995) Repair and recovery, following spinal cord injury in a neonatal marsupial (*Monodelphis domestica*). Clinical and Experimental Pharmacology and Physiology 22, 518–526

Saunders N R Kitchener P Knott G W Nicholls J G Potter A and Smith T J (1998) Development of walking, swimming and neuronal connections after spinal cord transection in the neonatal opossum, *Monodelphis domestica*. Journal of Neuroscience 18, 339–335

Shimizu I Oppenheim R W O'Brian M and Shneiderman A (1990) Anatomical and functional recovery following spinal cord transection in the chick embryo. Journal of Neurobiology 21, 918–937

Stelzner D J Ershler W B and Weber E D (1975) Effects of spinal transection in neonatal and weanling rats: survival of function. Experimental Neurology 46, 156–177

Stelzner D J Weber E B and Prendergast J N (1979) A comparison of the effect of mid-thoracic spinal hemisection in the neonatal or weanling rat on the distribution and density of dorsal root axons in the lumbosacral spinal cord of the adult. Brain Research 172, 407–426

Treherne J M Woodward S K A Varga Z M Ritchie J M and Nicholls J G (1992) Restoration of conduction and growth of axons through injured spinal cord of neonatal opossum in culture. Proceedings of the National Academy of Sciences USA 89, 431–434

Varga Z M Fernandez J Blackshaw S Martin A R Muller K J Adams W B and Nicholls J G (1996) Neurite outgrowth through lesions of neonatal opossum spinal cord in culture. Journal of Comparative Neurology 366, 600–612

Wang X M Terman J R and Martin G F (1996) Evidence for growth of supraspinal axons through the lesion after transection of the thoracic spinal cord in the developing opossum *Didelphis virginiana*. Journal of Comparative Neurology 371, 104–115

Wang X M Qin Y Q Terman J R and Martin G F (1997) Early development and developmental plasticity of the fasciculus gracilis in the North American opossum *Didelphis virginiana*. Developmental Brain Research 98, 151–163

Wang X M Basso D M Terman J R Bresnahan J C and Martin G F (1998a) Adult opossums (*Didelphis virginiana*) demonstrate near normal locomotion after spinal cord transection as neonates. Experimental Neurology 151, 50–69

Wang X M Terman J R and Martin G F (1998b) Regeneration of supraspinal axons after transection of the spinal cord in the developing opossum, *Didelphis virginiana*. Journal of Comparative Neurology 398, 83–97

Weber E D and Stelzner D J (1977) Behavioural effects of spinal cord transection in the developing rat. Brain Research 125, 241–255

Windle W F (1955) Regeneration in the Central Nervous System. Springfield: Charles C Thomas Publisher

Xu X M and Martin G F (1991) Evidence for new growth and regeneration of cut axons in developmental plasticity of the rubrospinal tract in the North American opossum. Journal of Comparative Neurology 313, 103–112

3

Intrinsic determinants of differential axonal regeneration by adult mammalian central nervous system neurons

Patrick N Anderson and A Robert Lieberman

Introduction

Traumatic injuries to the spinal cord cannot be repaired and have devastating and protracted effects on the lives of many thousands of people, mainly because axons in the injured CNS, unlike those in peripheral nerves, do not normally regenerate. Consequently, one of the most important goals of neurobiology is to understand the causes of the failure of axonal regeneration in the CNS so that treatments for the injured spinal cord can be developed. For many years, however, it was widely believed that most CNS neurons were intrinsically unable to regrow their axons following injury. This view was dramatically changed, and the study of CNS regeneration was reinvigorated by the experiments of Aguayo and colleagues in Montreal, and Berry in London, with peripheral nerve grafts implanted into the CNS (eg Benfey and Aguayo, 1982) or attached to severed optic nerves (eg Berry *et al.*, 1986), which showed that injured CNS axons could regenerate under some circumstances. Aguayo's experiments were not entirely novel; it has been known for a century that peripheral nerve grafts in the brain and spinal cord are invaded by axons, but the interpretation of such results was confused by the difficulty of identifying the source of the invading axons (which could have been from sensory or autonomic neurons with cell bodies outside the CNS). The considerable impact of the early studies by Aguayo's group was, therefore, partly the result of their careful exploitation of recently developed axonal tracing technology which allowed the categorical demonstration that neurons inside the CNS had regenerated their axons. Nerve grafting experiments added considerable weight to the hypothesis that CNS neurons are capable of axonal regeneration, but that the environment surrounding the injured axons is normally non-permissive or inadequate for axonal growth.

Almost all the information available about long distance regeneration of CNS axons in adult mammals comes from studies of peripheral nerve grafts. Nerve grafts are the only form of intervention which has allowed the unambiguous reconstitution of a physiological pathway within the CNS by the regrowth of the damaged axons

(Thanos, 1992; Thanos *et al.*, 1997). It is particularly easy to use retrograde tracing to identify the cell bodies from which axons regenerating into peripheral nerve grafts

Key Points

- There has been a long standing belief that most CNS neurons are intrisically unable to regrow axons following injury.
- Experiments using peripheral nerve implanted into CNS lesions in the last 20 years by Aguayo (eg Benfey and Aguayo, 1982) have transformed this view.
- Exploitation of newer pathway tracing methods, has allowed the categorical demonstration that neurons inside the CNS had regenerated their axons, peripheral nerve graft experiments suggested that CNS neurons are intrisically capable of regeneration but that the local environment is normally non-permissive or inadequate for axonal growth.

originate. This has allowed the demonstration of axonal regeneration from many sites in the CNS into which pieces of peripheral nerve have been transplanted and has allowed both the identity and the cellular physiology of regenerating CNS neurons to be investigated. The peripheral nerve graft model thus offers a consistent, unambiguous and consequently very powerful demonstration of the latent capacity for long distance regrowth and re-establishment of function of damaged CNS axons. Furthermore, this model also provides a satisfactory general explanation for the success of regeneration associated with such grafts and failure without them in terms of differences in the environments provided for injured axons in the CNS and PNS. It is not surprising, therefore, that one of the less well recognised inferences from studies of peripheral nerve grafts is that different classes of CNS neurons display dramatic differences in their regenerative potential. Not all CNS neurons regenerate well: some have to be axotomized very close to their cell bodies if they are to regenerate their axons and others fail to express growth-related genes and will not regenerate into peripheral nerve grafts at all. In this chapter we examine such observations, consider how they can be interpreted and explained, and touch on the implications for strategies for the repair of human spinal cord and brain, of intrinsic differences among CNS neurons in their regenerative potential.

The ability of CNS axons to regenerate into grafts depends on how far from the cell body the axon is injured

It was obvious even in the early studies (reviewed by Aguayo, 1985) that most CNS axons in the vicinity of a peripheral nerve graft had not regenerated, or had not regenerated far enough into the graft to become retrogradely labelled. Axons whose cell bodies were more than a few millimetres away from the graft did not usually regenerate their axons, so that most labelled neuronal cell bodies were found close to the graft tip. However, the maximum distance that cell bodies could be situated from a graft and still regenerate an axon into it depended on the type of neuron. For example, retinal ganglion cells regenerated successfully when a graft was attached to the optic nerve within 2mm of the eye, but poorly if a graft was attached to the intracranial optic nerve, 7mm from the eye (Richardson *et al.*, 1982); most intrinsic spinal neurons which regenerated into grafts in the spinal cord were found within 3–4mm from the graft tip (David & Aguayo, 1981); rubrospinal neurons regenerated into grafts placed in the cervical region of the spinal cord but not into grafts in the thoracic region (Richardson *et al.*, 1984). The most likely explanation of these results is that grafts must be implanted in a site that results in neurons being axotomized close enough to the cell bodies to initiate a strong cell body response: the closer that the site of axotomy is to the cell body the more profound the cell body response (Lieberman, 1974). Furthermore, some neurons must be axotomized closer to their cell bodies than others if they are to mount a response. The strength of the intrinsic responses of neurons to axotomy has been shown to be an important determinant of regenerative ability in studies of the regeneration of the central axons of dorsal root ganglion neurons into grafts in the spinal cord or attached to injured dorsal roots: a peripheral nerve injury, which produces a much greater cell body response than dorsal root injury, enhanced the number of such axons regenerating into the grafts a hundred fold even though the cells then had to regenerate both their peripheral and their central processes (Richardson & Issa, 1984; Chong *et al.*, 1996). Recently, evidence has emerged that in some CNS neurons (Purkinje cells, whose response to peripheral nerve grafts is discussed below) the ability to mount some aspects of the cell body response to axotomy is under the control of retrogradely transported inhibitory signals produced by a myelin-associated protein (Zagrebelsky *et al.*, 1998). This hypothesis is based on a mixture of *in vivo* and *in vitro* experiments using the IN-1 antibody to block the function of a myelin growth-cone collapsing factor and/or colchicine to prevent axonal transport. For example, a very proximal, but not a distal, axotomy induced the upregulation of c-jun by Purkinje cells, and this effect was mimicked by treatment with IN-1 or colchicine. This was interpreted as suggesting that the length of the myelinated portion of the proximal axonal stump may be one critical factor determining the regenerative response. Presumably the 'Nogo' molecule recognised by IN-1 produces a signal in the axon which is retrogradely transported to the neuronal cell bodies to suppress the expression of growth-related genes. However, the role of the myelinated segment of the proximal stump in blocking the expression of growth-related genes remains tentative, not least because colchicine has well known toxic effects and the 'Nogo' protein detected by IN-1 may be expressed by cells other than oligodendrocytes (Huber *et al.*, 1998).

Only some types of CNS neuron regenerate vigorously into peripheral nerve grafts

The regeneration of CNS axons into grafts in the thalamus

Studies of nerve grafts in the thalamus of adult rodents provided the first categorical evidence that some types of neuron are intrinsically much more able to regenerate axons into nerve grafts than are others. Although several types of neuron, including thalamo-cortical projection cells, neurons in the thalamic reticular nucleus (TRN), and corticothalamic projection cells, are axotomized when segments of peripheral nerve are implanted into the thalamus, Benfey *et al.* (1985) reported that the great majority of the axons which regenerated into such grafts came from the TRN, a thin sheet of GABAergic neurons covering the dorsolateral and rostral surfaces of the dorsal thalamus,

Key Points

- Maximum distance that cell bodies can be situated from a peripheral nerve graft and still regenerate an axon into it depends on the type of neuron.
- Strength of intrinsic responses of neurons to axotomy appears to be an important determinant of regenerative ability.
- In some CNS neurons (eg Purkinje cells) ability to mount some aspects of cell body response to axotomy is controlled by retrogradely transported inhibitory signals produced by a myelin-associated protein.
- Only some types of neuron regenerate vigorously into peripheral nerve grafts.
- Axons of neurons that regenerate well into peripheral nerve grafts are: thalamic reticular nucleus, deep cerebellar nuclei, substantia nigra pars compacta, the rubrospinal tract, vestibulospinal tract, and reticulospinal tracts and optic nerve.
- Those which regenerate poorly or not at all into grafts are: thalamocortical, cerebellar cortical, corticospinal, neostriatal and corticostriatal projection neurons.

with axons that project to the ipsilateral dorsal thalamus (Ohara & Lieberman, 1985). Very few thalamocortical projection neurons regenerated their axons. Cossu *et al.* (1987) reported apparently different results in animals with additional cortical lesions, but our studies have confirmed the findings of Benfey *et al.* (1985) and also showed that the pattern of differential regeneration into the graft was crucially dependent on the size of the graft (Morrow *et al.*, 1993). TRN neurons regenerated well into tibial nerve grafts but very poorly into peroneal nerve grafts, which are about half the diameter of tibial grafts. Grafts comprising two lengths of peroneal nerve sutured side by side were, however, able to support TRN axonal regeneration as well as tibial grafts. Presumably, concentrations of a factor, derived from the graft and necessary for TRN regeneration, are below a threshold value following implantation of a single peroneal nerve. We also found that very few corticothalamic projection neurons regenerated axons into the grafts, which with the data from grafts in the corpus striatum and spinal cord (see below) adds to the evidence that cortical projection neurons generally do not produce a vigorous regenerative response to peripheral nerve tissue.

Overall more than 90% of the CNS neurons whose axons could be shown to have regenerated through peripheral nerve grafts in the thalamus of adult rats were found in the TRN, but neurons in or near part of the medial geniculate nucleus also appeared to have relatively good powers of regeneration into grafts placed in a sufficiently posterior part of the rat thalamus. Thus grafts in all parts of the thalamus were invaded by regenerating axons of GABAergic TRN neurons and those in the posterior thalamus also by axons of one group of thalamocortical projection neurons of unknown transmitter phenotype, but on the basis of an extensive literature on such neurons (e.g. Kharazia & Weinberg, 1994), were likely to be glutamatergic.

Grafts in the cerebellum

Peripheral nerve grafts in the cerebellum also provoke differential regenerative responses. Dooley & Aguayo (1982) reported that neurons of the cerebellar cortex were never retrogradely labelled by tracers applied to such grafts, although neurons in the deep cerebellar nuclei did regenerate successfully into the same grafts. The apparently total inability of cerebellar cortical neurons to regenerate into nerve grafts and the good regenerative response of deep cerebellar neurons were confirmed by work in our laboratory (Vaudano *et al.*, 1993, 1998; Zhang *et al.*, 1997a) and it was additionally noted that precerebellar nuclei in the brainstem also regenerated axons into the grafts. The results cannot be explained by the anatomical organization of the cerebellar cortex. Although granule cells, with axons extending only into the most superficial layers of the cerebellar cortex, may be poorly placed to grow into grafts deep in the cerebellum, they are never retrogradely labelled even when the graft tip is found in the cerebellar cortex. Furthermore, Purkinje cells, whose axons extend from the cerebellar cortex to the deep nuclei, provide perhaps the best example of CNS neurons with a very low, or possibly non-existent, capability of regenerating axons into peripheral nerve grafts. In contrast to granule cells, Purkinje cell axons are well placed to respond to implanted peripheral nerve in these experiments. Thus many Purkinje cell axons must be both injured and directly exposed to the graft tip in most experiments when peripheral nerve segments are implanted into the cerebellum, yet they have never been shown to regenerate axons into the grafts, even though the great majority of Purkinje cells survive for long periods and possibly indefinitely following axotomy.

Grafts in the corpus striatum

The implantation of peripheral nerve segments into the neostriatum of adult rats injures the axons from several different classes of neurons including neostriatal neurons, such as the GABAergic projection cells, which comprise 90% of the total population of neostriatal neurons, and striatal interneurons including the large cholinergic cells (about 2% of the total population), corticostriatal projection cells, and neurons in the substantia nigra pars compacta. In a retrograde labelling study of peripheral nerve grafts in the neostriatum we found (in 22 experiments) that approximately six times as many neurons (mean = 23) in the substantia nigra pars compacta as in the neostriatum (mean = 4) had grown axons into the grafts (Woolhead *et al.*, 1998). The retrogradely labelled neurons in the neostriatum displayed none of the morphological features of projection cells (with the exception of a single neuron in the 22 experiments). Furthermore, in an additional group of animals it was shown that many of the striatal neurons which regenerated their axons into the grafts displayed cholinergic markers. Retrogradely labelled cortical neurons were rare. If the graft tip or injury tract abutted onto the globus pallidus there were usually relatively large numbers of pallidal neurons retrogradely labelled. Many of these neurons could be shown to be cholinergic and therefore likely to be part of a displaced population of basal forebrain neurons projecting to the cerebral cortex. The results of these experiments suggest that neostriatal projection neurons and corticostriatal projection neurons are very poor at regenerating their axons into peripheral nerve grafts, whereas neurons in the substantia nigra pars compacta have a relatively high regenerative capacity. Even substantia nigra pars compacta neurons are less successful at regenerating into grafts than are TRN neurons, several hundred of which grow axons into grafts in the thalamus (Morrow *et al.*, 1993).

In a retrograde labelling study of grafts implanted into the corpus striatum of 8 adult rats, Benfey & Aguayo (1982) found a mean of 25 striatal neurons and 3 nigral neurons which had regenerated axons into the grafts. The reason for the apparent discrepancy between their findings and ours is not clear. Benfey & Aguayo used unconjugated horseradish peroxidase (HRP) as a tracer, which would be expected to result in fewer retrogradely labelled cells compared with the HRP-lectin conjugates used in our study. Another explanation of the apparent discrepancy with regard to the extent of regeneration of striatal neurons lies in the size of the grafts used. Benfey & Aguayo (1982) used sciatic nerve, while in our study only the tibial nerve was implanted into the striatum. The larger size of the grafts used by Benfey & Aguayo (1982) would explain the greater number of retrogradely labelled striatal neurons found in their study, but not the lower numbers of substantia nigra pars compacta neurons. Benfey & Aguayo (1982) also used a different, lateral approach to the striatum, compared to the dorsal approach for implantation used in our experiments but it is not clear how this could have influenced the pattern of axonal regeneration from neurons in the neostriatum or substantia nigra.

Grafts on the optic nerve

The axons of retinal ganglion cells, whether in the retina or in the optic nerves of adult mammals, show a marked propensity to produce regenerative sprouts following injury (McConnell & Berry, 1982; Hall & Berry, 1989; Zeng *et al.*, 1994) and to regenerate into peripheral nerve grafts (Berry *et al.*, 1986; Carter *et al.*, 1989; Thanos, 1992). It has been shown that when peripheral nerve bridge grafts are used to connect the optic nerve just behind the optic disc with the superior colliculus, the axons of retinal ganglion cells grow through the entire graft and re-innervate the tectum around the distal end of the graft, forming morphologically normal (Carter *et al.*, 1989) and functional (Keirstead *et al.*, 1989) synapses. If the distal end of the graft is implanted into the pretectal area, restoration of the pupillary light reflex may result (Thanos, 1992) and if it is implanted near to the lateral geniculate nucleus, the restoration of pattern vision has been reported (Thanos *et al.*, 1997). None the less, most of the retinal ganglion cell population dies following axotomy, unlike corticospinal neurons (Barron *et al.*, 1988) or Purkinje cells, so that the neurons which grow axons into the graft are a minority (less than 10%) of the original retinal ganglion cell population. It is consequently interesting to ask whether there is differential survival and/or differential regeneration in response to the nerve graft. Peripheral nerve grafts certainly promote the survival of a population of retinal ganglion cells (eg Bähr *et al.*, 1992). When nerve grafts are attached to severed optic nerves in cats the large alpha cells appear to preferentially regenerate into the grafts (Watanabe *et al.*, 1993), but the classes of neurons which survive in long term experiments in rats seem to depend on their ability to contact particular target areas in the brain (Thanos & Mey, 1995).

Grafts into the corticospinal tract

One of the most consistent and interesting findings in studies of peripheral nerve grafts in the spinal cord and brainstem is the absence of regeneration of corticospinal axons. This tract plays an important part in the control of voluntary motor activity, particularly in primates including man, and its fibres are one of the few types of CNS axon which have been convincingly demonstrated to have some powers of regeneration in the spinal cord even in the absence of a nerve graft.

In rats, corticospinal tract axons originate mainly from neurons in layer Vb of the sensorimotor cortex (Miller, 1987; Barron *et al.*, 1988). When segments of peripheral nerve are implanted into the brainstem or spinal cord, axons from several descending tracts, including the rubrospinal tract, vestibulospinal tract, and reticulospinal tracts, regenerate into the grafts (Richardson *et al.*, 1980, 1984). In contrast, several studies have confirmed that corticospinal axons do not regenerate into such grafts in rats (Richardson *et al.*, 1980,1984; Horvat *et al.*, 1989; Houle, 1991; Ye & Houle, 1997), although there has been one notable claim to the contrary (Cheng *et al.*, 1996). A lack of regeneration of injured corticospinal axons into peripheral nerve grafts has also been shown in cats (Sceats *et al.*, 1986). When nerve grafts are inserted into the cerebrum, cortical neurons in the layers and regions which give rise to the corticospinal tracts do regenerate axons into the grafts, although in lesser numbers than axons from neurons in other areas (Benfey & Aguayo, 1982; Horvat & Aguayo, 1985) and it has not been demonstrated conclusively that the retrogradely labelled cells gave rise to corticospinal axons. It is possible that these findings reflect the inability of corticospinal neurons to mount an appropriate cell body response when axotomised distally by a perpheral nerve graft. Indeed, corticospinal neurons in animals with cord injuries do not appear to upregulate the regeneration-related molecules (eg GAP-43) associated with axonal growth from other cells. However, corticospinal tract axons contain moderately high levels of GAP-43 even in the intact animal and they can regenerate into collagen gel implants after spinal cord lesions (Joosten *et al.*, 1995a,b) and regenerate for several millimetres if the cords are treated with antibodies against myelin proteins (Schnell & Schwab, 1993) and/or the animals are treated with neurotrophin-3, NT-3 (Schnell *et al.*, 1994; Grill *et al.*, 1997). It has not been reported whether such regeneration is accompanied by the enhanced expression of growth-associated molecules. None the less it appears that corticospinal neurons can make a regenerative response, even to a distal axotomy, if there is an appropriate stimulus, which cannot be provided by a peripheral nerve graft. The pattern of corticospinal tract regeneration following injury varies interestingly in the different classes of experiments. The studies from Joosten's laboratory show that the axons grow into, but not beyond, collagen gel implants at the lesion site (Joosten *et al.*, 1995a,b), whereas those from Schwab's group (Schnell & Schwab, 1993) using antibodies to myelin proteins, and by Grill *et al.* (1997), using implants of cells transfected so that they secrete NT-3, show regenerating corticospinal axons growing in the grey matter deep to the lesion sites. Furthermore, olfactory glia injected into the lesioned cord migrate into the degenerated tract caudal to the lesion and apparently allow corticospinal axons to regenerate, become myelinated and produce some functional recovery (Li *et al.*, 1997). It is notable, however, that very few studies of corticospinal tract regeneration have used complete transection of the spinal cord, and there is consequently the possibility that the lesions were incomplete and that a neuroprotective effect rather than regeneration was being demonstrated. Furthermore, since corticospinal fibres terminate widely in the grey matter of the cord there is also the possibility that the "regeneration" demonstrated was in fact sprouting of intact fibres which previously terminated rostral to the lesion.

Do neuroanatomical or neurochemical features explain the high regenerative potential of some CNS neurons?

CNS neurons whose axons regenerate readily into nerve grafts include those in the TRN, deep cerebellar nuclei and substantia nigra pars compacta, retinal ganglion cells, cholinergic interneurons in the neostriatum and cholinergic neurons in the globus pallidus. Since these represent a mixture of projection neurons with both diffuse and highly localized terminal fields and interneurons, simple neuroanatomical features cannot readily explain the ability to regenerate axons. Similarly, the type of transmitter utilized by the cells does not generally correlate well with regenerative ability: GABAergic TRN neurons regenerate well but GABAergic striatal projection neurons and Purkinje cells regenerate very poorly. A cholinergic phenotype is, however, often associated with successful regeneration, as in striatal interneurons and pallidal neurons, but whether this is a causal relationship, or a function of cholinergic CNS neurons all responding to nerve growth factor (NGF) (via trkA, see below) produced by the grafts is not clear. Although it would be expected that the myelination of axons might limit regeneration into nerve grafts because of the well established inhibitory effects of CNS myelin proteins (eg Schwab & Caroni, 1988), there is no evidence that this is the case: TRN neurons and those of the deep cerebellar nuclei have myelinated axons, but nigrostriatal axons and those from the cholinergic pallidal cells are unmyelinated, yet all regenerate well.

Molecular correlates of high regenerative potential

For CNS neurons to regenerate their axons into peripheral nerve grafts they presumably need to express both molecules necessary for axonal growth and, since elongation takes place on the surfaces of Schwann cells, also molecules necessary for the interactions of growth cones with Schwann cells (Figure 1). It is implicit to the discussion which follows that some of the molecular correlates of regenerative ability into nerve grafts may be necessary for regeneration involving interactions with Schwann cells but not necessary for regeneration entirely within the CNS. The study of peripheral nerve regeneration provides some clues to the nature of regeneration-related molecules. All types of axons in damaged peripheral nerve trunks can regenerate, although the success of functional reinnervation of targets varies (Swett *et al.*, 1991). Following injury a co-ordinated sequence of changes in gene expression has been shown to take place in motor and sensory neurons. These changes include a reduction in the expression of some genes including those for neurofilament proteins, proteins involved in neurotransmission and some peptides, and an increase in the expression of other genes including some tubulin isoforms, GAP-43 (Hoffman, 1989), other peptides, and the transcription factor c-jun (Jenkins *et al.*, 1993). Our studies of CNS regeneration have concentrated on GAP-43, c-jun and cell recognition molecules (see Table 1).

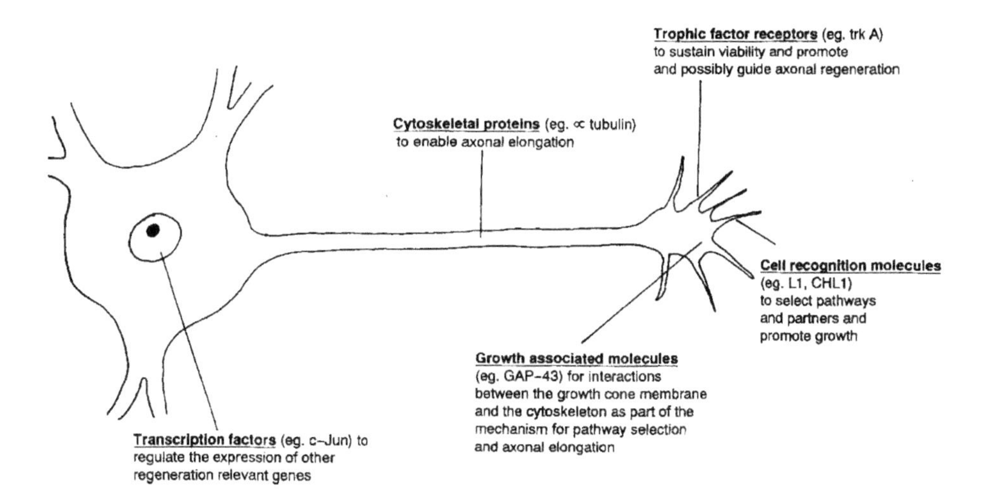

Figure 1 Some classes of molecules which neurons may have to express in order to regenerate an axon

The growth-associated protein, GAP-43, is highly concentrated in growth cones and although it is thought to play a role in the outgrowth of neurites (Benowitz & Routtenberg, 1997) and their subsequent guidance in response to environmental cues (Aigner & Caroni, 1995), it is not essential for the gross development of the nervous system (Strittmatter *et al.*, 1995). Most CNS neurons upregulate GAP-43 expression when they are regenerating their axons. Thus, when peripheral nerve grafts are implanted into the thalamus of adult rats, TRN neurons, and a group of neurons in or near the medial geniculate nucleus, both of which regenerate axons, upregulate GAP-43 mRNA from a very low base and the levels of expression remain high for several weeks while their axons are growing. GAP-43 protein is found in axons regenerating into the grafts (Campbell *et al.*, 1991; Vaudano *et al.*, 1995), which suggests that the increase in mRNA levels reflects an increase in gene expression. In contrast, most thalamocortical projection neurons do not upregulate GAP-43 and do not regenerate their axons (Vaudano *et al.*, 1995). TRN neurons upregulate GAP-43 expression even following stab wounds to the thalamus or the implantation of acellular grafts which do not support axonal regeneration, but expression is found in fewer cells and soon declines (Vaudano *et al.*, 1995). Thus the effect of a living nerve graft in the thalamus (presumably the effects of neuronal interactions with Schwann cells) is to increase the number of neurons expressing GAP-43 and to prolong its expression, but only in a population of neurons some of which are intrinsically able to re-express it following injury. A similar pattern is seen following peripheral nerve implantation into the cerebellum, when neurons in the deep cerebellar nuclei and brainstem precerebellar nuclei upregulate GAP-43 mRNA and regenerate their axons but Purkinje cells do neither (Vaudano *et al.*, 1993, 1998). Retinal ganglion cells regenerating into peripheral nerve

Table 1 The data summarised in Table 1 (page 63) are from work in our laboratory and derive almost entirely from *in situ* hybridization studies using synthetic oligonucleotide or cRNA probes labelled with S^{35}, alkaline phosphatase or digoxygenin; some of the c-jun data are based on immunohistochemistry using antibodies against the Jun protein. Data are shown for grafts implanted in three sites, thalamus, cerebellum, striatum and specifically for certain neuronal populations at, or projecting to, the implantation site. The regenerative potential of each of these populations is indicated as very low, low, high or very high on the basis of retrograde labelling studies with HRP or HRP-cholera toxin. In the main body of the Table are the expression pattern data for the mRNAs listed in the left hand column. In all cases the lower (or sole) symbol represents the level of expression of message in that neuronal population in unoperated animals: ❑ indicates that message is undetectable or barely detectable; ■ shows a detectable level of constitutive expression of the message in the population, the height of the bar providing an indication of the level of expression. The arrow (head)s comprising the upper symbols (where included) show the change in expression following implantation of a peripheral nerve graft, the height of the arrow giving an indication of the magnitude of the increase in expression. The absence of an upper symbol indicates that a change in hybridization signal could not be detected following peripheral nerve graft implantation. The most striking feature of the Table is the positive correlation between regenerative potential and either strong upregulation of expression of the mRNAs for GAP-43, L1, CHL1 and c-jun (in neurons of the thalamin reticular nuclers (TRN) and deep cerebellar nuclei (DCN) or a moderate to high level of constitutive expression of these same mRNAs. The data for putative cholinergic interneurons in the striatum (Put. chol.) are in parentheses because the evidence that the expression data reflect this subpopulation alone is not strong and in many cases is indirect. It should be noted that data summaries in this form are incapable of conveying the time course of the responses or the many subtleties of the findings. For example, whereas in some cases in which a modest increase in expression is indicated, the increase was a diffuse, generalised phenomenon involving many of the cells in the population (eg N-CAM in dorsal thalamus), in other cases the same symbol represents a much stronger upregulation, but in a small minority of the population (eg GAP-43 in dorsal thalamus). And in the case of cells showing modest increases in c-jun expression (eg. Purkinje cells) the affected cells may be those severely traumatized by graft insertion and undergoing degenerative changes. For further details and the original data, the following publications should be consulted: Vaudano *et al.* (1995, 1998); Zhang *et al.* (1995a, b); Woolhead *et al.* (1998).
SNpc: neurons of the substantia nigra pars compacta.

grafts also upregulate GAP-43 mRNA but in addition they accumulate GAP-43 protein in their perikarya (Schaden *et al.*, 1994), unlike most other regenerating CNS neurons. Cells in the substantia nigra pars compacta differ from the other types of successfully regenerating CNS neurons in that they do not appear to upregulate GAP-43 mRNA, at least as can be judged by *in situ* hybridization with non-radioactive probes (Woolhead *et al.*, 1998), but these neurons express moderately high levels of the mRNA even in ungrafted animals. However, in the same experiments a few large striatal neurons near the grafts, presumably those cells which regenerate axons, clearly upregulated GAP-43 mRNA, even though they too show moderately high constitutive expression of the molecule. In summary, GAP-43 mRNA is upregulated in most types of CNS neuron

Location of graft tip →	THALAMUS		CEREBELLUM		STRIATUM		
Neuronal populations →	Dorsal Thalamus	TRN	Purkinje Cells	DCN	Most neostriatal neurons	Put. Chol. neurons	SNpc
Regen. potential →	LOW	V.HIGH	V.LOW	HIGH	V.LOW	[HIGH]	HIGH

GAP-43

N-CAM

L1

CHL1

c-jun

NGFR (p75)

Tenascin-C

which regenerate their axons into nerve grafts, but some neurons which express the mRNA prior to axotomy can regenerate axons without further increasing its expression. No neurons have yet been shown to regenerate their axons without expressing GAP-43 mRNA. Because GAP-43 protein does not usually accumulate in the cell bodies of CNS neurons (other than in retinal ganglion cells), it is difficult to prove that the upregulation of its mRNA always reflects an increased expression of protein. However, immunohistochemical studies have demonstrated that CNS axons regenerating through peripheral nerve grafts contain GAP-43 protein (Campbell *et al.*, 1991; Vaudano *et al.*, 1995; Berry *et al.*, 1996) even though it is not detectable in the axonal shafts before they are injured. The role of GAP-43 in axonal pathfinding suggests that it probably

Key Points

Factors contributing to successful regeneration

- Cholinergic phenotype often associated with successful regeneration.
- Most CNS neurons upregulate GAP-43 expression when they are regenerating their axons.
- No neurons have yet been shown to regenerate their axons without expressing GAP-43 mRNA.
- c-jun is a promising candidate for gene regulation of regeneration related molecules.
- Differential expression by neurons of receptors for trophic factors produced by the grafts.
- Expression of cell recognition molecules (eg L1 and CHL1) potentially important for ability of neurons to extend their axons into nerve grafts.
- Upregulation of c-jun and GAP-43, L1and CHL1 or their constitutive expression are characteristics of CNS neurons regenerating into nerve grafts.

acts in growth cone exploratory activity into and within nerve grafts but whether GAP-43 is required for long distance axonal regeneration within the graft remains to be determined.

The most promising candidate for a gene capable of controlling the expression of several regeneration related molecules is c-jun. It encodes a transcription factor, c-Jun, which is induced in many types of neuron by axonal injury and forms part of the AP1 complex, comprising homodimers with other Jun molecules, or heterodimers with Fos and other molecules. It has been shown to have many important effects on cell physiology, and is essential for normal development (Johnson *et al.*, 1993). In neurons, c-jun expression has a curious, ambivalent association with both axonal regeneration and apoptosis (Jenkins *et al.*, 1993; Ham *et al.*, 1995). GAP-43 is among the genes known to have AP1 binding sites in their promoters (Eggen *et al.*, 1994). The control of c-jun activity occurs at both the transcriptional level and through its phosphorylation (Karin, 1996). Increased transcription is marked in regenerating neurons, but it is not known whether the phosphorylation state of c-jun changes. Motor neurons upregulate c-jun after peripheral nerve injury. Primary sensory (dorsal root ganglion) neurons massively upregulate c-jun after injury to the peripheral process of their axon, which regenerates vigorously, whereas there is little c-jun upregulation in response to injury of their central process which regenerates weakly (Jenkins *et al.*, 1993). Following peripheral axotomy, c-jun expression by dorsal root ganglion neurons persists until target tissues are re-innervated after which it is downregulated. There is thus a strong presumption that c-jun may play a major role in the control of gene expression during peripheral nerve regeneration. The upregulation of c-jun by CNS neurons is also one of the changes in gene expression most consistently correlated with the ability of neurons to regenerate their axons. When peripheral nerve grafts are placed in the thalamus, neurons in the TRN upregulate c-jun and regenerate their axons, but the great majority of thalamocortical projection neurons do neither (Vaudano *et al.*, 1998). Purkinje cells in the cerebellum do not normally express c-jun and most do not upregulate it following graft implantation. A few Purkinje cells near to the site at which the graft passes through the cerebellar cortex do upregulate c-jun, as detected by *in situ* hybridization and immunohistochemistry (Chaisuksunt *et al.*, 1998 and unpublished observations) but they are mainly shrunken cells and probable represent neurons which are dying after injury. Neurons in the deep cerebellar nuclei, which regenerate readily into the grafts, upregulate c-jun mRNA and protein (Chaisuksunt & Zhang, unpublished observations). Retinal ganglion cells regenerating axons through peripheral nerve grafts attached to the retinal stump of severed optic nerves also upregulate c-jun (Hull & Bähr, 1994; Robinson, 1994). On the other hand, when grafts are placed in the neostriatum there is only a transient upregulation of c-jun by neurons in the substantia nigra pars compacta (Vaudano, unpublished observations) which has declined by two weeks after injury, but a prolonged upregulation by a small number of neurons in the striatum (Chaisuksunt, unpublished observations). A general principle seems to be that c-jun is upregulated by most CNS neurons when they regenerate axons into nerve grafts, but also by other neurons injured by graft implantation. This presumably reflects the involvement of c-jun in both regeneration and cell death and suggests that other, unknown, factors also determine the regeneration potential of neurons.

Another way in which differential regeneration could be determined by the ability to interact with Schwann cells is by the differential expression by neurons of receptors for trophic factors produced by the grafts. The expression of the trk family of receptors for neurotrophins has been fairly well documented and there is some correlation with the regenerative ability of CNS neurons. Cholinergic neurons in the striatum and globus pallidus are better at regenerating into grafts than most surrounding cells and both express trkA, the high affinity receptor for NGF. Regeneration into a peripheral nerve graft is clearly not confined to neurons expressing trkA, however, and one population of neurons which express trkA, thalamic paraventricular neurons, are not good at regenerating (our unpublished observations). Many neurons in the substantia nigra pars compacta are sensitive to brain-derived neurotrophic factor (BDNF) (Mufson *et al.*, 1994) and express trkB, the receptor for BDNF/neurotrophin-4 (NT4). Regenerating axons of substantia nigra pars compacta neurons appear to grow more slowly into grafts in the striatum than do axons of cholinergic neurons (Woolhead *et al.*, 1998), which would be compatible with the slow upregulation of BDNF (relative to NGF) in injured peripheral nerve (Meyer *et al.*, 1992). None the less, many small and medium sized neurons in the striatum express trkB as well as trkC yet they are poor at regenerating axons into grafts. Purkinje cells in the cerebellum express trkC, but not trkA or trkB, but fail to regenerate axons into nerve grafts, whereas inferior olivary neurons express trkC strongly and trkB only weakly and are capable of regenerating their axons. Many neurons in the dorsal thalamus express moderate levels of trkB and some express trkC, but most do not readily regenerate axons into nerve grafts. On the other hand, TRN neurons express trkB and regenerate well. Another gene which may prove important for regeneration into grafts and which is expressed by TRN neurons (Trupp *et al.*, 1997) is c-ret, a receptor for glial derived neurotrophic factor (GDNF).

The cell recognition molecules L1 (NILE) and N-CAM are important for the growth of neurites on the surfaces of Schwann cells in culture (Kleitman *et al.*, 1988), acting principally via homophilic interactions between similar molecules on the adjacent surfaces of Schwann cells and axons. The expression of L1 and N-CAM by neurons is therefore potentially important for their ability to extend their axons into nerve grafts. We have studied the expression of these molecules and their mRNAs during axonal regeneration into nerve grafts in the thalamus, cerebellum and corpus striatum. The CNS neurons which regenerate into grafts in the first two sites (cells in the TRN and cerebellar deep nuclei) upregulate L1 mRNA (Zhang *et al.*, 1995a; unpublished observations) whereas thalamocortical projection cells and Purkinje cells which fail to regenerate their axons do not upregulate L1. As far as can be detected by *in situ* hybridization with digoxygenin labelled riboprobes, N-CAM mRNA is present in, but not upregulated by the regenerating neurons. However, both N-CAM and L1 are present on the regenerating axons and the N-CAM is polysialylated (see above), unlike most N-CAM in the adult brain (Zhang *et al.*, 1995a). This pattern of expression of CAMs would tend to promote the ability of regenerating CNS axons to elongate on the surfaces of Schwann cells. Neurons in the substantia nigra pars compacta and neostriatum do not appear to upregulate L1 mRNA when regenerating their axons into grafts, but they express moderately high levels of the molecule even in the unoperated animal. Retinal ganglion cells also express moderately high levels of L1 but do not apparently upregulate it when regenerating their axons into nerve grafts (Jung *et al.*, 1997). Although all the CNS neurons regenerating axons in our studies appear to express high levels of L1 mRNA, the changes in expression of cell L1 mRNA during CNS axonal regeneration do not follow the events which take place during peripheral nerve regeneration so accurately as the expression of GAP-43. The mRNA for another cell recognition molecule CHL1, a close homologue of L1, is upregulated by motor neurons when they regenerate their axons (Roslan *et al.*, 1998 and in preparation) and we have recently studied its expression during CNS axonal regeneration (Chaisuksunt *et al.*, 1998 and in preparation). When nerve grafts are placed in the thalamus, striatum or cerebellum the results are similar: the classes of neurons which regenerate their axons upregulate CHL1 mRNA and those which do not regenerate fail to upregulate CHL1 mRNA. Once again, the neurons in the substantia nigra pars compacta are an exception in that they do not upregulate CHL1 mRNA but express moderate to high levels even in the unoperated state. CHL1 is a powerful promoter of neurite outgrowth *in vitro* whether in solution or presented as a substrate, but its ligands are unknown and it is not clear how CHL1 expression by neurons can influence regeneration, although its involvement in Schwann cell/axon interactions or autocrine or paracrine effects on growth cones are possibilities.

Tenascin-C is an extracellular matrix protein which is strongly upregulated in segments of injured peripheral nerve during axonal regeneration, and which has powerful effects on axonal growth *in vitro* (Taylor *et al.*, 1993) and may enhance outgrowth when presented as a continuous substrate or act as a barrier to neurite outgrowth when it forms part of a discontinuous substrate. Clearly, tenascin-C does not prevent axonal regeneration in peripheral nerves, and may play a role in promoting regeneration (Martini, 1994) but the regenerating axons do not seem to bind tenascin-C at their surface. CNS axons regenerating into peripheral nerve grafts in the thalamus seem to more strongly express a tenascin-C receptor, since they are coated with graft-derived tenascin-C as they elongate through the Schwann cell columns (Zhang *et al.*, 1995b) and it is possible that tenascin-C may also promote their regeneration. The upregulation of an unknown tenascin-C receptor may, therefore, be part of the regenerative response of CNS neurons which are able to grow axons into grafts. Tenascin-C is upregulated in some parts of the CNS following injury, mainly by astrocytes (Zhang *et al.*, 1995c, 1997b) but to a much lesser extent than in injured nerves.

In general, the upregulation of c-jun and the upregulation or constitutive expression of GAP-43, L1and CHL1 are characteristics of those CNS neurons which can grow axons into nerve grafts. It is tempting to speculate that c-jun is responsible for orchestrating at least some of the many changes in gene expression which enable these processes to take place. It is not the case that all CNS neurons need to upregulate GAP-43 or L1 in order to regenerate axons. Some neurons, such as those of the substantia nigra pars compacta, have a pattern of growth-associated gene expression even when uninjured which seems preadapted for axonal regeneration, and they consequently have to make fewer changes in gene expression than others in order to grow a new axon. There is no simple correlation between expression of trk family receptors and regenerative ability, but neurons expressing trkA often have high regenerative powers and those expressing only trkC poor regenerative powers at least into peripheral nerve grafts. TrkB-expressing cells include a complex mixture of neurons which regenerate well and those which do not. The poor response of neurons expressing only trkC may be the result of the downregulation of expression of NT-3 in injured peripheral nerve, in contrast to the elevation of NGF and BDNF. Presumably CNS neurons need to be able to obtain trophic support and tropic stimuli from grafts, navigate into and within the grafts and interact successfully with the Schwann cells if they are to regenerate their axons.

Why do corticospinal axons not regenerate into nerve grafts?

The importance of long tracts in the spinal cord and the effects of their destruction in spinal injury patients means that it is worthwhile giving special consideration to the responses to injury of the corticospinal tracts. As described above, the majority of studies with peripheral nerve grafts in the spinal cord and brainstem indicate that these axons will not regenerate into nerve grafts. This may be the result of the inadequate interactions of corticospinal axons with Schwann cells. Schwann cells appear less promising adjuvants of corticospinal tract regeneration than olfactory glia (Li *et al.*, 1997; Ramón-Cueto *et al.*, 1998): when Matrigel columns containing cultured Schwann cells are used to bridge the severed spinal cord (Xu *et al.*, 1995, 1997; Bunge, chapter 8 this volume) corticospinal axons do not enter the implant (cf the results of peripheral nerve grafting), although many other axons do. The evidence that corticospinal axons will grow into nerve grafts in the forebrain (limited to an abstract, Horvat & Aguayo, 1985) suggests that corticospinal axons are specifically unable to respond to the environment offered by peripheral nerve grafts, unless their cell bodies are stimulated by a very proximal axotomy. The axons can regenerate, however, given a sufficient trophic stimulus and/or the removal of inhibitory influences in the spinal cord (Schnell & Schwab, 1993; Joosten *et al.*, 1995a,b; Blesch & Tuszynski, 1997; Grill *et al.*, 1997). It may be germane that NT-3, to which corticospinal axons are sensitive, is not upregulated by the non-neuronal cells in injured peripheral nerves (Funakoshi *et al.*, 1993) unlike NGF and BDNF. Perhaps olfactory glia express higher quantities of NT-3 than do Schwann cells. However, an explanation of the lack of regeneration of corticospinal axons into nerve grafts based on the lack of NT-3 is confused by the observation that most corticospinal neurons also express trkB and can be protected from cell death by BDNF (Giehl *et al.*, 1998). Clearly graft-derived BDNF is insufficient to promote axonal regeneration by these neurons although other types of neurons responsive to BDNF (such as retinal ganglion cells) do grow axons through grafts. BDNF is upregulated slowly in injured peripheral nerves (Funakoshi *et al.*, 1993) and it may be that by the time it is fully expressed in the grafts, the corticospinal axons may no longer be able to regenerate, perhaps because of glial scar formation. Furthermore, recent evidence shows that even the transfection of the Schwann cells so that they express high levels of NT-3 does not allow corticospinal axons to grow into a peripheral nerve graft (personal communication from J. Verhaagen). Another possibility is that sprouts formed by injured corticospinal axons lack cell surface molecules, eg L1 or CHL1, necessary for optimal growth on Schwann cell surfaces.

Is the expression of growth associated molecules necessary for the regeneration of axons by CNS neurons and can it be manipulated?

A fundamental weakness in our knowledge of axonal regeneration is that correlations between the expression of particular molecules and the ability of neurons to regenerate their axons does not necessarily imply that such molecules are either essential for axonal regeneration or are critical factors capable of producing the regrowth of axons. Some molecules are presumably expressed because the axon is regenerating rather than being necessary for regeneration. Indeed, it is difficult to think of a growth-related gene whose product has been shown to be essential for axonal regeneration *in vivo*. The most likely approach to a solution of this problem is through transgenic technology, i.e. the creation of transgenic animals or genetic manipulation using viral or other vectors. It should be possible to demonstrate that regeneration is impaired in the absence of the molecules which have been found to be correlated with axonal regeneration and enhanced when they are expressed, if they are indeed functionally important or essential for regeneration. Transgenic animals with null mutations of growth-associated molecules offer an opportunity to study this problem. GAP-43 knockout mice have been generated (Strittmatter *et al.*, 1995) but nothing has been published on the ability of their neurons to regenerate axons, presumably because most of the animals die early in life. That the nervous system develops in these animals does not prove that GAP-43 is not required for axonal regeneration because axonal growth during development takes place over very much shorter distances than axonal regeneration in the adult and may not have the same requirements. None the less, in view of the many similarities in the patterns of gene expression during axonal growth in development and regeneration, the presumption must be that GAP-43 will not prove necessary, at least for regeneration over short distances. Tenascin-C knock-out mice have little if any phenotypic abnormalities (Saga *et al.*, 1992) but nothing has been published on axonal regeneration. Null mutations of c-jun have been created but the animals die during development from liver pathology (Johnson *et al.*, 1993) and are consequently not available for regeneration experiments. In any case, null mutation experiments have been criticised on the grounds of the presence of compensatory changes which may take place during development. In the case of c-jun, transgenic animals with a dominant-negative c-jun gene under a neuron-specific promoter offer the most hope of definitive experiments to establish whether its expression is necessary for regeneration.

Key Points

- Correlations between expression of particular molecules and ability of neurons to regenerate axons does not necessarily imply such molecules are essential for regeneration.
- Some molecules may be expressed because the axon is regenerating rather than being necessary for regeneration.
- Transgenic animals with null mutations of growth-associated molecules offer an opportunity to distinguish between these two possibilities.
- A limitation of this approach is that in null mutation experiments there may be compensatory changes during development.
- Overexpression of putative growth factors, particularly where regeneration does not normally occur, is an alternative approach.

If the lack of particular growth-associated molecules is the factor limiting axonal regeneration from those classes of CNS neurons which do not regenerate into peripheral nerve grafts, then inserting those genes into the neurons should enable them to grow axons after injury. Several strains of mice overexpressing GAP-43 in CNS neurons have been created. Work in our laboratory (by G. Campbell and M. Mason) has involved the insertion of tibial nerve grafts into the thalamus of mice (from Dr Pico Caroni's laboratory in Basel) which overexpress GAP-43 in the medial thalamic nuclei. In wild type animals these neurons do not regenerate axons into grafts and our experiments show that GAP-43 expression is not sufficient to enable them to regenerate in the transgenic mice. We have also placed grafts into the cerebellum of mice from Dr Joost Verhaagen's laboratory in Amsterdam, in which Purkinje cells overexpress GAP-43. The result was similar: GAP-43 may be necessary for axonal regeneration but it was not sufficient to allow the Purkinje cells to regenerate their axons. Other transgenic experiments to establish whether particular molecules, including L1 and c-jun, are sufficient to allow the regeneration of otherwise refractory

CNS neuron are in progress. Of course, it may be that studying the neurons which can regenerate axons into nerve grafts can give a misleading impression of the characteristic which neurons need in order to regenerate axons. In particular, such studies are bound to identify neurons capable of expressing those genes necessary for axonal growth on Schwann cells, which are not normally present in the CNS. The regenerative response of corticospinal axons indicates that this may indeed be so. None the less, it should be possible soon to establish that the expression of particular genes, or more likely, groups of genes, is both necessary and sufficient for CNS neurons to be able to regenerate their axons, in grafts or in the spinal cord. The question of how to manipulate gene expression in cells which do not normally mount a regenerative response will then arise. It may be possible to increase the expression of some regeneration-related genes by applying appropriate pharmacological stimuli to the injured axons or directly to the cell bodies, but the huge variability in the regenerative abilities of different classes of CNS neurons suggests that some may require the direct manipulation of neuronal gene expression with viral or other vectors to produce an adequate regenerative response.

Acknowledgements

This work was supported by grants from the MRC and the Wellcome Trust. We are immensely grateful to our colleagues in the laboratory, past and present, for allowing us to make use of their work in this chapter. We thank in particular Yi Zhang, Greg Campbell, Elizabetta Vaudano, Julia Winterbottom and Vip Chaisuksunt.

References

Aguayo A J (1985) Axonal regeneration from injured neurons in the adult mammalian central nervous system. In Synaptic Plasticity, edited by C.W. Cotman, pp.475–484. New York: The Guilford Press

Aigner L and Caroni P (1995) Absence of persistent spreading, branching, and adhesion in GAP-43-depleted growth cones. Journal of Cell Biology 128, 647–660

Bähr M Eschweiler G W and Wolburg H (1992) Precrushed sciatic nerve grafts enhance the survival and axonal regrowth of retinal ganglion cells in adult rats. Experimental Neurology 116, 13–22

Barron K D Dentinger M P Popp A J and Mankes R (1988) Neurons of layer Vb of rat sensorimotor cortex atrophy but do not die after thoracic cord transection. Journal of Neuropathology and Experimental Neurology 47, 62–74

Benfey M and Aguayo A J (1982) Extensive elongation of axons from rat brain into peripheral nerve grafts. Nature 296, 150–152

Benfey M Bunger U R Vidal-Sanz M Bray G M and Aguayo A J (1985) Axonal regeneration from GABAergic neurons in the adult rat thalamus. Journal of Neurocytology 14, 279–296

Benowitz L I and Routtenberg A (1997) GAP-43: an intrinsic determinant of neuronal development and plasticity. Trends in Neuroscience 20, 84–91

Berry M Carlile J and Hunter A (1996) Peripheral nerve explants grafted into the vitreous body of the eye promote the regeneration of retinal ganglion cell axons severed in the optic nerve. Journal of Neurocytology 25, 147–170

Berry M Rees L and Sievers J (1986) Unequivocal regeneration of rat optic nerve axons into sciatic nerve isografts. In Neural Transplantation and Regeneration, edited by G D Das and R B Wallace, pp. 63–79: New York: Springer-Verlag

Blesch A and Tuszynski M H (1997) Robust growth of chronically injured spinal cord axons induced by grafts of genetically modified NGF-secreting cells. Experimental Neurology 148, 444–52

Campbell G Anderson P N Lieberman A R and Turmaine M (1991) GAP-43 in the axons of mammalian CNS neurons regenerating into peripheral nerve grafts. Experimental Brain Research 87, 67–74

Carter D A Bray G M and Aguayo A J (1989) Regenerated retinal ganglion cell axons can form well-differentiated synapses in the superior colliculus of adult hamsters. Journal of Neuroscience 9, 4042–4050

Chaisuksunt V Zhang Y Schachner M Anderson P N and Lieberman A R (1998) Expression of the cell recognition molecule, CHL1, (close homologue of L1), in adult rat brain following peripheral nerve graft implantation. Society for Neuroscience Abstracts 28, 2009

Cheng H Cao Y and Olson L (1996) Spinal cord repair in adult paraplegic rats: partial restoration of hind limb function. Science 273, 510–513

Chong M S Woolf C J Turmaine M Emson P C and Anderson P N (1996) Intrinsic vs extrinsic factors in determining the regeneration of the central processes of rat dorsal root ganglion neurons: the influence of a peripheral nerve graft. Journal of Comparative Neurology 370, 97–104

Cossu M Martelli A Pau A Viale E S Siccardi D and Viale G L (1987) Axonal elongation into peripheral nerve grafts between thalamus and somatosensory cortex of the rat. An experimental model. Brain Research 415, 399–403

David S and Aguayo A J (1981) Axonal elongation into peripheral nervous system "bridges" after central nervous system injury in adult rats. Science 214, 931–933

Dooley J M and Aguayo A J (1982) Axonal elongation from cerebellum into peripheral nervous system grafts in the adult rat. Annals of Neurology 12, 221 (Abstract)

Eggen B J Nielander H B Rensen de Leeuw M G Schotman P Gispen W H and Schrama L H (1994) Identification of two promoter regions in the rat B-50/GAP-43 gene. Molecular Brain Research 23, 221–234

Funakoshi H Frisen J Barbany G Timmusk T Zachrisson O Verge V M and Persson H (1993) Differential expression of mRNAs for neurotrophins and their receptors after axotomy of the sciatic nerve. Journal of Cell Biology 123, 455–465

Giehl K M Schütte A Mestres P and Yan Q (1998) The survival-promoting effects of glial cell line-derived neurotrophic factor on axotomized corticospinal neurons *in vivo,* is mediated by an endogenous brain- derived neurotrophic factor mechanism. Journal of Neuroscience 18, 7351– 7360

Grill R Murai K Blesch A Gage F H and Tuszynski M H (1997) Cellular delivery of neurotrophin-3 promotes corticospinal axonal growth and partial functional recovery after spinal cord injury. Journal of Neuroscience 17, 5560–5572

Hall S and Berry M (1989) Electron microscopic study of the interaction of axons and glia at the site of anastomosis between the optic nerve and cellular or acellular sciatic nerve grafts. Journal of Neurocytology 18, 171–184

Ham J Babij C Whitfield J Pfarr C M Lallemand D Yaniv M and Rubin L L (1995) A c-Jun dominant negative mutant protects sympathetic neurons against programmed cell death. Neuron 14, 927–939

Hoffman P N (1989) Expression of GAP-43, a rapidly transported growth-associated protein, and class II β tubulin, a slowly transported cytoskeletal protein, are coordinated in regenerating neurons. Journal of Neuroscience 9, 893–897

Horvat J C and Aguayo A J (1985) Elongation of axons from adult rat motor cortex into PNS grafts. Society for Neuroscience Abstracts 11, 254

Horvat J C Pecot-Dechavassine M Mira J C and Davarpanah Y (1989) Formation of functional endplates by spinal axons regenerating through a peripheral nerve graft. A study in the adult rat. Brain Research Bulletin 22, 103–114

Houle J D (1991) Demonstration of the potential for chronically injured neurons to regenerate axons into intraspinal peripheral nerve grafts. Experimental Neurology 113, 1–9

Huber A B Chen M S van der Haar M E and Schwab M E (1998) Developmental expression pattern and functional analysis of NOGO (formerly NI-35/250), a major inhibitor of CNS regeneration. Society for Neuroscience Abstracts 24, 1559

Hüll M and Bähr M (1994) Regulation of immediate-early gene expression in rat retinal ganglion cells after axotomy and during regeneration through a peripheral nerve graft. Journal of Neurobiology 25, 92–105

Jenkins R McMahon S B Bond A B and Hunt S P (1993) Expression of c-jun as a response to dorsal root and peripheral nerve section in damaged and adjacent intact primary sensory neurons in the rat. European Journal of Neuroscience 5, 751–759

Johnson R S van-Lingen B Papaioannou V E and Spiegelman B M (1993) A null mutation at the c-jun locus causes embryonic lethality and retarded cell growth in culture. Genes and Development 7, 1309–1317

Joosten E A Bar P R and Gispen W H (1995a) Collagen implants and cortico-spinal axonal growth after mid-thoracic spinal cord lesion in the adult rat. Journal of Neuroscience Research 41, 481–490

Joosten E A Bar P R and Gispen W H (1995b) Directional regrowth of lesioned corticospinal tract axons in adult rat spinal cord. Neuroscience 69, 619–626

Jung M Petrausch B and Stuermer C A (1997) Axon-regenerating retinal ganglion cells in adult rats synthesize the cell adhesion molecule L1 but not TAG-1 or SC-1. Molecular and Cellular Neuroscience 9, 116–131

Karin M (1996) The regulation of AP-1 activity by mitogen-activated protein kinases. Philosophical Transactions of the Royal Society of London B 351, 127–134

Keirstead S A Rasminsky M Fukuda Y Carter D A Aguayo A J and Vidal-Sanz M (1989) Electrophysiologic responses in hamster superior colliculus evoked by regenerating retinal axons. Science 246, 255–25

Kharazia V N and Weinberg R J (1994) Glutamate in thalamic fibres terminating in layer IV of primary sensory cortex. Journal of Neuroscience 14, 6021– 6032

Kleitman N Simon D K Schachner M and Bunge R P (1988) growth of embryonic retinal neurites elicited by contact with Schwann cell surfaces is blocked by antibodoes to L1. Experimental Neurology 102, 298–306

Li Y Field P M and Raisman G (1997) Repair of adult rat corticospinal tract by transplants of olfactory ensheathing cells. Science 277, 2000–2002

Lieberman A R (1974) Some factors affecting retrograde neuronal responses to axonal lesions. In Essays on the nervous system, edited by R Bellairs and E G Gray, pp 71–105. Oxford: The Clarendon Press

Martini R (1994) Expression and functional roles of neural cell surface molecules and extracellular matrix components during development and regeneration of peripheral nerves. Journal of Neurocytology 23, 1–28

McConnell P and Berry M (1982) Regeneration of ganglion cell axons in the adult mouse retina. Brain Research 241, 362–365

Meyer M Matsuoka I Wetmore C Olson L and Thoenen H (1992) Enhanced synthesis of brain-derived neurotrophic factor in the lesioned peripheral nerve: different mechanisms are responsible for the regulation of BDNF and NGF mRNA. Journal of Cell Biology 119, 45–54

Miller M W (1987) The origin of the corticospinal projection neurons in the rat. Experimental Brain Research 67, 339–351

Morrow D R Campbell G Anderson P N and Lieberman A R (1993) Differential regenerative growth of CNS axons into tibial and peroneal nerve grafts in the thalamus of adult rats. Experimental Neurology 120, 60–69

Mufson E J Kroin J S Sobreviela T Burke M A Kordower J H Penn R D and Miller J A (1994) Intrastriatal infusions of brain-derived neurotrophic factor: retrograde transport and colocalization with dopamine containing substantia nigra neurons in rat. Experimental Neurology 129, 15–26

Ohara P T and Lieberman A R (1985) The thalamic reticular nucleus of the adult rat: experimental anatomical studies. Journal of Neurocytology 14, 365–411

Ramón-Cueto A Plant G W Avila J and Bunge M B (1998) Long-distance axonal regeneration in the transected adult rat spinal cord is promoted by olfactory ensheathing glia transplants. Journal of Neuroscience 18, 3803–3815

Richardson P M and Issa V M (1984) Peripheral injury enhances central regeneration of primary sensory neurones. Nature 309, 791–793

Richardson P M Issa V M and Aguayo A J (1984) Regeneration of long spinal axons in the rat. Journal of Neurocytology 13, 165–182

Richardson P M Issa V M and Shemie S (1982) Regeneration and retrograde degeneration of axons in the rat optic nerve. Journal of Neurocytology 11, 949–966

Richardson P M McGuinness U M and Aguayo A J (1980) Axons from CNS neurons regenerate into PNS grafts. Nature 284, 264–265

Robinson G A (1994) Immediate early gene expression in axotomized and regenerating retinal ganglion cells of the adult rat. Molecular Brain Research 24, 43–54

Roslan R J Schachner M Vaudano E and Anderson P N (1998) The cell recognition molecule, close homologue of L1, is expressed in motor neurons in the lumbar spinal cord following axotomy. Society for Neuroscience Abstracts 28, 1053

Saga Y Yagi T Ikawa Y Sakakura T and Aizawa S (1992) Mice develop normally without tenascin. Genes and Development 6, 1821–1831

Sceats D J Jr Friedman W A Sypert G W and Ballinger W E Jr (1986) Regeneration in peripheral nerve grafts to the cat spinal cord. Brain Research 362, 149–156

Schaden H Stuermer C A and Bähr M (1994) GAP-43 immunoreactivity and axon regeneration in retinal ganglion cells of the rat. Journal of Neurobiology 25, 1570–1578

Schnell L and Schwab M E (1993) Sprouting and regeneration of lesioned corticospinal tract fibres in the adult rat spinal cord. European Journal of Neuroscience 5, 1156–1171

Schnell L Schneider R Kolbeck R Barde Y A and Schwab M E (1994) Neurotrophin-3 enhances sprouting of corticospinal tract during development and after adult spinal cord lesion. Nature 367, 170–173

Schwab M E and Caroni P (1988) Oligodendrocytes and CNS myelin are nonpermissive substrates for neurite growth and fibroblast spreading *in vitro*. Journal of Neuroscience 8, 2381–2393

Strittmatter S M Fankhauser C Huang P L Mashimo H and Fishman M C (1995) Neuronal pathfinding is abnormal in mice lacking the neuronal growth cone protein GAP-43. Cell 80, 445–452

Swett J E Hong C Z and Miller P G (1991) All peroneal motoneurons of the rat survive crush injury but some fail to reinnervate their original targets. Journal of Comparative Neurology 304, 234–252

Taylor J Pesheva P and Schachner M (1993) Influence of janusin and tenascin on growth cone behavior *in vitro*. Journal of Neuroscience Research 35, 347–362

Thanos S (1992) Adult retinofugal axons regenerating through peripheral nerve grafts can restore the light-induced pupiloconstriction reflex. European Journal of Neuroscience 4, 491–699

Thanos S and Mey J (1995) Type-specific stabilization and target-dependent survival of regenerating ganglion cells in the retina of adult rats. Journal of Neuroscience 15, 1057–1079

Thanos S Naskar R and Heiduschka P (1997) Regenerating ganglion cell axons in the adult rat establish retinofugal topography and restore visual function. Experimental Brain Research 114, 483–491

Trupp M Belluardo N Funakoshi H and Ibanez C F (1997) Complementary and overlapping expression of glial cell line-derived neurotrophic factor (GDNF), c-ret proto-oncogene, and GDNF receptor-alpha indicates multiple mechanisms of trophic actions in the adult rat CNS. Journal of Neuroscience 17, 3554–356

Vaudano E Woolhead C Anderson P N Lieberman A R and Hunt S P (1993) Molecular changes in Purkinje cells (PC) and deep cerebellar nuclei (DCN) neurons after lesion or insertion of a peripheral nerve graft into the adult rat cerebellum. Society for Neuroscience Abstracts 19, 1510

Vaudano E Campbell G Anderson P N Davies A P Woolhead C and Lieberman A R (1995) The effects of a lesion or a peripheral nerve graft on GAP-43 upregulation in the adult rat brain: an *in situ* hybridization and immunocytochemical study. Journal of Neuroscience 15, 3594–3611

Vaudano E Campbell G Hunt S P and Lieberman A R (1998) Axonal injury and peripheral nerve grafting in the thalamus and cerebellum of the adult rat: Upregulation of c-jun and correlation with regenerative potential. European Journal of Neuroscience 10, 2644–2656

Watanabe M Sawai H and Fukuda Y (1993) Number, distribution, and morphology of retinal ganglion cells with axons regenerated into peripheral nerve graft in adult cats. Journal of Neuroscience 13, 2105–2117

Woolhead C Zhang Y Lieberman A R Schachner M Emson P C and Anderson P N (1998) Differential effects of autologous peripheral nerve grafts to the corpus striatum of adult rats on the regeneration of axons of striatal and nigral neurons and on the expression of GAP-43 and the cell adhesion molecules N-CAM and L1. Journal of Comparative Neurology 391, 259–273

Xu X M Guenard V Kleitman N Aebischer P and Bunge M B (1995) A combination of BDNF and NT-3 promotes supraspinal axonal regeneration into Schwann cell grafts in adult rat thoracic spinal cord. Experimental Neurology 134, 261–272

Xu X M Chen A Guenard V Kleitman N and Bunge M B (1997) Bridging Schwann cell transplants promote axonal regeneration from both the rostral and caudal stumps of transected adult rat spinal cord. Journal of Neurocytology 26, 1–16

Ye J H and Houle J D (1997) Treatment of the chronically injured spinal cord with neurotrophic factors can promote axonal regeneration from supraspinal neurons. Experimental Neurology 143, 70–81

Zagrebelsky M Buffo A Skerra A Schwab M E Strata P and Rossi F (1998) Retrograde regulation of growth-associated gene expression in adult rat Purkinje cells by myelin-associated neurite growth inhibitory proteins. Journal of Neuroscience 18, 7912–7929

Zeng B-Y Anderson P N Campbell G and Lieberman A R (1994) Regenerative and other responses to injury in the retinal stump of the optic nerve in adult albino rats: transection of the intraorbital optic nerve. Journal of Anatomy 185, 643–661

Zhang Y Campbell G Anderson P N Martini R Schachner M and Lieberman A R (1995a) Molecular basis of interactions between regenerating adult rat thalamic axons and Schwann cells in peripheral nerve grafts I. Neural cell adhesion molecules. Journal of Comparative Neurology 361, 193–209

Zhang Y Campbell G Anderson P N Martini R Schachner M and Lieberman A R (1995b) Molecular basis of interactions between regenerating adult rat thalamic axons and Schwann cells in peripheral nerve grafts. II. Tenascin-C. Journal of Comparative Neurology 361, 210–224

Zhang Y Anderson P N Campbell G Schachner M Mohajeri H and Lieberman A R (1995c) Tenascin-C expression by glial cells and neurons in the rat spinal cord; changes during postnatal development and after dorsal root or sciatic nerve injury. Journal of Neurocytology 24, 585–601

Zhang Y Campbell G Anderson P N Martini R Schachner M and Lieberman A R (1997a) Upregulation of L1 mRNA in adult rat cerebellar neurons is correlated with their ability to regenerate axons. Society for Neuroscience Abstracts 23, 1721

Zhang Y Winterbottom J K Schachner M Lieberman A R and Anderson P N (1997b) Tenascin-C expression and axonal sprouting following injury to the spinal dorsal columns in the adult rat. Journal of Neuroscience Research 49, 433–450

4

Inflammation and the glial scar: Factors at the site of injury that influence regeneration in the central nervous system

Michael T Fitch and Jerry Silver

Introduction

During the past century, many researchers have sought to uncover the biological principles that underlie the failure of regeneration within the adult mammalian central nervous system (CNS; reviewed in Fitch & Silver, 1999). Many theories have been proposed to explain this unfortunate response to trauma, including: the intrinsic inability of CNS neurons to regenerate, inhibitory influences of CNS myelin, mechanical obstruction created by the astrocytic glial scar, and local production of inhibitory molecules in the vicinity of a lesion. In addition to a myriad of theories which have attempted to explain regeneration failure, numerous therapeutic strategies have been proposed and tested in various models of spinal cord and brain injury. Despite occasional claims over the years of pioneering advances in the field of regeneration research, CNS trauma remains a difficult clinical problem and very few basic research findings have made their way into routine patient care (see also Brown, chapter 1 this volume).

This chapter will introduce and discuss three major topics in the modern field of CNS regeneration research: (i) the popular hypothesis that molecules present in the normal white matter environment are potent inhibitors of regeneration, and recent evidence to the contrary found in the surprisingly robust potential for the adult CNS environment to support long distance regeneration; (ii) the restriction of axon regenerative responses by upregulation of inhibitory molecules associated with the glial scar in the immediate vicinity of a CNS lesion; and (iii) the poorly characterized inflammatory responses to trauma in the brain and spinal cord that may contribute to regenerative failure in some situations and yet allow regeneration to occur in others. We suggest that future investigations into the failure of the adult mammalian central nervous system to regenerate should focus on modulating the inflammatory response and other early post-injury events immediately surrounding the area of trauma which lead to a glial scar with its accompanying inhibitory molecular components.

Does the normal undamaged CNS inhibit regeneration?

It has been recognized for decades that the responses of the CNS to injury are fundamentally different than the responses of the peripheral nervous system (PNS). It is a stark contrast that functional regeneration of axons to their target tissues is often achieved following PNS trauma, while the CNS regenerative response is quite poor and most often leads to permanent disability. It is this biological dichotomy that has led numerous researchers to suggest that a fundamental difference exists between the CNS environment which apparently does not support regeneration, and the PNS environment which facilitates long distance axon regrowth. One popular hypothesis has targeted the oligodendrocyte as the cell type present in the CNS (but not in the PNS) that is responsible for the failure of regeneration within the mammalian brain and spinal cord. A number of investigators have published reports over the past ten years detailing aspects of this proposed explanation for the failure of regeneration (for a review of the oligodendrocyte and myelin inhibition hypothesis, see Caroni & Schwab, 1993).

The early literature supporting this hypothesis (see Caroni & Schwab, 1988) detailed some of the initial experiments that supported the *in vitro* documentation of oligodendrocytes and myelin as nonpermissive growth substrates. In this study, which was performed exclusively in tissue culture, the authors expanded upon the pioneering observations that initially suggested a nonpermissive role for oligodendrocytes and myelin membranes in respect to the growth of neurites and spreading of fibroblasts *in vitro*. It was findings from this paper (Caroni & Schwab, 1988) that established some of the foundations for their later work on this hypothesis. The authors demonstrated qualitatively in a tissue culture model that monoclonal antibodies IN-1 and IN-2 (generated against previously isolated myelin membrane protein fractions of 250 kDa and 35 kDa) neutralized the nonpermissive nature of myelin membrane preparations and oligodendrocyte monolayers. They presented some quantitative data for an enhanced degree of fibroblast spreading (an index they suggested reflects the degree of permissiveness of a substrate) on antibody treated myelin and/or oligodendrocytes. However, it was unclear whether this particular *in vitro* fibroblast spreading effect was relevant to the suggested *in vivo* inhibitory aspects of myelin for regeneration. The paper did go on to qualitatively describe increased numbers of neurites from cultured neurons growing on nonpermissive myelin or oligodendrocytes substrates in the presence of the IN-1 antibody, a result interpreted by the authors as the antibody serving to functionally neutralize undefined myelin inhibitors that are preventing neurite growth. IN-1 antibodies (and IN-2 antibodies) were also demonstrated to bind preferentially to CNS myelin as well as the 250 kDa and 35 kDa myelin protein fractions, and important control experiments established that these neutralizing effects were specific and were not attained by other antibodies that recognize myelin antigens, at least in tissue culture. Finally, optic nerve explants, which do not normally support the ingrowth of neurites from cultured neurons, were induced to support neurite growth when injected with the IN-1 antibody but not with the control O_1 antibody which recognizes an unrelated myelin antigen. The authors suggested that the IN-1 antibody demonstrated this specific activity of enhancing neurite growth on previously nonpermissive substrates by blocking the potent inhibitory activity of myelin associated proteins.

This study (Caroni & Schwab, 1988) was one of several from the Schwab laboratory and others which reported inhibitory effects of oligodendrocytes and/or myelin membranes on fibroblast spreading and neurite outgrowth in two-dimensional tissue culture models. However, there are several published experiments which suggest that oligodendrocytes may not be universally inhibitory in tissue culture, such as the studies by Kobayashi *et al.* (1995) and Moorman & Gould (1997). In the series of experiments by Kobayashi *et al.* (1995), evidence was presented which demonstrated that neurites from retinal ganglion cell neurons are not inhibited by contact with oligodendrocytes, while neurites from dorsal root ganglion and sympathetic ganglion neurons were indeed inhibited by such contact (as was shown by Caroni & Schwab, 1988). This paper made an important contribution to the field of oligodendrocyte inhibition of neurite growth *in vitro,* as it demonstrated that not all neuron types are affected in tissue culture by the previously described oligodendrocyte-associated inhibitors. The recent publication by Moorman & Gould (1997) demonstrated that the developmental stage of oligodendrocytes is extremely important to consider when examining growth cone collapsing activity in culture. This paper reported that Stage 2 (pro-OL), Stage 3 (pre-GalC), and Stage 4 (Immature) oligodendrocytes cause growth cone collapse the majority of the time in culture, in agreement with other studies documenting the inhibitory interactions of growth cones and oligodendrocytes *in vitro.* However, these experiments also documented that Stage 1 (O2A) and Stage 5 (Mature) oligodendrocytes do not lead to significant numbers of growth cone collapses. It is particularly important to note that the Stage 5 oligodendrocytes express the 35 kDa putative inhibitory protein, myelin-associated glycoprotein (MAG), and myelin basic protein (MBP), and yet do not lead to significant growth cone collapse. The authors suggested that one explanation for this observation is that other molecules on the surface of mature oligodendrocytes may negate any inhibitory influences, and they go on to postulate that "When expression of putative inhibitory molecules is considered in the context of the changing array of molecules on the surface of an oligodendrocyte, they more likely function as modulators of neuronal growth cone motility and not as simple inhibitors." (Moorman & Gould, 1997).

These three *in vitro* studies by Caroni & Schwab (1988), Kobayashi *et al.* (1995) and Moorman & Gould (1997) along with numerous other published manuscripts, demonstrated the nonpermissiveness of certain stages of oligodendrocytes and/or myelin as tissue culture substrates for neurite outgrowth from certain specific embryonic neuron types. These observations have supported the widely discussed hypothesis that normal adult CNS myelin contains inhibitors that are responsible for the failure of regeneration following trauma to the brain and spinal cord. While this is perhaps a logical hypothesis to generate based on such *in vitro* data, it is important to recognize that the nonpermissive nature of myelin or oligodendrocytes as the primary substrates for growth in two dimensional tissue culture assays may not accurately reflect their support characteristics in three dimensions inside the living mammalian brain or spinal cord, especially considering the possible influences of other cell types that are found within these tissues *in vivo.* It is also important to consider the difference between a substrate that is "nonpermissive" (i.e., not preferred by neurites as a molecular substrate for their outgrowth), and a substrate that is actively "inhibitory" for neurite outgrowth. The data discussed in these three example papers do not conclusively distinguish between "nonpermissive" and "inhibitory" characteristics of the oligodendrocyte and myelin substrates, because the substrates were not presented in combination with other potentially growth supporting surfaces as would be encountered *in vivo*. Thus it is most correct to state that oligodendrocytes of particular stages and purified myelin have been shown to be poor substrates for the tissue culture growth of neurites from some neuronal cell types. That these substrates are poorly supportive for growth in tissue culture should not be considered definitive proof that the molecules in question play an active role in inhibiting axon regeneration without further experimental evidence, particularly from *in vivo* studies.

To address the ultimate question of whether antibodies directed against the putative inhibitory 250 kDa and 35 kDa myelin proteins would influence the growth of axons *in vivo* following an injury to the CNS, a number of studies have been conducted using IN-1 antibodies and various models of trauma (Schnell & Schwab, 1990; Bregman, *et al.*, 1995). In both of these studies, hybridoma cells secreting the IN-1 antibody were implanted into rat host brains in conjunction with experimental spinal cord injuries. As a control, an anti-horseradish peroxidase (anti-HRP) antibody secreting hybridoma mass was implanted into the brains of some animals followed by experimental spinal injury. The results of each of these studies demonstrated significant sprouting and regeneration of axons in the animals receiving the IN-1 antibody treatment, and the second study (Bregman *et al.*, 1995) went on to demonstrate a small but significant recovery of contact placing behavior and locomotion in the animals treated with the IN-1. Control animals (receiving anti-HRP antibodies) did not display histological or behavioral evidence of regeneration. These experiments suggested that some aspect of the IN-1 hybridoma treatment led to a CNS environment that supported at least limited regeneration, and the authors suggested that it was a specific effect mediated by neutralization of so-called "myelin associated inhibitors." However, it is interesting to note that several other studies (Guest *et al.*, 1997; Thallmair *et al.*, 1998; Z'Graggen *et al.*, 1998) utilizing IN-1 antibodies indicated that this treatment can perhaps lead to increased sprouting but not long distance regeneration through the lesion area under these experimental conditions.

While the histological and functional recovery demonstrated in some studies using the IN-1 antibodies *in vivo* suggests the importance of modifying myelin components following a CNS lesion, the specificity of this treatment as a therapeutic agent to encourage axon regeneration remains unclear. It is interesting that the investigators conducting the experiments using the IN-1 antibody *in vivo* chose to utilize control antibodies raised against horseradish peroxidase (Schnell & Schwab, 1990; Bregman *et al.*, 1995; Guest *et al.*, 1997; Thallmair *et al.*, 1998; Z'Graggen *et al.*, 1998), an antigen that is not present within the normal myelin structure where the IN-1 antigens are located. This is in contrast to the more appropriate control antibodies used in the previously discussed *in vitro* study (Caroni & Schwab, 1988), which used the antibody O_1 that recognizes antigens that are actually found within adult CNS myelin. This raises the possibility that the effects seen with the *in vivo* use of the IN-1 antibody may not be a specific effect of binding to a particular myelin-associated antigen, but that perhaps antibodies that recognize other myelin proteins may have similar beneficial effects as those demonstrated in these studies. These beneficial effects, therefore, may not be a direct consequence of blocking specific inhibitory components of myelin, but instead may be due to an indirect effect caused by physical disruption to the myelin structure. In fact, experiments by other investigators have demonstrated regeneration of axons and electrophysiological recovery following disruption of the myelin structure via *in vivo* treatments with Gal-C or O_4 antibodies (against antigens that are found in normal myelin) combined with complement (Keirstead *et al.*, 1995; see also Tetzlaff & Steeves, chapter 5 this volume). Such results demonstrate that less specific antibody-mediated disruption of myelin is beneficial to axon regeneration, and that a component of the normal inflammatory response can mediate such alterations in myelin structure. Importantly, studies using the IN-1 antibodies as therapeutic agents have not addressed the possibility that antibodies that bind to oligodendrocytes and myelin membranes may, in turn, create a persistent low grade inflammatory reaction that is secondary to antibody binding or myelin destruction. Such limited inflammatory responses have previously been associated with beneficial effects on regeneration *in vivo* (as discussed later in this chapter, and in Fitch & Silver, 1999), perhaps through the direct effects of secreted inflammatory products on axon growth or indirectly via cytokine modulation of reactive astrocytes. Another aspect of the possible immune system involvement in the *in vivo* IN-1 antibody experiments is the use of Cyclosporine A immunosuppressive therapy to prevent rejection of the implanted hybridoma cells in the studies that demonstrated behavioural recovery (Schnell & Schwab, 1990; Bregman *et al.*, 1995; Thallmair *et al.*, 1998; Z'Graggen *et al.*, 1998). A recent study (Palladini *et al.*, 1996) demonstrated histological and behavioural recovery after spinal injury with no antibody treatments to modify inhibitors and only Cyclosporine A treatment used to influence inflammation. This provides an additional suggestion that the modulation of the inflammatory response to injury may be an important aspect to consider in treating CNS trauma.

Despite the popularity of the myelin-inhibition hypothesis, recent *in vivo* experiments from our laboratory have demonstrated that adult CNS myelinated axon tracts are highly permissive environments for long distance regeneration of adult neurons without the need for myelin inhibitor neutralization (Davies *et al.*, 1997). In these experiments, adult neurons from the dorsal root ganglion of donor animals were gently transplanted directly into the myelinated tracts of the corpus callosum and fimbria. The neurons were microtransplanted using a micropipette (diameter <120 μm) and minute volumes (0.5 μL) pulsed gently with a picospritzer. This technique minimizes tissue trauma and eliminates the formation of a histologically apparent glial scar. The transplants were assayed for regeneration by examining a subset of dorsal root ganglion neurons that express the neurotransmitter calcitonin gene related peptide (CGRP), which is not normally present in axons of the corpus callosum and fimbria. These CGRP containing adult dorsal root ganglion neurons were shown in tissue culture to be "inhibited" by oligodendrocytes (i.e., oligodendrocytes were a nonpermissive substrate for neurite outgrowth), confirming the previous *in vitro* studies by Caroni & Schwab (1988) and others. Despite the prediction from these tissue cultures studies, it was remarkable to find upon *in vivo* transplantation that rapidly advancing streamlined growth cones were documented on regenerating axons that gained access to the host glial terrain. These regenerating axons grew at speeds of 1–2 mm per day all the way from the transplant location, across the midline, and into the contralateral brain hemisphere. These experiments have demonstrated for the first time that at least one type of adult neuron is capable of robust regeneration within myelinated white matter tracts of the adult CNS. Experiments in which dorsal root ganglion neurons were transplanted

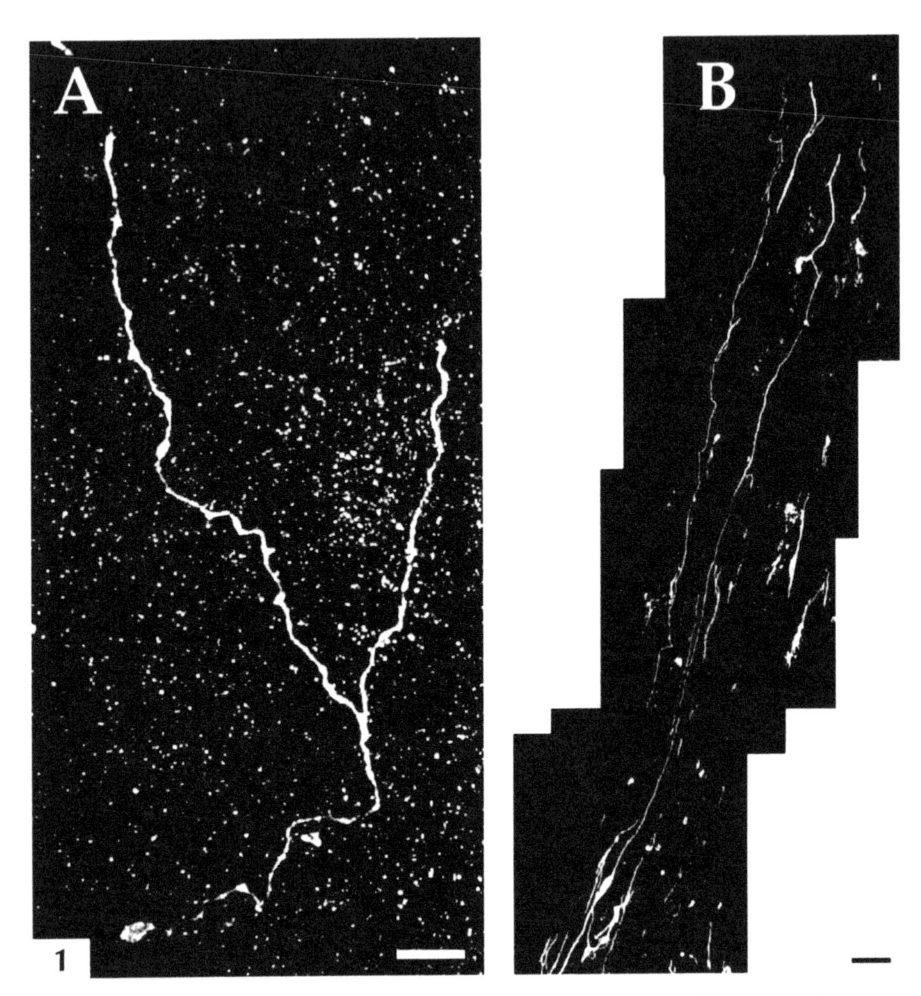

Figure 1 Adult rat dorsal root ganglion cells regenerating within the adult rat dorsal columns of the spinal cord. The neurons have been marked prior to transplantation with an adenovirus vector (a kind gift of K. Moriyoshi, see Neuron 16, 255–260, 1996.) and are expressing green fluorescent protein (GFP). (A) Single regenerating dorsal root ganglion neuron four days after transplantation. Scale bar, 9 μm. (B) Several regenerating dorsal root ganglion neurons extending processes through the dorsal columns eleven days after transplantation. Scale bar, 50 μm.

directly into the dorsal columns of the spinal cord (the normal pathway for these axons) have shown that adult dorsal root ganglion neurons are also capable of growth within their normal white matter tract in the adult CNS (see Figure 1) and can even regenerate in the presence of degenerating myelin (Davies *et al.*, 1999). These studies illustrate that the myelinated pathways of the adult CNS are actually highly permissive for axon regeneration, and suggest that any putative inhibitory effects of myelin that may exist *in vitro* may be relatively minor factors *in vivo* when compared to the growth supporting environment that is present in the CNS.

What molecules in the vicinity of the lesion can explain the failure of regeneration after injury?

Another important aspect of the study by Davies *et al.* (1997) was the demonstrated association of regenerative failure with the upregulation of glial cell extracellular matrix in the immediate vicinity of the lesion site. Those transplants of adult dorsal root ganglion neurons that failed to regenerate were surrounded by a rich extracellular matrix containing chondroitin sulphate proteoglycans, molecules that have been previously shown to be inhibitory to axon regeneration. This failure of regeneration is the most striking evidence to date for a direct relationship between increases in extracellular matrix molecules and boundaries of regeneration failure (for a review of extracellular matrix molecules as boundaries in development and regeneration, see Fitch & Silver, 1997b). Importantly, even in the absence of a physical barrier to regeneration, increases in inhibitory molecules such as proteoglycans as a part of the glial scar that forms in the immediate area of tissue damage may negatively influence regeneration (reviewed in Fitch & Silver, 1998).

Proteoglycans are a class of molecules produced by reactive astrocytes and have been suggested to play a role in the modulation of axon growth and regeneration. Proteoglycans consist of a protein core with attached sugar moieties called glycosaminoglycans and are characterized as chondroitin sulphate, heparan sulphate, keratan sulphate, and dermatan sulphate (for a review of nervous tissue proteoglycans see Margolis & Margolis, 1993). The upregulation of proteoglycans is found in many tissues throughout the body in pathological conditions, such as regenerating skeletal muscle, arterial injury, atherogenesis, and corneal injury. The nervous system is no exception, as increases in proteoglycans have also been demonstrated *in vivo* following trauma to the adult CNS. Chondroitin sulphate proteoglycans are one category of molecules that increase and persist in the extracellular matrix of the CNS following injury, including the spinal cord following dorsal root injury, in the fornix following transection, in the brain following stab wound, in explants of wounded striatum, and in the spinal cord following penetrating crush injury (for review see Fitch & Silver, 1997b). The presence of these putative inhibitory proteoglycans *in vivo* following injury suggests a role for these molecules in contributing to the nonpermissive environment encountered in the CNS around the immediate area of a lesion.

Proteoglycans associated with reactive astrocytes have been demonstrated by many different investigators to inhibit neurite outgrowth (for example see Snow *et al.*, 1990; and McKeon *et al.*, 1991). The experiments by Snow *et al.* (1990) were the first to illustrate that a purported boundary in the CNS which contained proteoglycans might actually function as a boundary as a direct result of its proteoglycan content. It was shown in culture that a chondroitin/keratan sulphate proteoglycan boundary was inhibitory to dorsal root ganglion neurite growth as growing neurites avoided and turned away when presented with an abrupt transition to this inhibitory substrate. McKeon *et al.* (1991) went on to demonstrate that the growth of axons on glial scar tissue harvested directly from the adult brain via implanted nitrocellulose is inhibited at least in part by the proteoglycans present in the extracellular matrix. This inhibition could be partially reversed by treatment of the scar tissue with enzymes that removed specific sugar epitopes from the proteoglycan molecules (McKeon *et al.*, 1995). While such studies provide compelling evidence for a direct effect of proteoglycans on neurite

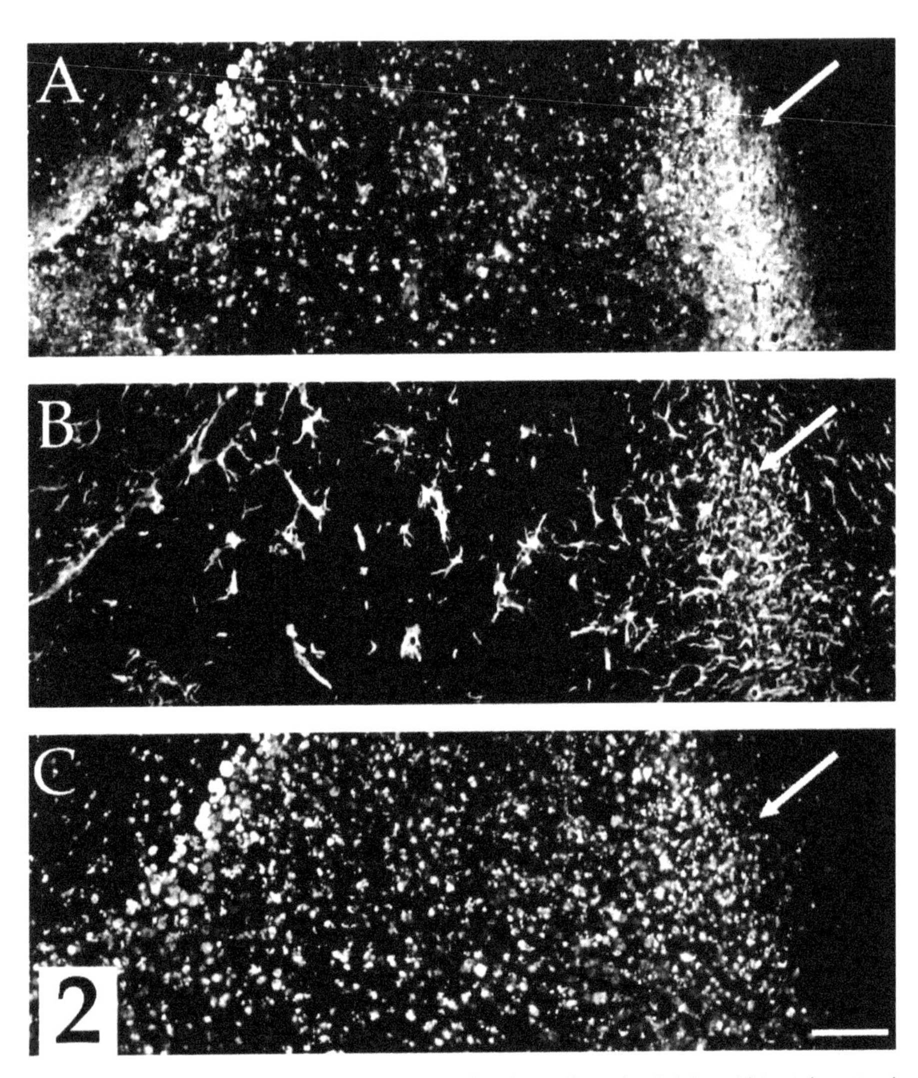

Figure 2 A developing cyst in the brain two weeks after unilateral stab injury. This triple stained section illustrates proteoglycan upregulation (A), GFAP positive astrocytes (B), and activated macrophages and microglial cells (C). Note the sparse population of astrocytes found within the developing cavity (B) and the dense macrophage infiltrate (C). Proteoglycan upregulation is found at the interface between activated macrophages that fill the developing cavity and the surrounding astrocytes (arrows). Scale bar, 100 μm. From Fitch and Silver (1997a), used with permission.

extension, a recent report suggests that in some situations chondroitin sulphate proteoglycans may regulate and organize other extracellular matrix associated molecules, perhaps by directly binding to various undefined growth promoting or inhibiting factors (Emerling & Lander, 1996). Such findings indicate that proteoglycan molecules present in glial scars may play direct and/or indirect roles in the lack of regeneration in the injured adult CNS, and that modifications of the inhibitory effects of post-injury CNS glial scar tissue may be possible.

Proteoglycans and other astrocyte extracellular matrix products have also been found closely associated with the formation of cystic cavities in the CNS. Cystic cavitation, which can be the result of progressive necrosis and secondary tissue damage, is an important problem in CNS injury as the size and extent of injuries can continue to increase from days to weeks after the primary event. Astrocytes normally produce a basal lamina (which contains proteoglycans) at the pial surface of the CNS, and following traumatic injury to the CNS astrocytes produce ectopic basal lamina components as part of the glial scar which borders the cut edges of the injured tissue. Chondroitin sulphate proteoglycans are examples of components that have been described by MacLaren (1996) and our laboratory (Fitch & Silver, 1997a) at the interface between developing cavities and the surrounding viable tissue. The study by MacLaren (1996) suggests that successful regrowth of ganglion cells through retinal lesions is due to the lack of inhibitory molecule upregulation in the retina as a direct contrast to the association of inhibitory proteoglycans with developing cavities in the cortex where regeneration does not occur. The study from our laboratory (Fitch & Silver, 1997a) demonstrated an association of proteoglycans with glial scarring and the borders of cystic cavities of both the brain and the spinal cord (see Figure 2). The function of extracellular matrix products surrounding a necrotic cavity of the CNS is open for speculation, and it is an intriguing possibility that glial scar components such as proteoglycans may play a role in "walling off" the injured tissue in an effort to protect the surrounding viable cellular environment from further damage. It is possible that the CNS uses proteoglycans as a molecular protectant of tissue destruction by degradative enzymes or secondary tissue damage by inflammatory cells following a traumatic injury.

One important question that so far has remained unanswered is the identity of the molecular triggers that control the production of astroglial inhibitory extracellular matrix after an injury. Using lesion models of the brain and spinal cord, our laboratory has suggested (see Fitch & Silver, 1997a) that the increases in inhibitory molecules are associated with a breakdown of the blood-brain barrier and infiltrating macrophages present within the lesion site (see Figure 3). These observations suggest that either leakage of serum proteins or infiltrating inflammatory cells (or both) influence the production of extracellular matrix molecules in the vicinity of a CNS wound. These results highlight the importance of inflammation in the complicated sequelae of a CNS injury, and point to modulation of these events as a potential therapeutic avenue.

Inflammation after CNS injury: good or bad for regeneration?

The inflammatory response by the CNS following injury is composed primarily of two components: activation of intrinsic microglial cells and recruitment of bone marrow derived inflammatory cells from the peripheral bloodstream (for review see Perry *et al.*, 1993). Chemical injuries to the brain appear to lead to a predominantly microglial cell inflammatory response, while direct stab wounds and injuries are composed mostly of peripheral monocytes. However, it is generally accepted that both microglia and peripherally derived macrophages respond to injury in various proportions depending on the type and severity of the lesion. This inflammatory response to injury is thought by some investigators to contribute to secondary tissue damage within the CNS. As an

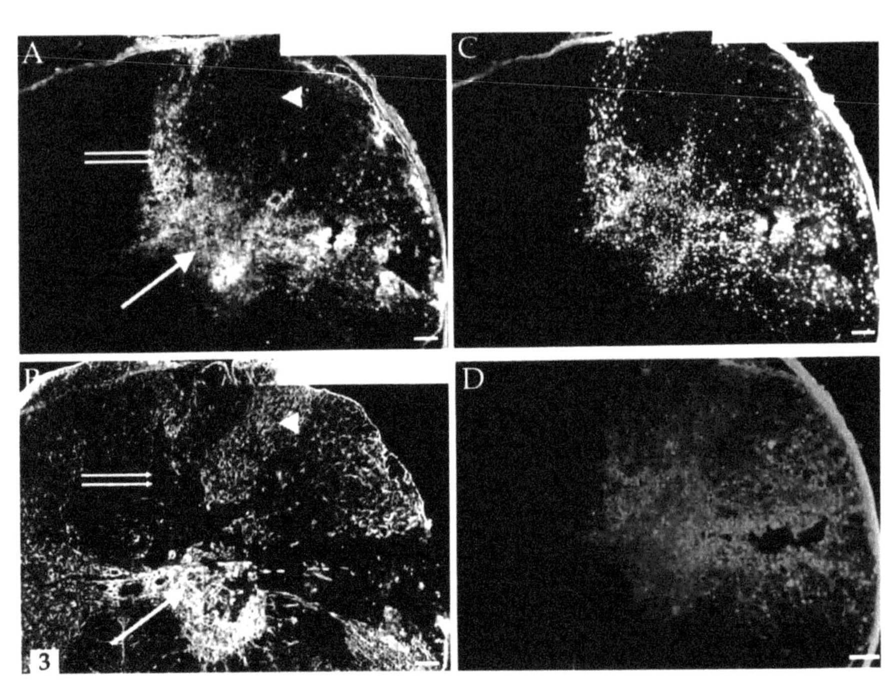

Figure 3 Upregulation of inhibitory proteoglycans is associated with inflammation and breakdown of the blood brain barrier. (A-C) Representative cross section of a spinal cord hemicrush lesion one week after injury, triple stained for (A) chondroitin sulphate proteoglycans, (B) GFAP to visualize astrocytes, and (C) macrophages / microglial cells. (D) An adjacent section stained for extravasated immunoglobulins demonstrates areas with compromised blood brain barrier. Large arrow (A-B) indicates a region of astrocyte gliosis associated with increased levels of proteoglycan. Arrowhead (A-B) indicates a region of astrocyte gliosis not associated with increased proteoglycan staining. Double arrows (A-B) indicate an area with no GFAP positive astrocytes that demonstrates increased proteoglycan. Note the close association between the distribution of proteoglycan, inflammation, and blood brain barrier breakdown (A, C, and D) that is not consistently associated with regions of astrocyte gliosis (B), suggesting that extravasation of serum components and/or products secreted by inflammatory cells are triggering the production of inhibitory extracellular matrix molecules (see Fitch & Silver, 1997a). Scale bar, 100 μm. From Fitch & Silver (1997a), used with permission.

excellent example, Blight (1994) utilized silica as a macrophage toxin to deplete animals of inflammatory cells prior to or immediately following experimental spinal crush injuries. The effect of the macrophage depletion was a delay in secondary onset of functional loss below the lesion and the number of axons remaining in the region of injury was higher, suggesting that macrophages in the non-depleted control group were contributing to the normal secondary tissue loss and onset of behavioral deficits. Giulian *et al.* (1993) demonstrated that activated microglial cells in tissue culture secrete a novel class of neuron-killing molecules that are toxic to neurons in culture but do not kill other brain cell types, and these cytotoxic effects by inflammatory cells that are normally present at the lesion site following trauma are consistent with the areas of expanding tissue damage often found after CNS injury.

A recent report in the field of Alzheimer's disease research, while not directly discussed by the authors in relation to CNS injury, presents compelling evidence that activated inflammatory cells can lead to significant damage to nearby tissue. This study by Weldon *et al.* (1998) demonstrates that fibrillar ß-amyloid (but not soluble ß-amyloid or vehicle) injected into the rat striatum leads to activation of inflammatory cells and subsequent loss of parvalbumin and nNOS (neuronal nitric oxide synthase) containing neurons outside the immediate area of injection, an effect that the authors suggest is secondarily due to inflammation and not direct toxicity of the ß-amyloid. These three examples, along with many others found in the literature and our recent study (Fitch *et al.*, 1999), suggest that inflammatory events accompanying trauma can have negative effects on the tissue response to injury and may in fact lead to more damage as a secondary effect.

However, there are also reports in the literature that do not support the hypothesis that inflammatory cells play a role in secondary tissue damage following CNS injury. Dusart & Schwab (1994) published a study in which the events occurring at spinal cord lesion sites from 1 hour to 3 months after injury were characterized. They found that recruitment of inflammatory cells into the lesion in their partial transection model lagged a day or two after an area of secondary cell death had already developed, and they hypothesized that this cell death was of a necrotic type. They present an argument, based on the timing of appearance of inflammatory cells in this model, that the infiltrating macrophages and neutrophils arrive too late into the CNS tissue to be solely responsible for the cell death. However, activated microglial cells are seen within 1 hour of the lesion which is considerably before the time of secondary cell death described after 12 hours and these cells could potentially play a role in secondary cell death, an issue that is acknowledged by the authors in this manuscript. The infiltrating macrophages do seem to play a role in the eventual development of cystic cavities after clearing damaged tissue. A recent report by Klusman & Schwab (1997) contradicts other recent reports which suggest that inflammatory cytokines may contribute to further tissue pathology after CNS trauma. This study examined the effects of exogenous administration of the pro-inflammatory cytokines IL-1ß, IL-6, and TNF-α to lesioned spinal cords of adult mice. Despite the finding that this cocktail of cytokines did indeed lead to large increases in activated macrophages and microglial cells within the spinal cord, it was surprising to learn that animals receiving the mixture actually had smaller lesions than those receiving control solutions. Under this particular set of experimental conditions, it appears that inflammation may actually be helpful in controlling the size of the spinal cord lesion. These two experiments directly contradict the hypothesis that inflammation always leads to increases in secondary tissue damage.

While it may seem a bit paradoxical, the study by Klusman & Schwab (1997) is not the first to suggest that the inflammatory response within the CNS may actually have positive effects on the healing of nervous system wounds. Several experiments that demonstrated limited regeneration of PNS axons into the CNS environment noted the presence of a mild inflammatory reaction at the site of axon entry into the CNS, suggesting a positive role for regeneration (Kliot *et al.*, 1990; Siegal *et al.*, 1990). In fact, two recent studies from different laboratories highlight a hypothesis that transplanting inflammatory cells activated in certain ways may actually contribute to some repair of CNS injuries (Lazarov-Spiegler *et al.*, 1996; Rabchevsky & Streit, 1997). Lazarov-Spiegler *et al.* (1996) suggest that transplanting activated macrophages that are preincubated with peripheral nerve will lead to an increase in regeneration of the transected optic nerve, and they document regeneration using anterograde and retrograde tracing. However, it is difficult to determine whether their transection model is a complete lesion and the possibility remains that the observed results could be accounted for by enhanced sparing or increased survival of spared axons. Rabchevsky & Streit (1997) documented increases in neurite growth into lesion cavities of the spinal cord in response to transplanted microglial cells from tissue culture and suggested that activated microglial cells may actually support axon regeneration via the production of growth promoting molecules such as laminin. The increased number of neurites growing into the region of spinal cord injury and microglial cell transplantation led the authors to suggest that these inflammatory cells were a growth promoting substrate for neurites. The authors hypothesized that microglial cells may counteract mechanisms that inhibit regeneration. This particular study only examined neurite growth into the microglial cell graft site and did not attempt to demonstrate any long distance regeneration out of the graft or document behavioral changes. Thus, it seems that transplanted inflammatory cells activated in specific ways may serve to exert positive influences on injured CNS tissue in as yet undefined ways, and future experiments will continue to shed light on how modifying inflammatory responses may enhance regenerative responses.

Modulation of the inflammatory response is one area of research that is being aggressively pursued towards finding a therapy for CNS injury. The current standard of care for human spinal cord injury is the use of rapid and high doses of methylprednisolone (Bracken *et al.*, 1990), a steroid which certainly influences the inflammatory cascade (see also Brown, chapter 1, this volume). Over the years a variety of "unusual" therapies have been proposed that lacked a clearly defined and rigorously tested mechanism for their reportedly beneficial effects on experimental CNS injuries (e.g. X-ray therapy, hypothermia, low power lasers, electromagnetic fields), but it is interesting to note that many of these reported treatment protocols probably modify the immune system and inflammatory responses to injury and may actually act to minimize damage or promote recovery and least partially based on this type of activity. In fact, many investigators have suggested that inflammatory modulators can be used therapeutically to treat CNS injury, with examples as long ago as when Windle *et al.* (1952) used Piromen, an inflammatory agent, to inhibit the formation of the glial scar and allow regeneration. Giulian *et al.* (1989) demonstrated that chloroquine or colchicine reduced inflammation after brain injury and limited subsequent gliosis and damage, and suggested that the use of drugs to modulate trauma induced inflammation may be useful in the treatment of acute brain injury. More recently, Guth *et al.* (1994) used bacterial lipopolysaccharide (LPS) as an inflammatory mediator (similar to Piromen) in combination with Indomethacin (to inhibit prostaglandin synthesis) and demonstrated histological and functional recovery from spinal cord crush injuries. Other anti-inflammatory agents were demonstrated in a later study (Zhang *et al.*, 1997) to attenuate the process of secondary necrosis and limit the lesion size in the spinal cord. These studies, along with others, highlight the benefits of a combination approach to treatment of CNS injury with agents that modulate the inflammatory response in different ways.

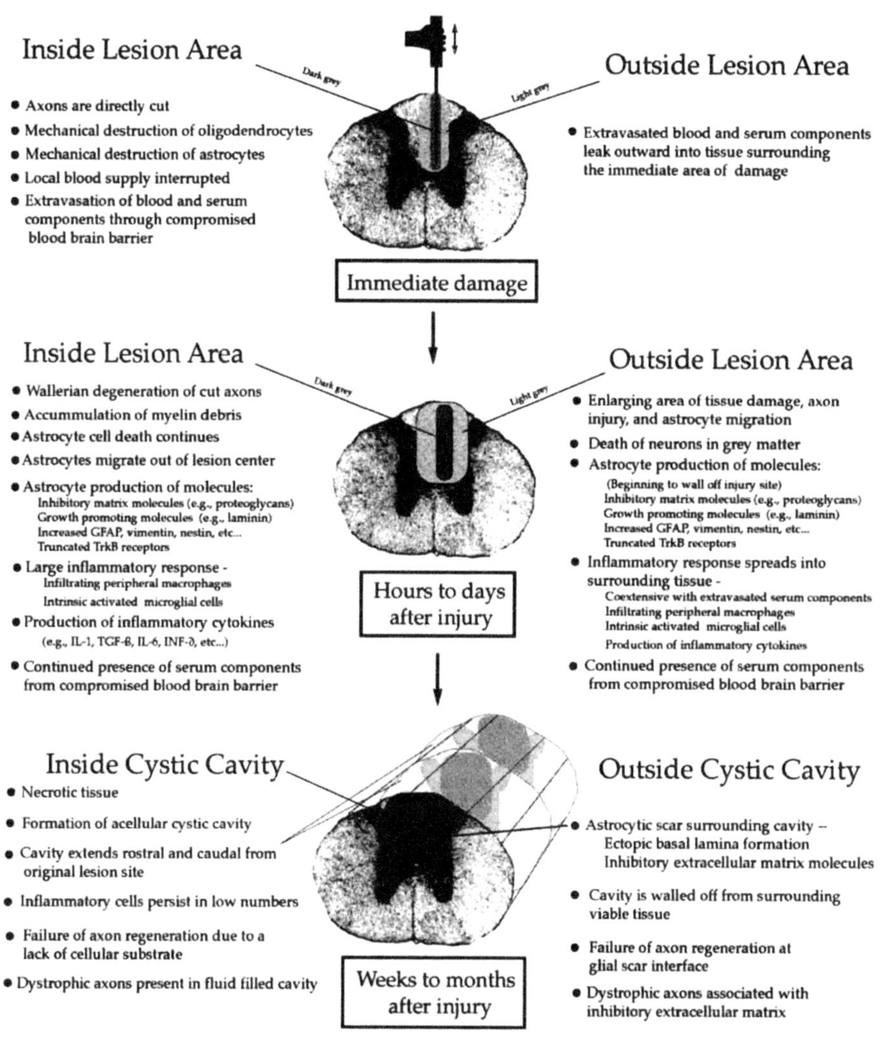

Figure 4 Cellular and molecular events following spinal cord injury that lead to progressive necrosis and cystic cavitation. Schematic illustration summarizing some of the major events that occur in three general phases of the response to spinal cord injury: immediate cellular and tissue damage, events occurring hours to days after injury, and the resolution of the wound weeks to months after injury. While this figure uses a spinal cord stab injury as a model for central nervous system wound healing, the general features of the tissue responses are relevant to trauma to both brain and spinal cord by various mechanisms. Note especially the enlarging acellular cavity that can extend laterally as well as rostral and caudal to the lesion site, and the astrocytic response to this progressive necrosis which includes the production of inhibitory molecules as a way to "wall off" the injury site. This figure demonstrates that leakage of serum factors through a compromised blood brain barrier and subsequent inflammatory infiltrates interact to influence the astrocyte reaction to CNS damage that ultimately leads to increased cell death and regenerative failure. From Fitch & Silver (1998), used with permission.

Future directions

This chapter has presented arguments and highlighted evidence demonstrating that the normal CNS environment can be highly permissive for regeneration, despite the *in vitro* predictions that molecules present in the normal brain and spinal cord are nonpermissive substrates. Therefore, one of the central keys to developing successful therapies for CNS injuries lies in the reactions of the tissue in and around the immediate site of a lesion and the subsequent glial scar formation. It is the complex cellular and molecular responses to injury in this defined area that lead to a local environment that is either not supportive or overtly inhibitory to long distance axon regeneration (see Figure 4). Inflammation, production of inhibitory extracellular matrix molecules in the glial scar, and secondary damage to surrounding tissue all contribute to the failure of regeneration in the mammalian CNS. Future research should focus on modulation of the inflammatory reactions and other events which occur in the immediate vicinity of a lesion as investigators continue to search for ways to modify these tissue responses to create a more supportive environment for regeneration in the central nervous system.

Key Points

- The normal adult central nervous system can be highly permissive for long distance axon regeneration from adult neurons

  Myelin may be a nonpermissive substrate for tissue culture, but it is not universally a potent inhibitor for all axon regeneration *in vivo*

- Proteoglycans and other extracellular matrix molecules that are upregulated in the region of the glial scar following central nervous system injury contribute to regeneration failure in the immediate vicinity of a lesion

- The inflammatory response to brain and spinal cord injury is complex and may have effects on regeneration

  Inflammatory cells can contribute to secondary tissue damage and progressive necrosis

  Inflammation may trigger responses from the tissue surrounding the lesion that are detrimental to axon growth

  Inflammation under some circumstances may actually be beneficial to regeneration

- Future research needs to focus on the modulation of inflammation as a key component of any therapy for central nervous system injury

References

Blight A R (1994) Effects of silica on the outcome from experimental spinal cord injury: implication of macrophages in secondary tissue damage. Neuroscience 60, 263–273

Bracken M B Shepard M J Collins W F Holford T R Young W Baskin D S Eisenberg H M Flamm E Leo S L Maroon J et al. (1990) A randomized, controlled trial of methylprednisolone or naloxone in the treatment of acute spinal-cord injury. Results of the Second National Acute Spinal Cord Injury Study. New England Journal of Medicine 322, 1405–1411

Bregman B S Kunkel B E Schnell L Dai H N Gao D and Schwab M E (1995) Recovery from spinal cord injury mediated by antibodies to neurite growth inhibitors. Nature 378, 498–501
Caroni P and Schwab M E (1993) Oligodendrocytes- and myelin-associated inhibitors of neurite growth in the adult nervous system. Advances in Neurology 61, 175–179
Caroni P and Schwab M E (1988) Antibody against myelin-associated inhibitor of neurite growth neutralizes nonpermissive substrate properties of CNS white matter. Neuron 1, 85–96
Davies S J A Fitch M T Memberg S P Hall A K Raisman G and Silver J (1997) Robust regeneration of adult axons in white matter tracts of the central nervous system. Nature 390, 680–683
Davies S J A Goucher D R Doller C and Silver J (1999) Robust regeneration of adult sensory axons in degenerating white matter of the adult rat spinal cord. Jounal of Neuroscience. 19, 5810–5822
Dusart I and Schwab M E (1994) Secondary cell death and the inflammatory reaction after dorsal hemisection of the rat spinal cord. European Journal of Neuroscience 6, 712–724
Emerling D E and Lander A D (1996) Inhibitors and promoters of thalamic neuron adhesion and outgrowth in embryonic neocortex: Functional association with chondroitin sulfate. Neuron 17, 1089–1100
Fitch M T Doller C Combs C K Landreth G E and Silver J (1999) Cellular and molecular mechanism of glial scarring and progressive cantation: *In vivo* and *in vitro* analysis of inflammation induced secondary injury following central nervous system trauma. Journal of Neuroscience. In press.
Fitch M T and Silver J (1997a) Activated macrophages and the blood brain barrier: Inflammation after CNS injury leads to increases in putative inhibitory molecules. Experimental Neurology 148, 587–603
Fitch M T and Silver J (1997b) Glial cell extracellular matrix: Boundaries for axon growth in development and regeneration. Cell and Tissue Research 290, 379–84
Fitch M T and Silver J (1999) Beyond the glial scar: Cellular and molecular mechanisms by which glial cells contribute to CNS regenerative failure. In CNS Regeneration: Basic Science and Clinical Advances, edited by M H Tuszynski and J H Kordower, pp 55–88. Academic Press
Guest J D Hesse D Schnell L Schwab M E Bunge M B and Bunge R P (1997) Influence of IN-1 antibody and acidic FGF-fibrin glue on the response of injured corticospinal tract axons to human Schwann cell grafts. Journal of Neuroscience Research 50, 888–905
Giulian D Chen J Ingeman J E George J K and Noponen M (1989) The role of mononuclear phagocytes in wound healing after traumatic injury to adult mammalian brain. Journal of Neuroscience 9, 4416–4429
Giulian D Vaca K and Corpuz M (1993) Brain glia release factors with opposing actions upon neuronal survival. Journal of Neuroscience 13, 29–37
Guth L Zhang Z DiProspero N A Joubin K and Fitch M T (1994) Spinal cord injury in the rat: treatment with bacterial lipopolysaccharide and indomethacin enhances cellular repair and locomotor function. Experimental Neurology 126, 76–87
Keirstead H S Dyer J K Sholomenko G N McGraw J Delaney K R and Steeves J D (1995) Axonal regeneration and physiological activity following transection and immunological disruption of myelin within the hatchling chick spinal cord. Journal of Neuroscience 15, 6963–6974
Kliot M Smith G M Siegal J D and Silver J (1990) Astrocyte-polymer implants promote regeneration of dorsal root fibers into the adult mammalian spinal cord. Experimental Neurology 109, 57–69
Klusman I and Schwab M E (1997) Effects of pro-inflammatory cytokines in experimental spinal cord injury. Brain Research 762, 173–184
Kobayashi H Watanabe E and Murikami F (1995) Growth cones of dorsal root ganglion but not retina collapse and avoid oligodendrocytes in culture. Developmental Biology 168, 383–394

Lazarov-Spiegler O Solomon A S Zeev-Brann A B Hirschberg D L Lavie V and Schwartz M (1996) Transplantation of activated macrophages overcomes central nervous system regrowth failure. FASEB Journal 10, 1296–1302

MacLaren R E (1996) Development and role of retinal glia in regeneration of ganglion cells following retinal injury. British Journal of Opthalmology 80, 458–464

Margolis R K and Margolis R U (1993) Nervous tissue proteoglycans. Experientia 49, 429–446

McKeon R J Hoke A and Silver J (1995) Injury-induced proteoglycans inhibit the potential for laminin mediated axon growth on astrocytic scars. Experimental Neurology 136, 32–43

McKeon R J Schreiber R C Rudge J S and Silver J (1991) Reduction of neurite outgrowth in a model of glial scarring following CNS injury is correlated with the expression of inhibitory molecules on reactive astrocytes. Journal of Neuroscience 11, 3398–3411

Moorman S J and Gould R M (1997) Differentiating oligodendrocytes inhibit neuronal growth cone motility in different ways. Journal of Neuroscience Research 50, 791–797

Palladini G Caronti B Pozzesser G Teichner A Buttarelli F R Morselli E Valle E Venturini G Fortuna A and Pontieri F E (1996) Treatment with cyclosporine A promotes axonal regeneration in rats submitted to transverse section of the spinal –II– Recovery of function. Journal für Hirnforschung 37, 145–153

Perry V H Andersson P B and Gordon S (1993) Macrophages and inflammation in the central nervous system. Trends in Neuroscience 16, 268–273

Rabchevsky A G and Streit W J (1997) Grafting of cultured microglial cells into the lesioned spinal cord of adult rats enhances neurite outgrowth. Journal of Neuroscience Research 47, 34–48

Schnell L and Schwab M E (1990) Axonal regeneration in the rat spinal cord produced by an antibody against myelin-associated neurite growth inhibitors. Nature 343, 269–272

Siegal J D Kliot M Smith G M and Silver J (1990) A comparison of the regeneration potential of dorsal root fibers into gray or white matter of the adult rat spinal cord. Experimental Neurology 109, 90–97

Snow D M Lemmon V Carrino D A Caplan A I and Silver J (1990) Sulfated proteoglycans in astroglial barriers inhibit neurite outgrowth *in vitro*. Experimental Neurology 109, 111–130

Thallmair M Metz G A S Z'Graggen W J Raineteau O Kartje G L and Schwab M E (1998) Neurite growth inhibitorys restrict plasticity and functional recovery following corticospinal tract lesions. Nature Neuroscience 1, 124–131

Windle W F Clemente C D and Chambers W W (1952) Inhibition of formation of a glial barrier as a means of permitting a peripheral nerve to grow into the brain. Journal of Comparative Neurology 96, 359–369

Z'Graggen W J Metz G A S Kartje G L Thallmair M and Schwab M E (1998) Functional recovery and enhanced corticofugal plasticity after unilateral pyramidal tract lesion and blockage of myelin-associated neurite growth inhibitors in adult rats. Journal of Neuroscience 18, 4744–4757

Zhang Z Krebs C J and Guth L (1997) Experimental analysis of progressive necrosis after spinal cord trauma in the rat: etiological role of the inflammatory response. Experimental Neurology 143, 141–152

5

Intrinsic neuronal and extrinsic glial determinants of axonal regeneration in the injured spinal cord

Wolfram Tetzlaff and John D Steeves

Introduction

It is likely that a combination of therapeutic interventions will be required to overcome spinal cord injury and promote functional regeneration in higher vertebrates (including humans) and these therapies will have to address: 1) the reduction of secondary cell damage, near the spinal cord injury site, to both neurons and glia that have survived the initial neurotrauma, 2) the bridging of any resulting tissue loss (i.e. cyst cavities) or physical scarring at the injury site, and 3) the promotion of functional axonal regeneration and/or sprouting of axons across the site of injury. Active rehabilitation therapy will also be essential to provide the necessary sensorimotor activity that will enhance plasticity within surviving circuits, as well as functionally consolidate any synaptic connections formed by regenerated axons (Muir & Steeves, 1997). In this chapter, we attempt to review the role played by some of the intrinsic neuronal and extrinsic glial determinants in axonal regeneration within the vertebrate spinal cord.

Key points

- Effective therapies for spinal cord injury will involve: 1) reduction of secondary cell damage near the spinal cord injury site, 2) bridging of any resulting tissue loss or physical scarring at the injury site, and 3) promotion of functional axonal regeneration and/or sprouting of axons across the site of injury.

Uncovering the neuronal determinants of CNS regeneration

Experiments in the early 1980s by Peter Richardson, Sam David and Albert Aguayo (David & Aguayo, 1981; Richardson *et al.*, 1980) extended the original findings of Tello and Ramón y Cajal (Ramón y Cajal, 1928) that CNS axons of higher vertebrates can regenerate into a permissive environment such as that provided within a peripheral nerve transplant. This demonstrated that at least some CNS neurons are not intrinsically unable to regenerate and possess many of the intrinsic programs for axonal regrowth. It also suggested that the extrinsic environment surrounding an adult CNS neuron may be responsible for the lack of axonal regeneration after injury (Richardson *et al.*, 1980). Since then, potential extrinsic axon growth inhibition has been identified in association with CNS myelin (Caroni & Schwab, 1988; Filbin, 1995; Keirstead *et al.*, 1992; McKerracher *et al.*, 1994), as well as subpopulations of astrocytes (Fitch & Silver, 1997). The roles of astrocytes and the problem of glial scar formation in spinal cord injury is the subject of another chapter in this volume (see chapter 4 by Fitch & Silver). In the second part of this review, we shall mainly focus on the possible inhibitory role of CNS myelin, but in the first part of this review, we will focus on the intrinsic neuronal determinants of axon growth within the injured brain and spinal cord (Caroni, 1997). A greater understanding of both the intrinsic neuronal mechanisms driving axonal regeneration and the extrinsic influences from the surrounding CNS environment is necessary to suggest the most appropriate path for promoting functional recovery after spinal cord injury.

One line of evidence which points towards the importance of an appropriate regenerative response within the neuronal cell body, after an axonal injury, comes from "cross-age" transplantation experiments. As a rule immature (e.g. fetal), but not mature (e.g. adult), neurons have a greater propensity to regenerate within the adult CNS, despite the apparently restrictive environment of the adult CNS. For example, embryonic neurons can regenerate axons after transplantation into the myelinated (white matter) CNS tracts of adult rodents (Davies *et al.*, 1994; Wictorin *et al.*, 1990). The argument could be made that these immature neurons have yet to express receptors to extrinsic CNS inhibitors (i.e. are unresponsive to negative signals) and/or still retain active intrinsic growth programs that overcome inhibitory influences (Figure 1). In other words, the relative balance of these two influences is more important to the regenerative outcome than the presence of an individual axonal growth promotor or inhibitor.

In vitro studies, by Schneider and colleagues, suggest that young neurons can respond to inhibitory myelin epitopes of the CNS environment, yet are able to overcome this inhibition. They used explant co-cultures of retina and midbrain tectum from hamsters (Chen *et al.*, 1995). In this example, the regeneration of retinal ganglion cell axons into the tectum appears to be dependent on the age of the retinal ganglion cell neurons, as opposed to the age of the tectum (i.e. axons from embryonic and early postnatal retina can regrow into adult tectum, but not vice versa).

The addition of the myelin-associated IN-1 antibodies (raised against a putative CNS myelin protein) enhanced the retinal ganglion cell regeneration. In an analogous fashion, Raisman and coworkers, (Li *et al.*, 1995) demonstrated that the age of the axons from the entorhinal cortex and not the age of the target dentate gyrus determines the success of *in vitro* axonal regeneration. The challenging and still unresolved question is what do these neurons have (or lack?) that alters their regenerative vigour?

It appears that at least some adult neurons, such as CGRP-positive dorsal root ganglion neurons retain this regenerative vigour after axonal injury, as evidenced by their axonal growth after transplantation into CNS white matter tracts (Davies *et al.*, 1997). However, the actual CNS lesion site must be restricted to minimize scar formation which would block axonal regrowth (Davies *et al.*, 1997). Moreover, successful axonal growth within the adult CNS has only been well documented for this subset of CGRP-positive dorsal root ganglion cells. The lack of spontaneous axonal regeneration by CNS neurons argues very persuasively that not all neuronal phenotypes respond

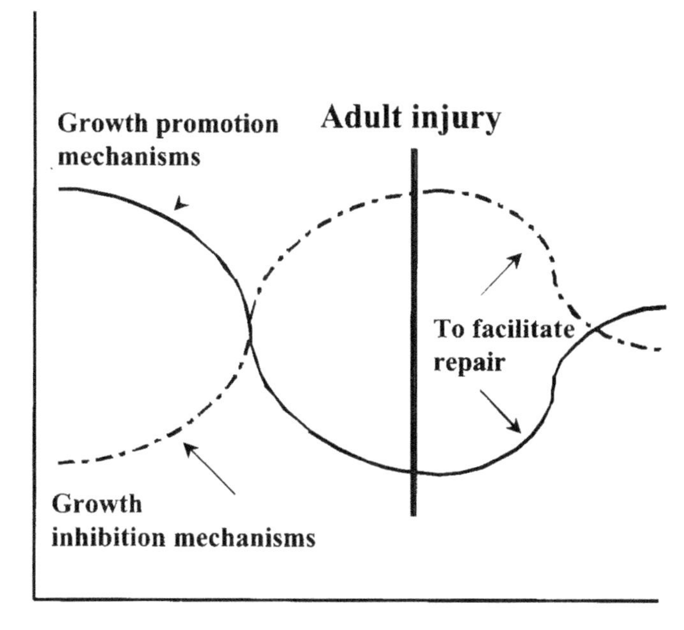

Figure 1

with equally vigorous axonal growth. Thus, what are the differences in the intrinsic neuronal growth programs within neurons which can regenerate after injury (e.g. dorsal root ganglia and motoneurons), as opposed to neurons which do not mount an appropriate regenerative response (e.g. most CNS neurons)?

A second line of evidence stressing the importance of intrinsic neuronal determinants is based on the limited success of CNS axon regeneration into supportive peripheral nerve transplants. CNS axons fail to regenerate when the peripheral nerve transplant is inserted some distance from the parent neuronal cell bodies (Richardson *et al.*, 1984). Moreover, there are considerable differences between the types of CNS neurons that are able to grow into a peripheral nerve transplant, e.g. Purkinje cells of the cerebellum are notorious for their failure to regenerate (Rossi *et al.*, 1997; see also chapter 3 by Anderson and Lieberman, this volume). In addition, the central axons of dorsal root ganglion neurons only regenerate into peripheral nerve transplants inserted into the dorsal columns of the cervical spinal cord when the dorsal root ganglion peripheral projections have also been cut previously (Richardson & Issa, 1984). This work laid the foundation for the concept that the failure of axonal regeneration could also be related to weak intrinsic neuronal growth activities, even in situations where a permissive growth environment is offered. We learned from later studies that peripheral, but not central, axotomy injury of dorsal root ganglion neurons reliably induces the appropriate axonal growth programs, as indicated by the expression of several proteins (e.g. GAP-43 Schreyer & Skene, 1993) or c-jun (Jenkins *et al.*, 1993a).

Is the activation of an appropriate neuronal response to axotomy by an adult peripheral neuron (which successfully regenerates) re-iterating its initial embryonic developmental programs or does it differ? Furthermore, is an appropriate intrinsic neuronal reaction sufficient to promote growth in an inhibitory CNS environment? The laboratory of Martin Berry recently provided evidence that the neuronal cell body response of retinal ganglion cells can be stimulated enough to promote the regeneration of their axons within the normally inhibitory environment of the adult optic nerve (Berry *et al.*, 1996). The insertion of a small peripheral nerve segment into the vitreous humor of the eye was the trigger for this successful regeneration, and presumably provided some, as yet uncharacterized, trophic support to the retinal ganglion cell bodies. In other words, some factor(s) offered at the level of the cell body enhance(s) the regenerative vigour of axotomized retinal ganglion cells. The challenge then resolves itself to the definition of these factors and the intrinsic neuronal mechanisms critical to the regeneration response.

Key points

- **CNS axons of higher vertebrates can regenerate into a permissive environment such as that provided within a peripheral nerve transplant.**
- **Extrinsic environment surrounding an adult CNS neuron may be responsible for the lack of axonal regeneration after injury.**
- **Extrinsic axon growth inhibition is associated with CNS myelin, as well as subpopulations of astrocytes.**
- **Age and type of neurons may affect their regenerative "vigour".**

The response to axotomy of peripheral neurons

Most peripheral neurons survive interruption from their targets and undergo characteristic and presumably favourable changes in gene expression, including:

(1) changes in transcription factors, (e.g. c-jun, NGFIA, junB, junD; reviewed by Herdegen *et al.*, 1997 and Bisby in chapter 12 of this volume),
(2) changes in cytoskeletal proteins, such as specific tubulin isotypes, actin, peripherin, neurofilaments (Bisby & Tetzlaff, 1992), as well as the axonal transport motors kinesin and dynein (Takemura *et al.*, 1996),

(3) changes in growth associated proteins (Benowitz & Routtenberg, 1997; Skene, 1989),
(4) changes in neurotransmitters and neuropeptides (for review see Zigmond *et al.*, 1996), as well as their receptors and ion channels (Dib-Hajj *et al.*, 1998),
(5) changes in cell adhesion and guidance molecules (Martini, 1994), and
(6) changes in trophic factors and their receptors (Ernfors *et al.*, 1993; Kobayashi *et al.*, 1996).

Significant differences between axotomized PNS versus CNS neurons have been observed in growth associated protein and trophic factor expression.

Transcription factors
Several transcription factors have been reported to change their expression after axonal injury, e.g. c-jun, junB (Herdegen *et al.*, 1997). One of the most consistent changes in gene expression within axotomized peripheral neurons is an increase in c-jun expression (Herdegen *et al.*, 1997). C-jun can form homodimers or heterodimers with other transcription factors (to form the activator protein-1, AP-1, complex) and the obvious next step is to determine c-jun's role in the regulation of downstream genes involved in peripheral nerve regeneration. One candidate gene might be GAP-43 which carries an AP-1 binding site in its promoter region (Weber & Skene, 1997). While there is correlated upregulation of both c-jun and GAP-43 expression in axotomized peripherally projecting neurons (Herdegen *et al.*, 1997), we have found that axotomized corticospinal neurons increase their expression of GAP-43 mRNA levels in the absence of detectable c-jun immunoreactivity (Giehl & Tetzlaff, in preparation). However, we do not know to what extent the mRNA levels for GAP-43 are regulated on the transcriptional versus post-transcriptional level (Perrone-Bizzozero *et al.*, 1993; Vanselow *et al.*, 1994).

Therefore, an important experiment would be to inhibit the expression of c-jun in axotomized peripheral neurons and test whether this affects the expression of GAP-43. Another function for c-jun may be an involvement in axotomy induced cell death (Herdegen *et al.*, 1997) which may be triggered, but ultimately aborted in adult peripheral neurons due to the availability of abundant trophic support. Consistent with this role is the rescue of axotomized neurons in newborns (which lack sufficient trophic support and consequently die more readily after axotomy) with jun kinase inhibitors (Maroney *et al.*, 1998), as well as the observed decrease in c-jun expression after the application of trophic factors (Blottner & Herdegen, 1998).

Key points

- Several transcription factors change their expression after axonal injury, eg c-jun increases.
- Axotomized neurons in newborns (which lack sufficient trophic support and consequently die more readily after axotomy) are rescued by jun kinase inhibitors.
- Decrease in c-jun expression occurs after application of trophic factors.

Growth associated protein (GAP)

Growth associated proteins (GAPs) were initially described in 2-dimensional gels of fast axonally transported proteins from regenerating peripheral nerves, as well as in regenerating CNS neurons of lower, but not higher, vertebrates (Skene, 1989). The most extensively studied, GAP-43, is involved in the transduction of extracellular signals regulating the actin cytoskeleton within the neuronal growth cone (Benowitz & Routtenberg, 1997). In the context of CNS regeneration it is interesting to note that *in vitro* over-expression of GAP-43 renders growth cones of dorsal root ganglion cells more resistant to myelin associated growth cone inhibitors (Aigner & Caroni, 1995). Conversely, the *in vitro* suppression of GAP-43 with antisense oligonucleotides reduces growth cone formation by dorsal root ganglia (Shea & Benowitz, 1995). *In vivo*, over-expression of GAP-43 in transgenic mice induces terminal sprouting at the neuromuscular junction, which is dramatically enhanced if a second GAP, GAP-23, is simultaneously over-expressed (Aigner *et al.*, 1995). The synergistic and potentially compensatory actions of GAPs might explain the relatively normal axonal development observed in the GAP-43 knockout mouse (Strittmatter *et al.*, 1995).

To the best of our knowledge, axonal regeneration in the GAP-43 knockout mice has yet to be reported. Several examples of axonal growth in the "absence of increased" (i.e. low) expression of GAP-43 have been reported. For example, the terminal sprouting of axons at the neuromuscular junction, in response to denervation of the adjacent muscle fibres, is that not associated with significant increases in GAP-43 (Bisby *et al.*, 1996). Another example is that some dorsal root ganglia fail to mount an increase in GAP-43 expression after axotomy of their central process. Nevertheless, their regeneration within the dorsal root proceeds, albeit at a much slower rate than their GAP-43 positive neighbours (Andersen & Schreyer, 1999).

All these observations are consistent with the suggestion that local sprouting or slow axonal growth can occur by utilizing the resident set of proteins within the terminal axon or growth cone, even in the absence of a sustained response by the parent cell body (Smith & Skene, 1997). In contrast, regenerative growth over long distances requires an effective cell body response, which typically includes the increased expression of GAP-43 and other regeneration associated genes.

Key points

- **Local sprouting may be possible in the absence of a sustained cell body response to injury.**
- **Regenerative growth over long distances requires an effective cell body response, including increased expression of GAP-43 and other regeneration associated genes.**

Cell adhesion molecules and axonal guidance molecules

Cell adhesion molecules and axonal guidance molecules play an important role in the development of the PNS and CNS (Walsh & Doherty, 1997). The expression of cell adhesion molecules changes considerably in regenerating peripheral nerves (Martini, 1994). L1, a member of the immunoglobulin superfamily, has been observed to increase in regenerating peripheral nerves (Bernhardt *et al.*, 1996) and more recently the increase in mRNA expression of an L1-like protein was described in axotomized peripheral neurons (Chaisuksunt *et al.*, 1998). More than one binding mechanism may account for the broad range of functions of L1: homophilic binding, binding to integrin or axogenin-1, as well as fibroblast growth factor (FGF)-receptor dimerization and activation (Brummendorf *et al.*, 1998). Antibodies to FGF receptors or dominant negative FGF receptors inhibit the L1 stimulation of neurite outgrowth (Walsh & Doherty, 1997). *In vivo,* L1 knockout mice show hypoplasia of the corticospinal tract which also fails to decussate properly at the pyramid (Dahme *et al.*, 1997). While the development of peripheral nerves in these mice shows only minor impairment in the ensheathment of axons by Schwann cells (Dahme *et al.*, 1997), it remains to be shown whether regeneration after peripheral nerve injury is hampered or whether other cell adhesion molecules compensate for a deficit in L1 expression. In spinal cord projection neurons of lower vertebrates, the expression of L1 corresponds well with their regenerative propensity (Becker *et al.*, 1998). For example, zebrafish brainstem-spinal neurons with high levels of L1, N-CAM and GAP-43 regenerate successfully whereas other descending cells (e.g. Mauthner cells), which lack one or more of these molecules, exhibit poor regeneration (Becker *et al.*, 1998). Thus, the combined expression of regeneration associated genes and growth promoting cell adhesion molecules may correlate with better regenerative success.

The functional distinction between the above mentioned cell adhesion molecules and the currently identified families of axonal guidance molecules (semaphorins, netrins and ephrins) may be somewhat artificial. Both play roles in axonal guidance during neural development. However, little is known about the latter group during axonal regeneration. Since semaphorins, netrins and ephrins can have axonal growth inhibiting properties, it is conceivable that their re-expression may also contribute to the failure of axonal regeneration in the CNS of higher vertebrates.

The Eph receptor family, named after the first member discovered in a erythropoietin-producing hepatocellular carcinoma cell line, is the largest known receptor tyrosine kinase family with over a dozen members (Flanagan & Vanderhaeghen, 1998). This family has been implicated in a variety of patterning events during embryonic development, particularly in the nervous system. Ephrins play a role in axonal guidance and topographic map formation of neural projections (Orioli & Klein, 1997). After denervation of the optic tectum some guidance properties are re-expressed, suggesting a role for this "family" in the reformation of the retino-tectal map (Drescher *et al.*, 1997). The gene deletion of Eph4 demonstrated an involvement of the Eph/ephrin-family in the formation of the corticospinal tract. Hence it is conceivable that these factors also play a role in spinal cord regeneration.

Key points

- **Expression of cell adhesion molecules changes considerably in regenerating peripheral nerves.**
- **Combined expression of regeneration associated genes and growth promoting cell adhesion molecules may correlate with better regenerative success.**

Peripherally projecting neurons express trophic factors after injury

Peripheral axonal injury disconnects the neuron from its target which is believed to provide a source of trophic support. In adult rodents, most peripheral neurons survive target disconnection indicating that they no longer critically depend for survival on these target derived factors (Oppenheim, 1996). Alternatively, other sources of trophic support may become available from the reactive Schwann cells of the proximal stump or the reactive glial cells surrounding the parent cell bodies and perhaps even within the axotomized neurons themselves. Axotomized dorsal root ganglia, as well as motoneurons, have been reported to express BDNF and FGF-2 after axonal injury (e.g. Huber *et al.*, 1997; Kobayashi *et al.*, 1996). Both neuron types express the appropriate receptors for these factors, trkB and FGFR-1 respectively (Ji *et al.*, 1996), and after motoneuron axotomy they are up-regulated appropriately (Kobayashi *et al.*, 1996). Interestingly, the increased expression of FGFR-1 may enhance the effects of L1 which the injured axons might encounter during their regeneration.

The functional significance of BDNF and FGF-2 may be hard to assess in axotomized PNS neurons since they are responsive to and exposed to a wide variety of trophic factors (Oppenheim, 1996). For example, an increasing number of cytokines are expressed in axotomized neurons and/or their perineuronal glial cells, including: Interleukin-6 (IL-6), transforming growth factor-β, TGF-β (Murphy *et al.*, 1995; Streit *et al.*, 1998), and ciliary neurotrophic factor (CNTF) (Tetzlaff, unpublished observation). As discussed below, this rich expression of trophic factors in and around axotomized motoneurons is not seen in CNS neurons (e.g. rubrospinal neurons). As yet, we do not know exactly why we do not observe similar changes or whether stimulation of a similar expression pattern would facilitate better CNS regeneration?

CNS axotomy

Distance dependent effects in the neuronal cell body response

The cell body response of axotomized CNS neurons differs from PNS neurons. Although many CNS neurons increase their expression of regeneration associated genes, it is usually only seen after an axotomy relatively close to the neuronal cell body (see aslo chapter 3 by Andersen & Lieberman, this volume). In the context of spinal cord injury, it is interesting to note that rubrospinal neurons increase their expression of c-jun, GAP-43 and Tα1-tubulin after axotomy within the cervical spinal cord (20 mm distance), but not after an injury at the low thoracic level (approximately 60 mm distance; Jenkins *et al.*, 1993b; Tetzlaff *et al.*, 1994). The observed increase in rubrospinal neuron gene expression, after a cervical injury, peaks 7 days after injury and then declines with an accompanying atrophy of the neuronal cell body (Tetzlaff *et al.*, 1991; Tetzlaff *et al.*, 1994). Corticospinal neurons, in contrast, do not mount an increase in GAP-43 mRNA expression unless axotomized as close as 200 μm from their cell bodies; a lesion that also results in 50% cell death within one week (Giehl & Tetzlaff, 1996; Tetzlaff *et al.*, 1994).

The reason for the distance dependency of the CNS neuronal cell response to axotomy is not well understood. It stands in contrast to the response in PNS neurons (Figure 2). There is evidence that target derived factors may suppress the expression of GAP-43 in some neurons, such as rat dorsal root ganglia (Liabotis & Schreyer, 1995), rat spinal motoneurons (Fernandes *et al.*, unpublished observations), or fish retinal

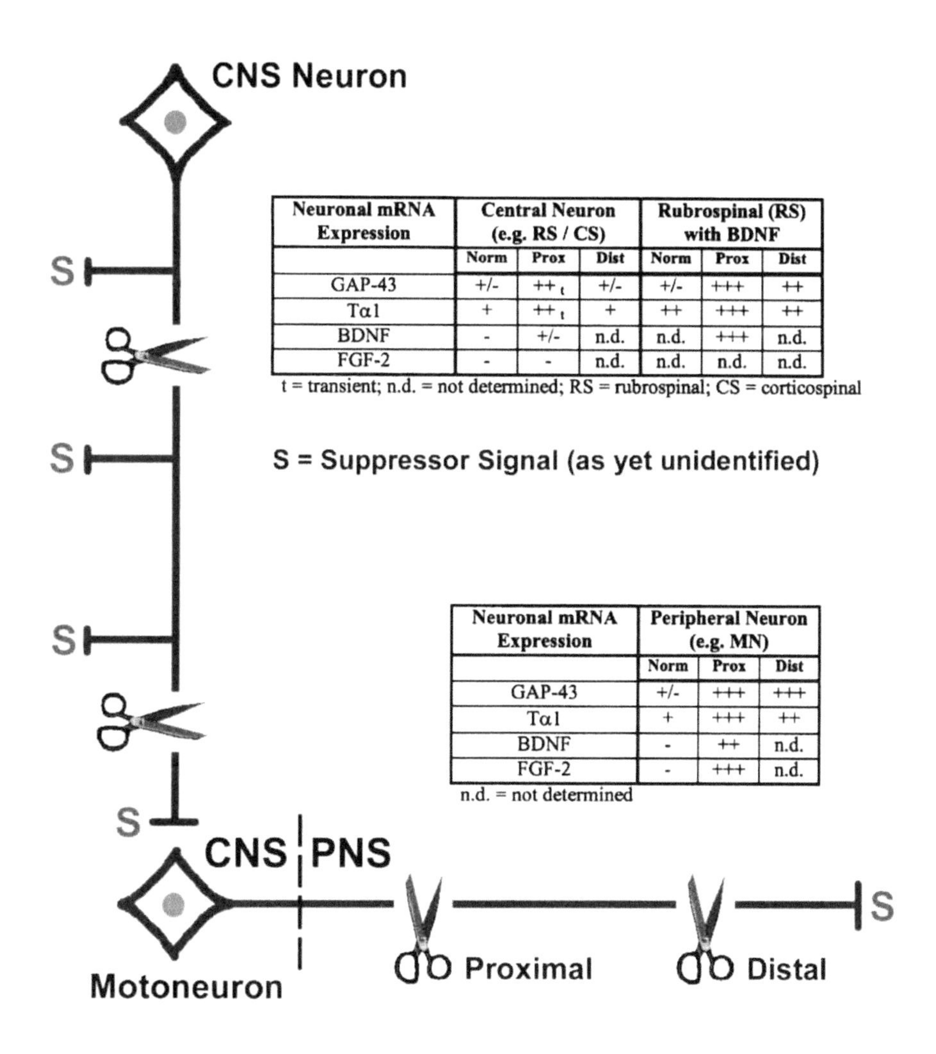

Neuronal mRNA Expression	Central Neuron (e.g. RS / CS)			Rubrospinal (RS) with BDNF		
	Norm	Prox	Dist	Norm	Prox	Dist
GAP-43	+/-	++ $_t$	+/-	+/-	+++	++
Tα1	+	++ $_t$	+	++	+++	++
BDNF	-	+/-	n.d.	n.d.	+++	n.d.
FGF-2	-	-	n.d.	n.d.	n.d.	n.d.

t = transient; n.d. = not determined; RS = rubrospinal; CS = corticospinal

Neuronal mRNA Expression	Peripheral Neuron (e.g. MN)		
	Norm	Prox	Dist
GAP-43	+/-	+++	+++
Tα1	+	+++	++
BDNF	-	++	n.d.
FGF-2	-	+++	n.d.

n.d. = not determined

Figure 2 Comparison of CNS and PNS neuron responses to injury.

ganglion cells (Bormann *et al.*, 1998). An axonal injury (i.e. the blockade of axonal transport) would prevent the "return" of these target derived suppressing signals to the neuronal cell body. A characteristic feature of these neurons is the paucity of collateral branches (i.e. a source of the potential suppressor signal) emanating from the primary axon. A lack of substantial collateral branches might explain why in case of peripheral projecting neurons the increase in GAP-43 expression is comparable after either a proximal or distal axotomy, since in both cases a target derived supply of "suppressor" is interrupted to the same extent.

If a similar retrograde signal is responsible for the suppression of GAP-43 expression within the more branched CNS neuron, one could imagine that uninjured axon collaterals might still supply the parent cell bodies with sufficient GAP-43 "suppressor" triggered by the surviving synaptic connections (Figure 2). Thus, in the case of a distal axotomy of a long projection CNS cell (e.g. brainstem-spinal neurons), those sustaining collaterals between the cell body and the axotomy site could still provide significant suppression signals. Conversely, a more proximal axonal injury of a CNS projection neuron would disconnect more axon collateral terminals from the parent cell body and remove more of the suppressing signals, thereby releasing a greater expression of GAP-43 and other regeneration associated genes.

GAP-43 expression can be regulated at the level of gene transcription, as well as by mRNA stabilization. GAP-43 transcription appears to be under the control of suppressing transcription factors (Chiaramello *et al.*, 1996) and axonal injury of peripheral neurons involves transcriptional activation of the GAP-43 gene (Vanselow *et al.*, 1994). Stabilization of GAP-43 mRNA levels (i.e. post-transcriptional control) can be achieved by the application of nerve growth factors (NGF) to cultured PC12 and dorsal root ganglion cells (Burry & Perrone-Bizzozero, 1993). Interestingly, the exogenous application of another neurotrophin family member, BDNF, to rubrospinal neurons stimulated and maintained the expression of GAP-43 mRNA after cervical axotomy and even after thoracic axotomy.

It would appear that the BDNF application has a primary effect on GAP-43 mRNA stabilization since BDNF application to intact rubrospinal neurons did not increase GAP-43 expression (Kobayashi, 1998). Thus, we favour the hypothesis that the neurotrophic factor stabilizes GAP-43 mRNA expression within rubrospinal neurons. Even though some transcriptional activation of GAP-43 occurs after a thoracic axotomy, it is less than after a cervical injury, perhaps because more "suppressor" is returned via the surviving collaterals. After cervical injury there is a significant transcriptional increase (due to the removal of most collateral derived suppressors), but due to the lack of trophic support (perhaps in part, because there are few surviving collaterals), the GAP-43 increase is not maintained. This hypothesis provides testable predictions and future studies will be required to prove its validity.

Trophic support for axotomized spinal cord projection neurons may be insufficient to support regeneration associated gene expression, but enough to support neuron survival

Preliminary work from our laboratory suggests that the availability of trophic factors after axonal injury is another difference between axotomized CNS and PNS neurons. In contrast to axotomized motoneurons, we failed to observe FGF-2 or BDNF expression

Key points

- Cell body response of axotomized CNS neurons differs from PNS neurons.
- Many CNS neurons increase expression of regeneration associated genes, but usually only after an axotomy relatively close to the neuronal cell body.
- Target derived factors may suppress the expression of GAP-43 in some neurons.

in injured or intact rubrospinal neurons as they underwent atrophy, but this atrophy could nevertheless be prevented by the exogenous application of BDNF (Kobayashi *et al.*, 1997). This lack of endogenous trophic support is also evident in injured corticospinal neurons, where proximal axotomy leads to 50% corticospinal cell death. BDNF is expressed in the cortex and cell death increases to 70% when function-blocking antibodies to BDNF are infused into the cortex (Giehl *et al.*, 1998). Conversely, the exogenous applications of BDNF or NT-3 prevent axotomy induced cell death (Giehl & Tetzlaff, 1996). Taken together, these data indicate the existence of autocrine/ paracrine trophic loops (BDNF), which however, are insufficient to support the entire population after axotomy close to the cell bodies.

After distal axotomy injuries in the spinal cord, corticospinal, rubrospinal and reticulospinal neurons undergo atrophy, but not death (Barron *et al.*, 1988; McBride & Steeves, unpublished observations). The survival of long projection CNS neurons after a distal axotomy has been suggested to be the consequence of sustaining support received via the numerous axon collaterals that exist between a distal axotomy site and the parent cell body. Although 35%–40% of rubrospinal neurons have been reported to die, within several months, after a relatively proximal axonal injury within the cervical spinal cord (Mori *et al.*, 1997), when BDNF is infused onto these chronically axotomized rubrospinal neurons, 6 months after injury, their pronounced atrophy is reversed and over 85% of the rubrospinal neurons appear to survive (Kobayashi, 1998).

After spinal cord injury, it appears that endogenous or exogenous trophic support of many long projection CNS neurons will promote their survival, yet it is insufficient to stimulate an effective neuronal response for axonal regeneration (e.g. upregulation of regeneration associated gene expression). Thus, the "good news" is that many of the candidate neurons (e.g. brainstem-spinal), essential for functional recovery after spinal cord injury, will survive and can respond, even to delayed therapeutic intervention. Nevertheless, the "unresolved" issue is how can we best generate the appropriate conditions for their effective regeneration.

The capacity of mammalian CNS neurons to regenerate their axons correlates with the expression of regeneration associated genes

The expression of regeneration associated genes appears to be necessary, but may not be sufficient for the regeneration of CNS neurons. The expression of regeneration associated genes has been reported in a variety of CNS neurons regenerating into peripheral nerve grafts or transplants of embryonic tissue. For example regenerating

Key points

- After spinal cord injury, endogenous or exogenous trophic support of many long projection CNS neurons appears to promote survival, but is insufficient to sustain a regenerative response.
- If neurons survive axotomy, how can we best generate appropriate conditions for their effective regeneration?

retinal ganglion cells show high levels of GAP-43 (Schaden *et al.*, 1994), developmental tubulin isoforms (McKerracher *et al.*, 1993), as well as L1 (Jung *et al.*, 1997). Similar observations were made for other mammalian CNS neurons as well (Woolhead *et al.*, 1998).

Likewise, regeneration of spinal cord projection neurons into peripheral nerve grafts is only observed after axonal injuries that also induce the expression of regeneration associated genes (e.g. GAP-43 and Tα1 tubulin). For example, rubrospinal neurons fail to regenerate into thoracic peripheral nerve grafts (Houle, 1991; Kobayashi, 1998; Richardson *et al.*, 1984) much like retinal ganglion cells fail to regenerate into intracranial peripheral nerve grafts (Vidal-Sanz *et al.*, 1991). Although there is variability between different types of neurons in the expression of GAP-43 and other regeneration associated genes after axotomy, it appears that high levels of GAP-43 expression correlate well with their endogenous capacity to sprout or regenerate within the CNS (Alonso *et al.*, 1997; Benowitz & Routtenberg, 1997). Purkinje cells, which do not express GAP-43 in development (Console-Bram *et al.*, 1996), fail to re-express GAP-43 after injury and fail to regenerate even in a permissive environment (Zagrebelsky *et al.*, 1998 and references herein). Nevertheless, the mere overexpression of GAP-43 is not sufficient to enhance the regeneration of Purkinje cells (Buffo *et al.*, 1997), emphasizing the need for a concerted action of several regeneration associated genes (as yet unidentified).

In short, the coordinated expression of regeneration associated genes, like c-jun, GAP-43, L1, and Tα1, appears to enhance the neuron's intrinsic regenerative capacity and may be a prerequisite for successful regeneration within the unfavourable environment of the adult CNS. Proof of this suggestion would require the specific induction of these regeneration associated genes and subsequent observation of substantial axonal regeneration. The *in vivo* analysis of the neuronal expression of regeneration associated genes after trophic factor treatment might serve as a predictor of the expected regeneration success.

Application of trophic factors to an injured neuronal cell body stimulates both RAG expression and axonal regeneration

Implantation of peripheral nerve transplants into the vitreous humour of the eye stimulated the expression of GAP-43 by retinal ganglion cells and significantly enhanced the number of retinal ganglion cells regenerating into peripheral nerve grafts (Ng *et al.*, 1995). This approach also stimulated retinal ganglion cell regeneration after an optic nerve crush (Berry *et al.*, 1996) which was somewhat unexpected given the inhibitory

Key points

Expression of regeneration associated genes appears to be necessary, but may not be sufficient for the regeneration of CNS neurons.

High levels of GAP-43 expression correlate well with their endogenous capacity to sprout or regenerate within the CNS.

Some neurons, eg Purkinje cells, do not express GAP-43 in development, fail to re-express GAP-43 after injury and fail to regenerate even in a permissive environment, also do not regenerate even when overerexpressing GAP-43.

Coordinated expression of regeneration associated genes, like c-jun, GAP-43, L1, and Tα1, appears to enhance neurons intrinsic regenerative capacity.

environment of the rat optic nerve (although this may also change after an optic nerve crush). The challenging questions become: 1) what is (are) the factor(s) that stimulate(s) this cell body and axonal response (Mey & Thanos, 1996), and 2) what are the molecular signaling mechanisms? One candidate factor is BDNF which is secreted by reactive Schwann cells (Meyer *et al.*, 1992); however, injection of BDNF into the eye failed to stimulate retinal ganglion cell regeneration (Mansour-Robaey *et al.*, 1994). It appears that BDNF stimulated only some of the identified regeneration associated genes (e.g. GAP-43, but not Tα1 expression, Fournier *et al.*, 1997).

It would appear that therapeutic approaches will have to target a broad range of changes in the neuron. In contrast to the eye, after proximal or distal axotomy injury, we found that exogenous BDNF application in the vicinity of rubrospinal and corticospinal neurons stimulated the expression of several regeneration associated genes (e.g. GAP-43, Tα1; Kobayashi, 1998; Kobayashi *et al.*, 1997). In addition, when we examined the regeneration of rubrospinal neurons into peripheral nerve grafts, we observed axonal regeneration into both cervical and thoracic placed peripheral nerve grafts (Kobayashi, 1998; Kobayashi *et al.*, 1997). As expected all regenerating neurons expressed high levels of GAP-43, Tα1 and c-jun (Fan & Tetzlaff, unpublished observations). It remains to be shown whether BDNF and a peripheral nerve graft will stimulate the regeneration of corticospinal neurons.

Interestingly, NT-3 failed to stimulate GAP-43 expression after exogenous application to corticospinal neurons (Giehl *et al.*, unpublished observations). Yet NT-3, in contrast to BDNF, was more effective in promoting re-growth of injured corticospinal neurons when administered at the spinal cord injury site in conjunction with antibodies to myelin associated growth inhibitors (Schnell *et al.*, 1994). This is consistent with the promotion of corticospinal regeneration by NT-3 producing encapsulated cells (Grill *et al.*, 1997).

The most parsimonious explanation is that NT-3 exerts a local effect on the growth cones which does not involve the neuronal cell body or that NT-3 may act indirectly via other cells (e.g. glia). In our hands, BDNF failed to promote regeneration when applied to the injured spinal cord (12 μg/day for 7 days, Kobayashi, 1998). This negative finding could be explained by the high levels of truncated trkB receptors in the spinal cord (Armanini *et al.*, 1995). It remains to be shown whether a combination of treatments to the cell body, e.g. BDNF, and to the injured spinal cord, e.g. NT-3, would further enhance axonal regeneration. Obviously, this neurotrophin approach would not stimulate those spinal cord projection neurons that are not BDNF/NT-3 responsive.

Thus, to address the variablity in the trophic requirements of other spinal projecting neurons we have used *in vitro* assays of brainstem-spinal neurons that were retrograde labeled prior to explantation (Pataky *et al.*, unpublished observations). For example, we have observed that FGF-2 effectively promoted both vestibulospinal survival and neurite outgrowth. In contrast, FGF-1 failed to promote vestibulospinal neuron survival, but was equally effective as FGF-2 in stimulating neurite outgrowth. Conversely, FGF-9 enhanced survival but not outgrowth (Pataky *et al.*, unpublished observations). Thus, as in other systems, the signaling pathways promoting survival appear to be distinct from those promoting outgrowth (Kaplan & Miller, 1997).

Most interesting in the context of the above *in vitro* findings was the recent report that FGF-1 application to the severed adult spinal cord facilitated increased brainstem-spinal and propriospinal regeneration, when combined with directed peripheral nerve bridges across the site of injury (Cheng *et al.*, 1996). More detailed knowledge of the trophic factor influences on a specific neuronal population may benefit the development of more effective *in vivo* therapies to promote repair of the damaged CNS. For example, application of NT-3 at the spinal cord injury site enhanced reticulospinal and rubro-spinal axon regrowth, but did not promote raphe-spinal or vestibulospinal regeneration (Ye & Houle, 1997).

The challenge therefore, is to not only identify the differences in the trophic requirements of PNS and CNS neurons, but also to characterize the specific signaling pathways responsible for axonal regeneration for each candidate spinal cord projection phenotype.

Key points

After proximal or distal axotomy injury, exogenous BDNF application in the vicinity of rubrospinal and corticospinal neurons stimulated the expression of several regeneration associated genes.

Different growth factors are effective in promoting growth of axons in different neuronal populations.

An important part of the challenge in devising methods of repair of injured spinal cord lies in characterizing the specific trophic factors and signaling pathways responsible for axonal regeneration for each candidate spinal cord projection phenotype.

Extrinsic determinants of axonal regeneration associated with CNS myelin

There is good evidence that the axonal regeneration failure within the higher vertebrate CNS is at least in part due to an unfavourable axonal growth environment. Here, we will focus on the axonal growth cone inhibitors that have been associated with CNS myelin, and their parent cell bodies, the oligodendrocytes. It is important to remember that during neural development, axonal outgrowth is always followed by the subsequent myelination of that axon and therefore any potential myelin inhibitors do not impede the initial formation of that axonal projection. A possible developmental role of CNS myelin, that persists throughout adult life, may be the stabilization of axonal fibre tracts (i.e. the suppression in the generation of spontaneous axon collaterals). As a consequence of this role, CNS myelin may present an obstacle to regeneration after an adult axonal injury (Keirstead & Steeves, 1998).

Functional spinal cord regeneration within the developing spinal cord
Strong correlative evidence for the inhibitory role of myelin comes from developmental studies in vertebrate embryos. In the developing chicken, we observed functional axonal regeneration after a spinal cord transection, but only when the injury occurs prior to the onset of myelination at embryonic day (E) 13 (Keirstead *et al.*, 1992). Transection of the spinal cord after E13 (e.g. E15) results in rapidly diminishing regeneration and no functional recovery (Hasan *et al.*, 1993). Thus, there is a permissive developmental period for functional spinal cord regeneration, followed by a restrictive period where repair is as impoverished as that within the adult cord. By experimentally delaying the onset of myelination, using an intraspinal immunological protocol, we extended axonal regeneration and functional recovery to later stages of development (Hasan *et al.*, 1993; Keirstead *et al.*, 1992). The application of this protocol to the mature injured avian spinal cord (Keirstead *et al.*, 1995) and rats (Dyer *et al.*, 1999) will be discussed below.

Axonal regeneration of the immature spinal cord has also been observed in mammalian embryos, with the opossum becoming a widely used model (see chapter 2 by Saunders & Dziegielewska, this volume). Again, the regenerative capacity of most fibre tracts is limited to the developmental period prior to myelination (Varga *et al.*, 1995). Likewise, the developmental period for spontaneous regeneration can be extended by the application of the IN-1 antibody. As in all developing systems, late outgrowing axons may also contribute to the documented functional recovery, but double-lable tract-tracing (i.e. before and after the lesion), as well as electrophysiological studies have confirmed that regenerating axons are also contributing (Hasan *et al.*, 1993; Varga *et al.*, 1995).

Key points

In developing chicken, functional axonal regeneration after a spinal cord transection occurs, only when injury is prior to onset of myelination at E13.

Period of axonal regeneration extended by delaying onset of myelination.

In neonatal opossum there is a similar relation between cessation of a permissive period for axon growth following injury and the onset of myelination.

CNS myelin and axonal regeneration within lower vertebrates
If the hypothesis is correct that myelin associated inhibitors are developmental stabilizers, one would expect that they might be absent within the CNS of "lower" vertebrates, e.g. fish, amphibians and reptiles, all of which maintain indeterminate (continuous) growth throughout their entire life. The lamprey is one primitive vertebrate where compact CNS myelin is absent and indeed adult lamprey exhibit robust functional CNS regeneration after spinal cord injury (for review, see chapter 7 by Conti & Selzer, in this volume and McClellan, 1998). However, the story is not so straightforward when we examine other members of the so-called "lower" vertebrates.

For example, indeterminate growth in adult (i.e. sexually mature) teleost fish provides for the continual generation of new retinal ganglion cells which project their axons through the optic nerve to the adult tectum. In the goldfish, studies of CNS myelin inhibition of retinal ganglion cell regeneration have yielded somewhat controversial data (Sivron & Schwartz, 1995). Stuermer and collaborators initially indicated that goldfish optic nerve myelin and oligodendrocytes are devoid of inhibitory activity based on the successful growth of fish retinal ganglion cells on fish myelin substrates, and the inhibition of retinal ganglion cell axonal growth on sections of rat optic nerves (Wanner *et al.*, 1995). However, Schwartz and collaborators found that sections of the fish optic nerve are non-permissive to CNS axonal growth and this inhibitory activity can be neutralized by the application of IN-1 antibodies (Sivron *et al.*, 1994). Unlike the rat optic nerve, the fish optic nerve undergoes certain changes after axonal injury that may provide enhanced support for the regeneration of adult neurons (Sivron *et al.*, 1994). As elaborated further below, these changes might be related to inflammation and macrophage invasion (Rapalino *et al.*, 1998).

It should also be mentioned that the organization of the goldfish optic nerve is fundamentally different from mammals, which may compensate for the inhibitory activity of axon growth on nerve sections. The goldfish optic nerve shows inter- and intrafascicular compartments. The former are formed by fibroblasts and stain positive for fibronectin and laminin. In the light of what we know about growth substrates in general, it is not surprising the retinal ganglion cell axons readily grow on fibroblasts or laminin *in vitro* or *in vivo*. The oligodendrocytes within the fascicles have the E587 antigen, an L1-like molecule, known to facilitate axonal growth (Ankerhold *et al.*, 1998; Bastmeyer *et al.*, 1995). It is therefore not surprising that fish optic nerves can act as a permissive substrate for the regenerative growth of mammalian retinal ganglion cells. Thus, the *in vivo* regeneration of fish retinal ganglion cells after axotomy may be attributed, in part to: 1) the presence of these permissive axon growth molecules within the fish optic nerve environment, 2) the absence or modification of one or more unidentified inhibitory myelin proteins, 3) the different responses at the site of injury (Hirsch *et al.*, 1995), and/or 4) the robust expression of growth associated proteins by axotomized fish retinal ganglion cells (Benowitz *et al.*, 1990). In addition, the fish optic nerve does not show the characteristic invasion of GFAP positive reactive astrocytes at the injury site until late (>10 days) after injury, suggesting the development of a "glial scar" is sufficiently delayed to pose no impediment to regeneration (Hirsch *et al.*, 1995).

But why is spinal cord regeneration relatively limited in fish and amphibians when compared to retinal ganglion cell regeneration? For example, very little spinal cord regeneration is seen in the post-metamorphic frog (Beattie *et al.*, 1990). Again, both components, intrinsic neuronal and extrinsic glial factors may play a role. Studies in *Xenopus* revealed prominent immunostaining for IN-1 in the spinal cord, but not in the visual system (Lang *et al.*, 1995). When spinal cord regeneration has been observed in the zebrafish, a closer examination of those regenerating brainstem-spinal neurons indicated high levels of GAP-43, and L1 expression (Becker *et al.*, 1998). In summary, it appears that myelin-associated inhibitors, especially those which bind to the IN-1 epitope, are not absent within CNS myelin of lower vertebrates, i.e. fish and amphibians. More importantly, the concomitant expression of the necessary growth promoting molecules to maintain indeterminate growth throughout adult life may compensate or over-ride any inhibitory myelin signals. For example, retinal ganglion cell axons of lizards show only a transient collapse after contact with mammalian CNS myelin (Lang *et al.*, 1998) which is reminiscent of the response of mammalian embryonic neurons (Shewan *et al.*, 1995).

Key points

Myelin-associated inhibitors, especially those binding to IN-1 epitope, are present in CNS myelin of lower vertebrates, i.e. fish and amphibians.

Concomitant expression of necessary growth promoting molecules to maintain growth throughout adult life may compensate or over-ride inhibitory myelin signals.

Suppression of myelin inhibitors and adult spinal cord regeneration within higher vertebrates

Several studies have observed axonal regeneration within the injured adult rat spinal cord following neutralization of myelin-associated neurite growth inhibitors. IN-1 antibody treatment of young adult rats with a transected corticospinal tract resulted in sprouting at the lesion site, and fine axons and fascicles were observed 7–11 mm caudal to the lesion site (Schnell & Schwab, 1990). Following IN-1 treatment of young adult rats with an over-hemisection of the spinal cord, serotonergic brainstem-spinal and corticospinal tract axons also regenerated and behavioural analysis suggested that the improved recovery of some reflexes might be due to the regrowth of these descending projections (Bregman *et al.*, 1995; Bregman *et al.*, 1997).

What are the molecules responsible for axonal growth inhibition? Several inhibitory components of myelin have been suggested for the failure of CNS regeneration in mammals and have been the subject of several reviews (Filbin, 1996; Keirstead & Steeves, 1998; Schwab & Bartholdi, 1996). Experiments in Schwab's laboratory led to the generation of antibodies directed against the myelin epitopes NI-35 and NI-250 (Caroni & Schwab, 1988). The *in vivo* application of these antibodies (IN-1), via transplantated hybridoma cells, into the forebrain promoted the sprouting and regeneration of several CNS neuronal systems, including corticospinal, brainstem-spinal and septohippocampal neurons (Schwab, 1996). The recent *in vivo* application of these IN-1 antibodies after corticospinal tract lesions at the level of the pyramidal decussation in the rat resulted in the facilitation of some corticospinal regeneration and in sprouting of axons from the contralateral tract (Z'Graggen *et al.*, 1998). A subsequent second transection of the regenerated axons did not jeopardize the functional recovery (Z'Graggen *et al.*, 1998) indicating that recovery of function was not due to regeneration alone, but may be equally dependent on sprouting or inherent spinal cord "plasticity". Thus, the extent of axonal regeneration produced by IN-1 may be minor, yet its major benefit may be the enhanced plasticity of the system. However, the exact target of the IN-1 action is presently unknown. Does it specifically bind to one epitope, and if so what is the receptor or does it bind to several antigenic sites?

There is evidence for several other myelin associated molecules acting as axonal growth inhibitors, but perhaps the most compelling argument can be made for myelin associated glycoprotein (MAG, McKerracher *et al.*, 1994; Tang *et al.*, 1997; but see also chapter 4 by Fitch & Silver, this volume). A soluble fragment of MAG, consisting of its extracellular domain, is released from myelin and can inhibit axonal growth (Filbin, 1995). However, the repulsive actions of MAG can be modulated by cAMP-dependent activities and alterations in the extracellular calcium concentrations (Song *et al.*, 1998).

How many potential myelin (or non-myelin) inhibitors are associated with the adult CNS, is still unknown and reminds us that we may need to adopt different therapeutic strategies if there are more than a few inhibitory molecules.

The experimental suppression of myelin inhibition in the developing CNS stimulated us to investigate whether similar strategies would facilitate functional regeneration in the injured adult spinal cord of birds and mammals (Steeves *et al.*, 1994). We observed that the immunologically-induced demyelination (or myelin suppression) within the mature avian cord extended rostrocaudally over several spinal segments on either side of the infusion site and appears to also involve activated macrophages or microglia (Keirstead *et al.*, 1995). More significantly, the immunological protocol in these precocial hatchling birds was initiated after (not prior to) transection of the thoracic spinal cord and the induced demyelination would thus have removed all "inhibitory" myelin molecules simultaneously. Neuroanatomical regeneration of avian brainstem-spinal projections was assessed 2-4 weeks after injury.

In the mature avian spinal cord, myelin suppression facilitated axonal regeneration by approximately 6-19% of the severed brainstem-spinal neurons, which was substantially less than that observed after a transection during the permissive developmental period (i.e. prior to E13). Nevertheless, the axonal regrowth within the myelin-suppressed mature avian cord was accompanied by functional synaptogenesis (Keirstead *et al.*, 1995). Although hatchling chicks that had undergone spinal transection and subsequent myelin disruption did not show any voluntary signs of locomotor recovery, focal electrical stimulation of a brainstem locomotor region (in a decerebrate animal preparation) evoked rhythmic motor activity in the right and left leg (Keirstead *et al.*, 1995). In comparison to the myelin-suppressed developing spinal cord, it is easy to suggest that the reduced axonal regeneration observed after spinal cord injury to the mature (myelin-suppressed) cord is due to significant alterations in either intrinsic neuronal properties and/or the extrinsic neuronal environment. Unfortunately, we only have tantalizing hints as to what some of those mechanisms might be.

More recent studies of adult spinal cord injuries indicated that transient focal demyelination of the rat spinal cord using serum complement proteins and myelin-specific, complement-fixing (IgG) antibodies (e.g. GalC) also facilitates axonal regeneration (Dyer *et al.*, 1999; Keirstead *et al.*, 1998). Ultrastructural analysis also revealed regenerating growth cones that were restricted to demyelinated regions of the injured cord in numbers that correlated to the number of axons severed (Keirstead *et al.*, 1998). Using retrograde tract tracing techniques, transient immunological demyelination after a spinal cord injury facilitates some rubrospinal axon regeneration when compared to control treated animals (Dyer *et al.*, 1999). Preliminary evidence also suggests that delaying immunological demyelination treatment for as much as 2 months after a spinal injury can still facilitate regeneration of some (but not all) brainstem-spinal projections (Dyer & Steeves, unpublished observations). Once again, this underscores the need to understand the specific intrinsic and extrinsic regeneration requirements for each particular neuronal phenotype affected by a spinal injury.

The potential importance of activated macrophages or microglia to the beneficial effects of the immunological demyelination cannot be ignored (Dyer *et al.*, 1999; Rapalino *et al.*, 1998). Recent results from Michal Schwartz's group in Israel have suggested that transplantation of peripherally activated macrophages has impressive beneficial effects to functional spinal cord regeneration (Rapalino *et al.*, 1998).

The observations of myelin suppression contributing to both regeneration and sprouting is in agreement with the possible role of myelin associated growth inhibitors as "stabilizers" of axonal fibre morphology during the latter stages of any CNS tract's development, which is then maintained throughout adult life (Keirstead & Steeves, 1998). Interestingly, those CNS projections which retain a high degree of plasticity (e.g. raphe-spinal or coeruleo-spinal) are either unmyelinated or relatively thinly myelinated throughout adult life and usually display an increased ability for inherent plasticity or regeneration after axotomy. Such regeneration is also accompanied by the expression of growth associated proteins (McNamara & Lenox, 1997).

Key points

Myelin-associated neurite growth inhibitor, eg IN-1 antibody, treatment of young adult rats with a transected corticospinal tract resulted in sprouting at lesion site.

A soluble fragment of MAG, consisting of its extracellular domain, is released from myelin and can inhibit axonal growth.

In the mature avian spinal cord, myelin suppression facilitated axonal regeneration by approximately 6-19% of the severed brainstem-spinal neurons, which is appreciably less than after cord transection in development.

Conclusions

Most spinal cord injury investigators and clinicians agree it is unlikely any single therapeutic intervention will facilitate complete functional repair whether it promotes neuron survival, stimulates axonal outgrowth or transiently blocks inhibitory signals. Perhaps an equally promising path of investigation, that we have promoted in this review, is the idea that delicately altering the balance between those intrinsic neuronal factors, promoting regeneration, with those endogenous glial mechanisms, inhibiting adult CNS repair, may be sufficient to drive recovery in a functionally positive direction (Figure 1). If we are to design an integrated approach to neurotrauma recovery, we also need to determine the spatial and temporal organization of any interventions.

Acknowledgements

We would like to thank the Medical Research Council of Canada, the Natural Sciences and Engineering Research Council of Canada, the Canadian Neuroscience Network, The Rick Hansen Institute at UBC, the BC Neurotrauma Fund, and the Paralyzed Veterans Administration of the United States for their support. We are also indebted to our colleagues at CORD for their input and constructive advice on this chapter.

References

Aigner L Arber S Kapfhammer J P Laux T Schneider C Botteri F Brenner H R and Caroni P (1995) Overexpression of the neural growth-associated protein GAP-43 induces nerve sprouting in the adult nervous system of transgenic mice. Cell 83, 269–278

Aigner L and Caroni P (1995) Absence of persistent spreading, branching, and adhesion in GAP-43-depleted growth cones. Journal of Cell Biology 128, 647–660

Alonso G Prieto M Legrand A and Chauvet N (1997) PSA-NCAM and B-50/GAP-43 are coexpressed by specific neuronal systems of the adult rat mediobasal hypothalamus that exhibit remarkable capacities for morphological plasticity. Journal of Comparative Neurology 384, 181–199

Andersen L B and Schreyer D J (1999) Constitutive expression of GAP-43 correlates with rapid, but not slow regrowth of injured dorsal root axons in the adult rat. Experimental Neurology 155, 157–164

Ankerhold R Leppert C A Bastmeyer M and Stuermer C A (1998) E587 antigen is upregulated by goldfish oligodendrocytes after optic nerve lesion and supports retinal axon regeneration. Glia 23, 257–270

Armanini M P McMahon S B Sutherland J Shelton D L and Phillips H S (1995) Truncated and catalytic isoforms of trkB are co-expressed in neurons of rat and mouse CNS. European Journal of Neuroscience 7, 1403–1409

Barron K D Dentinger M P Popp A J and Mankes R (1988) Neurons of layer Vb of rat sensorimotor cortex atrophy but do not die after thoracic cord transection. Journal of Neuropathology & Experimental Neurology 47, 62–74

Bastmeyer M Ott H Leppert C A and Stuermer C A (1995) Fish E587 glycoprotein, a member of the L1 family of cell adhesion molecules, participates in axonal fasciculation and the age-related order of ganglion cell axons in the goldfish retina. Journal of Cell Biology 130, 969–976

Beattie M S Bresnahan J C and Lopate G (1990) Metamorphosis alters the response to spinal cord transection in *Xenopus laevis* frogs. Journal of Neurobiology 21, 1108–1122

Becker T Bernhardt R R Reinhard E Wullimann M F Tongiorgi E and Schachner M (1998) Readiness of zebrafish brain neurons to regenerate a spinal axon correlates with differential expression of specific cell recognition molecules. Journal of Neuroscience 18, 5789–5803

Benowitz L I Perrone-Bizzozero N I Neve R L and Rodriguez W (1990) GAP-43 as a marker for structural plasticity in the mature CNS. Progress in Brain Research 86, 309–320

Benowitz L I and Routtenberg A (1997) GAP-43: an intrinsic determinant of neuronal development and plasticity. Trends in Neurosciences 20, 84–91

Bernhardt R R Tongiorgi E Anzini P and Schachner M (1996) Increased expression of specific recognition molecules by retinal ganglion cells and by optic pathway glia accompanies the successful regeneration of retinal axons in adult zebrafish. Journal of Comparative Neurology 376, 253–264

Berry M Carlile J and Hunter A (1996) Peripheral nerve explants grafted into the vitreous body of the eye promote the regeneration of retinal ganglion cell axons severed in the optic nerve. Journal of Neurocytology 25, 147–170

Bisby M A and Tetzlaff W (1992) Changes in cytoskeletal protein synthesis following axon injury and during axon regeneration. Molecular Neurobiology 6, 107–123

Bisby M A Tetzlaff W and Brown M C (1996) GAP-43 mRNA in mouse motoneurons undergoing axonal sprouting in response to muscle paralysis of partial denervation. European Journal of Neuroscience 8, 1240–1248

Blottner D and Herdegen T (1998) Neuroprotective Fibroblast Growth Factor Type-2 Down-Regulates the C-Jun Transcription Factor in Axotomized Sympathetic Preganglionic Neurons of Adult Rat. Neuroscience 82, 283–292

Bormann P Zumsteg V M Roth L W and Reinhard E (1998) Target contact regulates GAP-43 and alpha-tubulin mRNA levels in regenerating retinal ganglion cells. Journal of Neuroscience Research 52, 405–419

Bregman B S Kunkel-Bagden E Schnell L Dai H N Gao D and Schwab M E (1995) Recovery from spinal cord injury mediated by antibodies to neurite growth inhibitors [see comments]. Nature 378, 498–501

Bregman B S McAtee M Dai H N and Kuhn P L (1997) Neurotrophic factors increase axonal growth after spinal cord injury and transplantation in the adult rat. Experimental Neurology 148, 475–494
Brummendorf T Kenwrick S and Rathjen F G (1998) Neural cell recognition molecule L1: from cell biology to human hereditary brain malformations. Current Opinion in Neurobiology 8, 87–97
Buffo A Holtmaat A J Savio T Verbeek J S Oberdick J Oestreicher A B Gispen W H Verhaagen J Rossi F and Strata P (1997) Targeted overexpression of the neurite growth-associated protein B-50/GAP-43 in cerebellar Purkinje cells induces sprouting after axotomy but not axon regeneration into growth-permissive transplants. Journal of Neuroscience 17, 8778–8791
Burry B W and Perrone-Bizzozero N I (1993) Nerve growth factor stimulates GAP-43 expression in PC12 cell clones independently of neurite outgrowth. Journal of Neuroscience Research 36, 241–251
Caroni P (1997) Intrinsic neuronal determinants that promote axonal sprouting and elongation. Bioessays 19, 767–775
Caroni P and Schwab M E (1988) Antibody against myelin-associated inhibitor of neurite growth neutralizes nonpermissive substrate properties of CNS white matter. Neuron 1, 85–96
Chaisuksunt V Zhang Y Schachner M Anderson P N and Lieberman A R (1998) Expression of the cell recognition molecule CHL1 (close homologue of L1), in adult rat brain following peripheral nerve graft implantation. Society for Neuroscience Abstracts 24, 2009
Chen D F Jhaveri S and Schneider G E (1995) Intrinsic changes in developing retinal neurons result in regenerative failure of their axons. Proceedings of the National Academy of Sciences USA 92, 7287–7291
Cheng H Cao Y and Olson L (1996) Spinal cord repair in adult paraplegic rats: partial restoration of hind limb function [see comments]. Science 273, 510–513
Chiaramello A Neuman T Peavy D R and Zuber M X (1996) The GAP-43 gene is a direct downstream target of the basic helix-loop-helix transcription factors. Journal of Biological Chemistry 271, 22035–22043
Console-Bram L M Fitzpatrick-McElligott S G and McElligott J G (1996) Distribution of GAP-43 mRNA in the immature and adult cerebellum: a role for GAP-43 in cerebellar development and neuroplasticity. Developmental Brain Research 95, 97–106
Dahme M Bartsch U Martini R Anliker B Schachner M and Mantei N (1997) Disruption of the mouse L1 gene leads to malformations of the nervous system. Nature Genetics 17, 346–349
David S and Aguayo A J (1981) Axonal elongation into peripheral nervous system "bridges" after central nervous system injury in adult rats. Science 214, 931–933
Davies S J Field P M and Raisman G (1994) Long interfascicular axon growth from embryonic neurons transplanted into adult myelinated tracts. Journal of Neuroscience 14, 1596–1612
Davies S J A Fitch M T Memberg S P Hall A K Raisman G and Silver J (1997) Regeneration of adult axons in white matter tracts of the central nervous system. Nature 390, 680–683
Dib-Hajj S D Black J A Cummins T R Kenney A M Kocsis J D and Waxman S G (1998) Rescue of α-SNS sodium channel expression in small dorsal root ganglion neurons after axotomy by nerve growth factor *in vivo*. Journal of Neurophysiology 79, 2668–2676
Drescher U Bonhoeffer F and Muller B K (1997) The Eph family in retinal axon guidance. Current Opinion in Neurobiology 7, 75–80
Dyer J K Bourque J A and Steeves J D (1998) Regeneration of brainstem-spinal axons after lesion and immunological disruption in adult rat. Experimental Neurology, 154, 12–22
Filbin M T (1995) Myelin-associated glycoprotein: a role in myelination and in the inhibition of axonal regeneration? Current Opinion in Neurobiology 5, 588–595
Filbin M T (1996) The muddle with MAG. Molecular & Cellular Neurosciences 8, 84–92

Fitch M T and Silver J (1997) Glial cell extracellular matrix: boundaries for axon growth in development and regeneration. Cell & Tissue Research 290, 379–384

Flanagan J G and Vanderhaeghen P (1998) The ephrins and Eph receptors in neural development. Annual Review of Neuroscience 21, 309–345

Fournier A E Beer J Arregui C O Essagian C Aguayo A J and McKerracher L (1997) Brain-derived neurotrophic factor modulates GAP-43 but not Tα1 expression in injured retinal ganglion cells of adult rats. Journal of Neuroscience Research 47, 561–572

Giehl K M Schutte A Mestres P and Yan Q (1998) The survival-promoting effect of glial cell line-derived neurotrophic factor on axotomized corticospinal neurons *in vivo* is mediated by an endogenous brain-derived neurotrophic factor mechanism. Journal of Neuroscience 18, 7351–7360

Giehl K M and Tetzlaff W (1996) BDNF and NT-3, but not NGF, prevent axotomy-induced death of rat corticospinal neurons *in vivo*. European Journal of Neuroscience 8, 1167–1175

Grill R Murai K Blesch A Gage F H and Tuszynski M H (1997) Cellular delivery of neurotrophin-3 promotes corticospinal axonal growth and partial functional recovery after spinal cord injury. Journal of Neuroscience 17, 5560–5572

Hasan S J Keirstead H S Muir G D and Steeves J D (1993) Axonal regeneration contributes to repair of injured brainstem-spinal neurons in embryonic chick. Journal of Neuroscience 13, 492–507

Herdegen T Skene P and Bähr M (1997) The c-Jun transcription factor–bipotential mediator of neuronal death, survival and regeneration. Trends in Neurosciences 20, 227–231

Hirsch S Cahill M A and Stuermer C A (1995) Fibroblasts at the transection site of the injured goldfish optic nerve and their potential role during retinal axonal regeneration. Journal of Comparative Neurology 360, 599–611

Houle J D (1991) Demonstration of the potential for chronically injured neurons to regenerate axons into intraspinal peripheral nerve grafts. Experimental Neurology 113, 1–9

Huber K Meisinger C and Grothe C (1997) Expression of fibroblast growth factor-2 in hypoglossal motoneurons is stimulated by peripheral nerve injury. Journal of Comparative Neurology 382, 189–198

Jenkins R McMahon S B Bond A B and Hunt S P (1993a) Expression of c-Jun as a response to dorsal root and peripheral nerve section in damaged and adjacent intact primary sensory neurons in the rat. European Journal of Neuroscience 5, 751–759

Jenkins R Tetzlaff W and Hunt S P (1993b) Differential expression of immediate early genes in rubrospinal neurons following axotomy in rat. European Journal of Neuroscience 5, 203–209

Ji R R Zhang Q Pettersson R F and Hokfelt T (1996) aFGF, bFGF and NGF differentially regulate neuropeptide expression in dorsal root ganglia after axotomy and induce autotomy. Regulatory Peptides 66, 179–189

Jung M Petrausch B and Stuermer C A (1997) Axon-regenerating retinal ganglion cells in adult rats synthesize the cell adhesion molecule L1 but not TAG-1 or SC-1. Molecular & Cellular Neurosciences 9, 116–131

Kaplan D R and Miller F D (1997) Signal transduction by the neurotrophin receptors. Current Opinion in Cell Biology 9, 213–221

Keirstead H S Dyer J K Sholomenko G N McGraw J Delaney K R and Steeves J D (1995) Axonal regeneration and physiological activity following transection and immunological disruption of myelin within the hatchling chick spinal cord. Journal of Neuroscience 15, 6963–6974

Keirstead H S Hasan S J Muir G D and Steeves J D (1992) Suppression of the onset of myelination extends the permissive period for the functional repair of embryonic spinal cord. Proceedings of the National Academy of Sciences USA 89, 11664–11668

Keirstead H S Hughes H C and Blakemore W F (1998) A quantifiable model of axonal regeneration in the demyelinated adult rat spinal cord. Experimental Neurology 151, 303–313

Keirstead H S and Steeves J D (1998) CNS myelin: Does a stabilizing role in neurodevelopment result in inhibition of neuronal repair after injury. The Neuroscientist 4, 273–284

Kobayashi N (1998) Neurotrophins and the neuronal response to axotomy. Ph.D. Thesis, University of British Columbia

Kobayashi N R Bedard A M Hincke M T and Tetzlaff W (1996) Increased expression of BDNF and trkB mRNA in rat facial motoneurons after axotomy. European Journal of Neuroscience 8, 1018–1029

Kobayashi N R Fan D P Giehl K M Bedard A M Wiegand S J and Tetzlaff W (1997) BDNF and NT-4/5 prevent atrophy of rat rubrospinal neurons after cervical axotomy, stimulate GAP-43 and Tα1-tubulin mRNA expression, and promote axonal regeneration. Journal of Neuroscience 17, 9583–9595

Lang D M Monzon-Mayor M Bandtlow C E and Stuermer C A (1998) Retinal axon regeneration in the lizard *Gallotia galloti* in the presence of CNS myelin and oligodendrocytes. Glia 23, 61–74

Lang D M Rubin B P Schwab M E and Stuermer C A (1995) CNS myelin and oligodendrocytes of the *Xenopus* spinal cord–but not optic nerve–are nonpermissive for axon growth. Journal of Neuroscience 15, 99–109

Li D Field P M and Raisman G (1995) Failure of axon regeneration in postnatal rat entorhinohippocampal slice coculture is due to maturation of the axon, not that of the pathway or target. European Journal of Neuroscience 7, 1164–1171

Liabotis S and Schreyer D J (1995) Magnitude of GAP-43 induction following peripheral axotomy of adult rat dorsal root ganglion neurons is independent of lesion distance. Experimental Neurology 135, 28–35

Mansour-Robaey S Clarke D B Wang Y C Bray G M and Aguayo A J (1994) Effects of ocular injury and administration of brain-derived neurotrophic factor on survival and regrowth of axotomized retinal ganglion cells. Proceedings of the National Academy of Sciences USA 91, 1632–1636

Maroney A C Glicksman M A Basma A N Walton K M Knight E Jr Murphy C A Bartlett B A Finn J P Angeles T Matsuda Y Neff N T and Dionne C A (1998) Motoneuron apoptosis is blocked by CEP-1347 (KT 7515) A novel inhibitor of the JNK signaling pathway. Journal of Neuroscience 18, 104–111

Martini R (1994) Expression and functional roles of neural cell surface molecules and extracellular matrix components during development and regeneration of peripheral nerves. Journal of Neurocytology 23, 1–28

McClellan A D (1998) Spinal cord injury: lessons from locomotor recovery and axonal regeneration in lower vertebrates. The Neuroscientist 4, 250–263

McKerracher L David S Jackson D L Kottis V Dunn R J and Braun P E (1994) Identification of myelin-associated glycoprotein as a major myelin-derived inhibitor of neurite growth. Neuron 13, 805–811

McKerracher L Essagian C and Aguayo A J (1993) Temporal changes in beta-tubulin and neurofilament mRNA levels after transection of adult rat retinal ganglion cell axons in the optic nerve. Journal of Neuroscience 13, 2617–2626

McNamara R K and Lenox R H (1997) Comparative distribution of myristoylated alanine-rich C kinase substrate (MARCKS) and F1/GAP-43 gene expression in the adult rat brain. Journal of Comparative Neurology 379, 48–71

Mey J and Thanos S (1996) Functional and biochemical analysis of CNS-relevant neurotrophic activity in the lesioned sciatic nerve of adult rats. Journal fur Hirnforschung 37, 25–50

Meyer M Matsuoka I Wetmore C Olson L and Thoenen H (1992) Enhanced synthesis of brain-derived neurotrophic factor in the lesioned peripheral nerve: different mechanisms are responsible for the regulation of BDNF and NGF mRNA. Journal of Cell Biology 119, 45–54

Mori F Himes B T Kowada M Murray M and Tessler A (1997) Fetal spinal cord transplants rescue some axotomized rubrospinal neurons from retrograde cell death in adult rats. Experimental Neurology 143, 45–60

Muir G D and Steeves J D (1997) Sensorimotor stimulation to improve locomotor recovery after spinal cord injury [see comments]. Trends in Neurosciences 20, 72–77

Murphy P G Grondin J Altares M and Richardson P M (1995) Induction of interleukin-6 in axotomized sensory neurons. Journal of Neuroscience 15, 5130–5138

Ng T F So K F and Chung S K (1995) Influence of peripheral nerve grafts on the expression of GAP-43 in regenerating retinal ganglion cells in adult hamsters. Journal of Neurocytology 24, 487–496

Oppenheim R W (1996) Neurotrophic survival molecules for motoneurons: an embarrassment of riches. Neuron 17, 195–197

Orioli D and Klein R (1997) The Eph receptor family: axonal guidance by contact repulsion. Trends in Genetics 13, 354–359

Perrone-Bizzozero N I Cansino V V and Kohn D T (1993) Posttranscriptional regulation of GAP-43 gene expression in PC12 cells through protein kinase C-dependent stabilization of the mRNA. Journal of Cell Biology 120, 1263–1270

Ramón y Cajal S (1928) Degeneration and Regeneration of the Nervous System New York: Oxford University Press (1991)

Rapalino O Lazarov-Spiegler O Agranov E Velan G J Yoles E Fraidakis M Solomon A Gepstein R Katz A Belkin M Hadani M and Schwartz M (1998) Implantation of stimulated homologous macrophages results in partial recovery of paraplegic rats. Nature Medicine 4, 814–821

Richardson P M and Issa V M (1984) Peripheral injury enhances central regeneration of primary sensory neurones. Nature 309, 791–793

Richardson P M Issa V M and Aguayo A J (1984) Regeneration of long spinal axons in the rat. Journal of Neurocytology 13, 165–182

Richardson P M McGuinness U M and Aguayo A J (1980) Axons from CNS neurons regenerate into PNS grafts. Nature 284, 264–265

Rossi F Bravin M Buffo A Fronte M Savio T and Strata P (1997) Intrinsic properties and environmental factors in the regeneration of adult cerebellar axons. Progress in Brain Research 114, 283–296

Schaden H Stuermer C A and Bähr M (1994) GAP-43 immunoreactivity and axon regeneration in retinal ganglion cells of the rat. Journal of Neurobiology 25, 1570–1578

Schnell L Schneider R Kolbeck R Barde Y A and Schwab M E (1994) Neurotrophin-3 enhances sprouting of corticospinal tract during development and after adult spinal cord lesion [see comments]. Nature 367, 170–173

Schnell L and Schwab M E (1990) Axonal regeneration in the rat spinal cord produced by an antibody against myelin-associated neurite growth inhibitors. Nature, 343, 269–272

Schreyer D J and Skene J H (1993) Injury-associated induction of GAP-43 expression displays axon branch specificity in rat dorsal root ganglion neurons. Journal of Neurobiology 24, 959–970

Schwab M E (1996) Structural plasticity of the adult CNS. Negative control by neurite growth inhibitory signals. International Journal of Developmental Neuroscience 14, 379–385

Schwab M E and Bartholdi D (1996) Degeneration and regeneration of axons in the lesioned spinal cord. Physiological Reviews 76, 319–370

Shea T B and Benowitz L I (1995) Inhibition of neurite outgrowth following intracellular delivery of anti-GAP-43 antibodies depends upon culture conditions and method of neurite induction. Journal of Neuroscience Research 41, 347–354

Shewan D Berry M and Cohen J (1995) Extensive regeneration *in vitro* by early embryonic neurons on immature and adult CNS tissue. Journal of Neuroscience 15, 2057–2062

Sivron T Schwab M E and Schwartz M (1994) Presence of growth inhibitors in fish optic nerve myelin: postinjury changes. Journal of Comparative Neurology 343, 237–246

Sivron T and Schwartz M (1995) Glial cell types, lineages, and response to injury in rat and fish: implications for regeneration. Glia 13, 157–165

Skene J H (1989) Axonal growth-associated proteins. Annual Review of Neuroscience 12, 127–156

Smith D S and Skene J H P (1997) A transcription-dependent switch controls competence of adult neurons for distinct modes of axon growth. Journal of Neuroscience 17, 646–658

Steeves J D Keirstead H S Ethell D W Hasan S J Muir G D Pataky D M McBride C B Petrausch B and Zwimpfer T J (1994) Permissive and restrictive periods for brainstem-spinal regeneration in the chick. Progress in Brain Research 103, 243–262

Streit W J Semple-Rowland S L Hurley S D Miller R C Popovich P G and Stokes B T (1998) Cytokine mRNA profiles in contused spinal cord and axotomized facial nucleus suggest a beneficial role for inflammation and gliosis. Experimental Neurology 152, 74–87

Strittmatter S M Fankhauser C Huang P L Mashimo H and Fishman M C (1995) Neuronal pathfinding is abnormal in mice lacking the neuronal growth cone protein GAP-43. Cell 80, 445–452

Takemura R Nakata T Okada Y Yamazaki H Zhang Z and Hirokawa N (1996) mRNA expression of KIF1A, KIF1B, KIF2, KIF3A, KIF3B, KIF4, KIF5, and cytoplasmic dynein during axonal regeneration. Journal of Neuroscience 16, 31–35

Tang S Woodhall R W Shen Y J deBellard M E Saffell J L Doherty P Walsh F S and Filbin M T (1997) Soluble myelin-associated glycoprotein (MAG) found *in vivo* inhibits axonal regeneration. Molecular & Cellular Neurosciences 9, 333–346

Tetzlaff W Alexander S W Miller F D and Bisby M A (1991) Response of facial and rubrospinal neurons to axotomy: changes in mRNA expression for cytoskeletal proteins and GAP-43. Journal of Neuroscience 11, 2528–2544

Tetzlaff W Kobayashi N R Giehl K M Tsui B J Cassar S L and Bedard A M (1994) Response of rubrospinal and corticospinal neurons to injury and neurotrophins. Progress in Brain Research 103, 271–286

Vanselow J Grabczyk E Ping J Baetscher M Teng S and Fishman M C (1994) GAP-43 transgenic mice: dispersed genomic sequences confer a GAP-43–like expression pattern during development and regeneration. Journal of Neuroscience 14, 499–510

Varga Z M Schwab M E and Nicholls J G (1995) Myelin-associated neurite growth-inhibitory proteins and suppression of regeneration of immature mammalian spinal cord in culture. Proceedings of the National Academy of Sciences USA 92, 10959–10963

Vidal-Sanz M Bray G M and Aguayo A J (1991) Regenerated synapses persist in the superior colliculus after the regrowth of retinal ganglion cell axons [published erratum appears in J Neurocytol 1992 21:234]. Journal of Neurocytology 20, 940–952

Walsh F S and Doherty P (1997) Neural cell adhesion molecules of the immunoglobulin syperfamily – Role in axon growth and guidance. Annual Review of Cell & Developmental Biology 13, 425–456

Wanner M Lang D M Bandtlow C E Schwab M E Bastmeyer M and Stuermer C A (1995) Reevaluation of the growth-permissive substrate properties of goldfish optic nerve myelin and myelin proteins. Journal of Neuroscience 15, 7500–7508

Weber J R M and Skene J H P (1997) Identification of a novel repressive element that contributes to neuron-specific gene expression. Journal of Neuroscience 17, 7583–7593

Wictorin K Brundin P Gustavii B Lindvall O and Bjorklund A (1990) Reformation of long axon pathways in adult rat central nervous system by human forebrain neuroblasts. Nature 347, 556–558

Woolhead C L Zhang Y Lieberman A R Schachner M Emson P C and Anderson P N (1998) Differential effects of autologous peripheral nerve grafts to the corpus striatum of adult rats on the regeneration of axons of striatal and nigral neurons and on the expression of GAP-43 and the cell adhesion molecules N-CAM and L1. Journal of Comparative Neurology 391, 259–273

Ye J H and Houle J D (1997) Treatment of the chronically injured spinal cord with neurotrophic factors can promote axonal regeneration from supraspinal neurons. Experimental Neurology 143, 70–81

Zagrebelsky M Buffo A Skerra A Schwab M E Strata P and Rossi F (1998) Retrograde regulation of growth-associated gene expression in adult rat Purkinje cells by myelin-associated neurite growth inhibitory proteins. Journal of Neuroscience 18, 7912–7929

Z'Graggen W J Metz G A Kartje G L Thallmair M and Schwab M E (1998) Functional recovery and enhanced corticofugal plasticity after unilateral pyramidal tract lesion and blockade of myelin-associated neurite growth inhibitors in adult rats. Journal of Neuroscience 18, 4744–4757

Zigmond R E Hyatt-Sachs H Mohney R P Schreiber R C Shadiack A M Sun Y and Vaccariello S A (1996) Changes in neuropeptide phenotype after axotomy of adult peripheral neurons and the role of leukemia inhibitory factor. Perspectives on Developmental Neurobiology 4, 75–90

6

Evolutionary hierarchy of optic nerve regeneration: Implications for cell survival, axon outgrowth and map formation

Lyn D Beazley and Sarah A Dunlop

Background

The vertebrate optic nerve has proved a popular model for studies of central nerve degeneration and regeneration. One valuable aspect is that the interpretation of the results of axotomy are simplified by the fact that the optic nerve represents a nerve tract composed entirely or almost entirely of axons of one origin, namely optic axons arising from retinal ganglion cells. Another advantage is that the parent cell somata form a well demarcated population, within the innermost cellular layer of the retina. Optic axons converge to form the optic nerve and on reaching the brain, traverse the optic tracts to terminate in contralateral, and usually to a lesser extent, in ipsilateral visual brain centres. The main visual centre in non-mammals is the optic tectum. In mammals there are two prominent primary visual centres, namely the superior colliculus, the homologue of the optic tectum, and the lateral geniculate nucleus. In the major visual centres, axons terminate in a topographically ordered fashion, related to the distribution of their somata within the retina. Many anatomical, physiological, pharmacological and behavioural studies describe the normal visual projection, providing a basis for studies of axonal degeneration and regeneration (Jacobson, 1991).

This review is concerned with optic nerve regeneration and it is therefore pertinent to describe the various lesion procedures followed. However, it must be conceded that some studies suffer from inadequate descriptions of either the mode and/or the site of lesion. The optic nerve can be severed by cutting, a procedure which induces an extensive zone of injury including macrophage invasion and neovascularisation (Berkelaar *et al.*, 1994) and may allow regenerating axons to escape into surrounding non-neural tissue. However, the more commonly adopted method, termed a nerve crush, involves severing the optic axons whilst leaving the tough dural nerve sheath intact as a conduit for regenerating axons. The task is achieved either by repeatedly squeezing the nerve with forceps until there is a distinct gap in the nerve parenchyma or by gaining access to the interior of the nerve via a slit in the sheath. Less frequently used procedures include cryosurgery. Care is taken to avoid compromising the blood supply by either avoiding ophthalmic vessels within the nerve or by locating the lesion at a site more distal to the eye, where the nerve and ophthalmic vessels are no longer associated. The location of the lesion, as well as its type, may also be relevant. Studies in both frogs and mammals indicate that the regenerative response differs when the lesion is close to the eye as compared to more distal sites (Humphrey & Beazley, 1982, 1983; Berkelaar *et al.*, 1994).

One reason for the continued popularity of the optic nerve for studies of nerve regeneration is the gradation of responses to axotomy across the vertebrate phylum. Insights can be gained as to the conditions required for sufficient regeneration to provide useful function. Regenerative responses range from almost completely successful in fish to abortive in mammals. However, using a technique pioneered by Aguayo and his colleagues, the brief and usually abortive axonal sprouting near the lesion site in mammals can be mobilised to trigger limited axonal regeneration. To do so, one end of an autologous peripheral nerve graft is anastomosed to the optic nerve and the other end inserted into a visual brain centre (Keirstead *et al.*, 1989).

Moreover, optic nerve regeneration, as we discuss here, can be considered to be a series of responses, each of which must have a successful outcome to restore vision. One major component is the response of the axotomised retinal ganglion cell themselves. Here we consider both their somata and dendrites, outgrowth of their axons to traverse the crush site and reach the brain and finally, the establishment and maintenance of appropriate reinnervation of visual centres, thereby mediating visual behaviour. The other component involves changes which extend beyond the ganglion cells themselves and include modifications to the macroglia, the vascular system, the status of the blood-brain barrier, as well as the removal of the distal segments of axons disconnected by axotomy. Here we consider both components, describing studies of the visual system and, where relevant, other sysrems. Aspects of our review are summarised in Figures 1 and 2 and Table 1.

Key points

- **The optic nerve is usually lesioned by a cut or a crush; the former causes greater damage to the nerve. The regenerative response varies from extensive in fish and amphibians to abortive in mammals.**

The retinal response

For optic nerve regeneration to be successful, retinal ganglion cell bodies must survive axotomy. The cells must switch their metabolism to support axonal regrowth and stimulate appropriate changes in intra-retinal wiring.

Survival of axotomised ganglion cells and the time course of cell death

The extent of ganglion cell survival after axotomy has been assessed by counts of ganglion cell somata and/or optic axons. Counts of ganglion cells require that these cells be distinguished from displaced amacrine cells which also lie in the retinal ganglion cell layer; morphological criteria may be used for cell identification but retrograde labelling with an axonal tracer is probably preferable. Since regenerating axons are unmyelinated, at least initially, axon numbers are best assessed by electron microscopy.

Ganglion cell survival during optic nerve regeneration is greatest in fish, at over 90% (Murray *et al.*, 1982). Rates are intermediate at 20–70% in frog (Humphrey & Beazley, 1985; Scalia *et al.*, 1985; Stelzner & Strauss, 1986 and Beazley *et al.*, 1986) and a mean of 65% in the lizard *Ctenophorus ornatus* (Beazley *et al.*, 1997) but minimal in mammals (Quigley *et al.*, 1995). However, as we describe below, increased ganglion cell survival can be achieved by suitable interventions in mammals. In each case, the surviving ganglion cells are distributed across the retina, offering the opportunity to regain vision throughout the visual field.

There are two modes of cell death, apoptotic and necrotic. In each vertebrate class, the axotomised retinal ganglion cells that die appear to do so by apoptosis. Their histologically degenerate appearance, with pyknotic and fragmenting chromatin, is compatible with this mode of death. The conclusion is supported in mammals by DNA electrophoresis and DNA fragmentation using the TdT-dVTP terminal nick end labelling (TUNEL) procedure (Berkelaar *et al.*, 1994; Quigley *et al.*, 1995). The apoptotic cells are engulfed by their neighbours or by invading macrophages.

In most tissues, degeneration is usually a trigger for neovascularisation. It is unclear whether this process occurs in mammalian retina during the cell death resulting from optic nerve lesion. However, retinal neovascularisation has been visualised in frog by perfusion of India ink. The time course of neovascularization correlated with that for ganglion cell death. Moreover, the transient vessels were most numerous in areas with the greatest number of dying cells, namely the area centralis of high ganglion cell density, and of dying axons, namely the optic nerve head (Tennant *et al.*, 1993a).

The time course of ganglion cell death after axotomy, usually assessed from counts of ganglion cell somata in retinal wholemounts, seems to extend over several weeks in all vertebrates studied. This is despite differences in metabolic rates between the cold and warm blooded classes. As examples, for frogs most axotomised ganglion cells die 2-10 weeks after nerve crush in *Litoria moorei* (Humphrey & Beazley, 1985) and 8-12 weeks in *Rana pipiens* (Beazley *et al.*, 1986, Humphrey, 1987). The time course of ganglion cell death has yet to be established for lizard but in rat after intraorbital nerve crush, it appears to be composed on an early rapid phase between 5–14 days with a more protracted second phase (Villegas-Perez *et al.*, 1993). In rabbit, intraorbital nerve crush leads to ganglion cell death mostly between one and four weeks; in monkey, cell death has commenced by 2 weeks and is approaching completion by 10 and 12 weeks (Quigley *et al.*, 1995). The time course of cell death, particularly its second phase, is markedly delayed the more distal the lesion from the eye (Berkelaar *et al.*, 1994).

Key points

- Most ganglion cells survive axotomy in fish, 20-70% do so in frog and 65% in lizard; all ganglion cells normally die after axotomy in mammals.
- Axotomised ganglion cells die by apoptosis. Most ganglion cell death takes place over 2 months in frog, and by 2, 4 and 12 weeks in rat, rabbit and monkey respectively.

Reasons for the death of axotomised ganglion cells

A key question in this field is why ganglion cells die after axotomy. There are at least three possibilities that are not mutually exclusive. The axotomised ganglion cells may die because axotomy directly or indirectly changes their metabolic status, initiating an apoptotic chain of events rather than one leading to axonal regeneration. In this scenario, the ganglion cell does not attempt to initiate axonal regeneration. The death of some mammalian β ganglion cells seems to fall into this category, being brought about by excitotoxicity to glutamate (Silveira *et al.*, 1994). In accord with this suggestion, the application of the N-methyl-D-aspartate (NMDA) antagonist MK-801 has been found to decrease ganglion cell death (Hahn *et al.*, 1988).

A second reason for ganglion cell death after axotomy is that the cells attempt to regrow an axon and survive if they do so, but die if the attempt fails. This seems to be the case in frog (Dunlop *et al.*, unpublished). The time course of ganglion cell death, along with counts of retrogradely labelled ganglion cells and of axons on either side of the lesion, indicate that ganglion cells survive whilst their axons attempt to cross the lesion. Indeed it is possible that the greater survival of axotomised ganglion cells in the tree frog *L. moorei* as compared to the leopard frog *R. pipiens* can be explained in this way. In the tree frog, optic axons do not withdraw back towards the eye to any discernible extent, presumably increasing their chance of crossing the lesion site and of survival. In the leopard frog, the evidence for axonal sprouting between the eye and crush site is compatible with axons withdrawing back towards the eye before starting to regrow their axons, thus presumably reducing their chances of survival.

In mammals also, at least some ganglion cells seem to die because optic axons fail in their attempt to cross the lesion site. As evidence, a minority survive if their axons can regenerate into an autologous peripheral nerve graft, even in the absence of other manipulations to delay or prevent ganglion cell death (Keirstead *et al.*, 1989). On the other hand, it seems that many optic axons retract into the retina after optic nerve lesion, their axons never re-exit the eye and the parent ganglion cell dies. It appears that there is a hierarchy of sites at which optic axons form additional processes after axotomy. If the cut end of the axon remains at the lesion site, then axonal regrowth into a peripheral nerve graft takes place preferentially at this site and the ganglion cell survives. If the axon retracts into the retina, then aberrant intra-retinal growth results and the ganglion cell dies. The extent of such aberrant growth, demonstrated using intracellular tracers, is increased by the application to the retina of exogenous neurotrophins including brain-derived neurotrophic factor (BDNF) or neurotrophin-4/5 (NT-4/5) (Sawai *et al.*, 1996).

A third possible reason for ganglion cell death during optic nerve regeneration is that optic axons regain visual centres but then compete via an activity dependent mechanism for terminal space in a fashion mimicking that seen during development (Jacobson, 1991). However, there is little support for this concept. The possibility has so far been tested only in frog. Retinal application of tetrodotoxin throughout optic nerve regeneration to prevent sodium-mediated action potentials, does not change the extent of ganglion cell death as assessed from counts of retinal ganglion cell somata (Sheard & Beazley, 1988). By contrast, in developing mammals, such treatment prevents some ganglion cell death (Jacobson, 1991). Until more extensive axonal regrowth can be induced after axotomy adult mammals, the possibility of central competition playing a role in ganglion cell death during optic nerve regeneration remains untested.

Key points

- In frog, ganglion cells die if their axons do not cross the lesion site but survive if they do so.
- In mammals, triggers of ganglion cell death are probably multi-factorial and include excitotoxicity and a failure of axons to negotiate the lesion site or distal nerve.
- Some ganglion cells survive if their axons regenerate into a supportive environment such as a peripheral nerve graft.

Prevention of ganglion cell death

Irrespective of the immediate triggers for ganglion cell death after axotomy, ultimately the death of many ganglion cells, as with neurons in other parts of the nervous system, can usually be linked to a lack of neurotrophic support. This support is normally retrogradely transported from target cells (Jacobson, 1991). In the interval between axotomy and regaining access to the brain, ganglion cells presumably survive by either a temporary resilience to a lack of neurotrophic support or by receiving such support from cells within the visual pathway. Presumably one or both these conditions apply to fish in which virtually all ganglion cells survive axotomy and to frog, in terms of those ganglion cells with axons that cross the lesion site. Surprisingly, there is now evidence that ganglion cells may receive adequate neurotrophic support even if their axons are not appropriately connected in the brain. As an example, in lizard, axotomised ganglion cells survive for at least a year despite their axons failing to form retinotopically ordered projections within visual centres (Beazley *et al.*, 1997); moreover, they form persistent projections to inappropriate targets such as the opposite optic nerve, the olfactory nerve, hypothalamus and tectal and poterior commissures (Dunlop *et al.*, in press). Similarly, in rat, ganglion cells survive well beyond the usual period of cell death even if the axons reinnervate a non-visual region, namely the cerebellum (Zwimpfer *et al.*, 1992).

Several neurotrophins have been applied in attempts to rescue axotomised retinal ganglion cells. When applied to the severed optic nerve in frog, nerve growth factor (NGF) does not diminish ganglion cell losses (Humphrey, 1987); however, it is unclear whether the NGF gained access to the ganglion cell somata. In rat, repeated intraocular application of NGF significantly delays the death of axotomised ganglion cells (Carmignoto *et al.*, 1989). Similarly, transient sparing is seen after application of BDNF (Berkelaar *et al.*, 1994), a result compatible with the localisation of the tyrosine kinase B (trkB) receptor for BDNF to rat ganglion cells (Jelsma *et al.*, 1983). Factors such as microglial inhibitory factor, fibroblast growth factor and ciliary neurotrophic factor also offer protection to some axotomised ganglion cells (Thanos *et al.*, 1993; Berry *et al.*, 1996; Peinado-Ramon *et al.*, 1996). Indeed, insertion of lengths or extracts from peripheral nerve (Ng *et al.*, 1995) or of viable Schwann cells into the vitreous humour of the eye are also effective (Maffei *et al.*, 1990). Although the protection by such exogenous neurotrophins is transient, it may prove adequate to allow other procedures to ensure a sufficient number of ganglion cells undergo axonal regeneration and thus regain endogenous neurotrophic support.

Another avenue to protect axotomised ganglion cells is based on defining the molecular program of apoptosis, irrespective of the factors leading to cell suicide, and then endeavouring to block the cascade. Several genes change their expression during apoptosis, with the gene bcl-x_L predominating in adult rat retina. The gene is down-regulated during ganglion cell apoptosis, suggesting that if it could be up-regulated, cell death might be prevented (Levin *et al.*, 1997). In other words, bcl-x_L may represent an 'antideath' gene for axotomised ganglion cells. The result is compatible with the finding that up 60% of axotomised ganglion cells survive for several months in mice transgenic for the bcl-2 gene (Cenni *et al.*, 1996). However, it is important to bear in mind that the ganglion cell population is already highly abnormal, as are many other aspects of such animals. Cells in other parts of the nervous system, including sympathetic neurons (Garcia *et al.*, 1992) that normally would have been removed by natural processes of cell death are presumably still present; such cells may themselves endow the animal with an atypical profile of both neurotrophic and other factors. Another promising avenue to rescue ganglion cells might be to block the cell death cascade by inhibiting lipid peroxidation (Levin *et al.*, 1996); an alternative strategy might be to up-regulate gene expression for the ferroxidase ceruloplasmin, which prevents the formation of certain reactive oxygen species (Levin & Geszvain, 1998).

Key points

- **In mammals, exogenous neurotrophins delay the death of a proportion of axotomised ganglion cells.**
- **Strategies to modify the expression of genes involved in apoptosis may be valuable in preventing ganglion cell death.**

Metabolic responses of axotomised ganglion cells

If a ganglion cell is to survive the challenge of axotomy, it must mobilise its metabolism to avoid apoptosis and initiate a regenerative response. Such changes manifest themselves as chromatolysis, an appearance associated with up-regulation of protein synthesis. Shortly after axotomy, all ganglion cell somata in fish (Barron *et al.*, 1985) and frog exhibit chromatolysis (Humphrey, 1988); morphometric analyses confirm that both somal and nucleolar enlargement take place. Moreover, it is clear that, although all ganglion cells undergo chromatolysis in frog, some will not complete axonal regeneration and will die. The observation is compatible with our concept that all ganglion cells in frog attempt to regenerate but some die if they fail to cross the lesion site. The chromatolytic appearance persists for weeks or months until optic axon regeneration is complete and stable central connections are re-established. By contrast, in mammals, the failure of optic nerve regeneration is compatible with an absence of chromatolysis and with ganglion cells themselves undergoing atrophy. The lizard represents the other extreme in that ganglion cells exhibit intense chromatolysis for at least a year after nerve lesion (Beazley *et al.*, 1997). The implication from our work mapping projections to the optic tectum in lizard (see "Retinotopicity of Regenerated Projections" below) is that the ganglion cells are held at the penultimate stage of axonal regeneration. The axons must constantly grow new membranes as their terminals search in vain for appropriate post-synaptic partners.

Many studies have defined proteins that are up-regulated during successful optic nerve regeneration (Quitsche & Schechter, 1983). Autoradiographic procedures indicated increased incorporation of the RNA precursor uridine in fish (Barron *et al.*, 1985). Gel electrophoresis of radioactively labelled transported proteins was used to demonstrate increased levels of GAP-43 (growth associated protein) in goldfish (Perry *et al.*, 1987), a result compatible with the presence of this molecule in growth cones and axon terminals. Similarly, the expression of GAP-43 is increased after axotomy in rat but only if the lesion is close to the eye (Doster *et al.*, 1991). The relationship between GAP-43 expression and the location of the nerve lesion matches the greater ability of optic axons to regenerate into a peripheral nerve graft anastomosed close behind the eye, rather than at a more distant location (So & Aguayo, 1985). Levels of GAP-43 are also elevated in those mammalian ganglion cells with axons regenerating into a peripheral nerve graft (Ng *et al.*, 1995).

Using specific polyclonal antibodies it has been shown that two other intermediate filament proteins, plasticin and gefiltin (formerly designated as ON_1), are also up-regulated during optic nerve regeneration in fish (Glasgow *et al.*, 1994). In mammals, in the absence of a peripheral nerve graft, expression of ß-tubulin mRNA transiently rises and then falls, presumably reflecting the early short-lived attempts at axonal regeneration; neurofilament subunit (NF-M) mRNA levels fall throughout (McKerracher *et al.*, 1993). However, when severed axons regenerate to the brain via a peripheral nerve graft, expression of ß-tubulin mRNA is increased in retinal "hot spots". By retrograde labelling, these spots were shown to correspond to the somata of axotomised ganglion cells with regenerated axons (McKerracher *et al.*, 1993). Furthermore, the sustained nature of the ß-tubulin mRNA synthesis is compatible with the continued growth of regenerated optic axon terminals within the visual brain centres (Carter *et al.*, 1994).

Key points

- **Synthesis of cytoskeletal molecules is up-regulated in ganglion cells during optic nerve regeneration in fish and amphibians and in mammals if optic axons regenerate via a peripheral nerve graft.**

Survival of different classess of ganglion cell

Retinal ganglion cells have been classified according to their functional types, characteristics that correlate with their morphology. For example, ganglion cells in frog are defined as detectors of movement, edges and changing light levels (Maturana *et al.*, 1960). In mammals (Saito, 1983), the major ganglion cell classes are referred to as α, β, δ, and γ; classes I, II and III in rat may be equivalent to α, β and γ in cat. Ganglion cells can be further defined as either 'on' or 'off' cells, according to their centre/surround response properties. Moreover, the various ganglion cell classes have characteristic central projections. Distinct immuno-reactivities are also seen amongst ganglion cells although their relationship to physiologically and/or morphologically defined classes is not yet fully documented.

Evidence suggests that all ganglion cell classes can survive axotomy in non-mammalian vertebrates and also in mammals if interventions allow axonal regeneration.

In frog, both physiological recording from the optic tectum (Keating & Gaze, 1970) and tectal immunoreactive profiles, revealed by both monoclonal and polyclonal antibodies (Humphrey *et al.*, 1995), indicate that some ganglion cells of each class survive optic nerve section. A previous report (Kuljis & Karten, 1985) suggesting that only ganglion cells with substance P-like immunoreactivity survive regeneration did not allow for the protracted survival of the disconnected distal segments of severed optic axons (Humphrey *et al.*, 1992). However, throughout these studies in frog, it has remained unclear whether extents of survival are comparable between ganglion cell classes.

For mammals, evidence of the survival and axonal regeneration of different classes of ganglion cells has come from work in both rats and cats. In rats, exogenous neurotrophic factors injected intravitreally were found to allow the preferential survival of large ganglion cells (Mey & Thanos, 1993). The cat is particularly suited to such studies because their ganglion cell classes are well understood and are morphologically distinct. Fukuda & Watanabe (1996) have shown that the large α cells are disproportionately represented amongst surviving ganglion cells, with the numbers of β and other cell classes unchanged or decreased. As mentioned above (see "Reasons for the death of axotomised ganglion cells"), it has been suggested that the preferential loss of β as compared to α cells may be explained by glutamate toxicity. The β cells are innervated solely by bipolar cells, a cell type which employs glutamate as their transmitter (Silveira *et al.*, 1994). An intriguing but paradoxical finding in cats is that physiological recording indicated an overwhelming preponderance of ganglion cells of the 'on' type. The result was gained despite most cells examined anatomically having retained their dendritic trees in the sub-lamina of the inner plexiform normally occupied by dendrites of 'off' cells.

In rats, as in cats, it is the large type I and type III ganglion cells that preferentially regenerate into a peripheral nerve graft (Mey & Thanos, 1993). Moreover, when rat optic axons regenerate via a peripheral nerve graft, the survival of ganglion cells seems to be target specific. Thus more large type I cells survive if the axons reinnervate the pretectum but smaller type II cells predominate if target is the superior colliculus (Thanos & Mey, 1995).

In both frog and in rat with a peripheral nerve graft, displaced ganglion cells are represented amongst the population of surviving ganglion cells (Dunlop *et al.*, 1992; Thanos & Mey, 1995).

Key points

- **All classes of ganglion cell survive axotomy in amphibians and in mammals with peripheral nerve grafts although large cells are probably disproportionately represented.**

Retinal re-wiring

Ganglion cell death associated with axotomy necessarily changes the wiring diagram of the retina. We assume that in fish, with a near complete survival of ganglion cells, the existing retinal circuitry is maintained. By contrast, the substantial depletion of the ganglion cell population during optic nerve regeneration in frog, lizard and mammals must disturb retinal wiring. In frog and presumably also in reptiles, birds and mammals, there is no compensatory generation of ganglion cells (Beazley *et al.*, 1998). Moreover, in frog, there is no discernible change in the number of cells in the inner nuclear layer of the retina, cells that innervate ganglion cells (Darby *et al.*, 1990). Hypertrophy of the dendritic trees of the ganglion cells remaining after optic nerve regeneration would be necessary to optimise the input to ganglion cells. The possibility is supported by a study reporting dendritic hypertrophy of large ganglion cells after optic nerve regeneration in the South African clawed frog *Xenopus laevis* (Straznicky, 1988). In lizard, it is not yet known whether remodelling takes place or if the inner nuclear layer retains its cell complement after optic nerve lesion.

In contrast to the findings for frog, retrograde degeneration has been reported within the inner nuclear layer of mammals when ganglion cells die after axotomy (Gills & Wadsworth, 1967). However, the extent of transneuronal cell death has yet to be analysed when partial optic nerve regeneration takes place via a peripheral nerve graft. In accord with a loss of cells from the inner nuclear layer, and in contrast to the findings for amphibians, hypertrophy has not been found amongst the dendrites of axotomised mammalian ganglion cells. Thus in cats with peripheral nerve grafts (Fukuda & Watanabe, 1996) single cell injection of tracers showed retention of many regenerating α ganglion cells with appropriate dendritic morphologies and sizes; compatible with these anatomical findings, receptive field properties were normal. However, there were also many unclassifiable cells, which may have represented α cells with atrophied dendritic trees. The β ganglion cells also had dendritic trees of normal size but physiologically their receptive fields were enlarged. The enlargement may reflect plastic changes resulting in an abnormal convergence of bipolar cell input onto surviving ganglion cells; alternatively, some loss of inhibition from surrounding cells may have led to the change.

For rodents, labelling with fluorescent tracers revealed that terminal and preterminal dendrites degenerate after optic nerve lesion (Thanos, 1988). However, the reduction in dendritic dimensions after axotomy was found to be less pronounced if axons regenerate into a peripheral nerve graft (Thanos & Mey, 1995). The study also reported that the pruning of dendritic trees is reduced by retinal application of microglial inhibitory factor; the response may be a direct one or indirect by encouraging cell survival and axonal regrowth.

Changes in retinal wiring may extend not only to dendrites but also to axons, again with similarities across the vertebrate phylum. In goldfish, single cell injection revealed that a minority of ganglion cells develop a second axon from the cell soma after optic nerve section (Becker & Cook, 1990). It is unclear whether the additional axon, which approaches the optic nerve head, survives in the long term and gains access to the visual centres. Ultrastructural evidence indicates that, in the frog *X. laevis*, section of the optic nerve or tract, results in the formation of a permanent swirl of neuritic processes around the optic nerve head (Bohn & Reier, 1982). It is uncertain whether these processes represent aberrant axonal growth; their functional impact is also unknown.

Silver staining and intracellular injection of fluorescent tracers have revealed the formation of ectopic axon-like processes in the rodent retina after ganglion cell axotomy, especially if the retina is treated with neurotrophins (Sawai *et al.*, 1996). Thus, neurotrophic agents may be valuable to prevent ganglion cell death after axotomy; they have the disadvantage, however, that they may encourage intra-retinal axonal growth, a response which is clearly inappropriate for a ganglion cell attempting to regenerate along the optic pathway. As we have described earlier (see "Reasons for the death of axotomised ganglion cells"), there may be a hierarchy in terms of the site of axonal sprouting, with the intra-retinal one being favoured only if the axon no longer retains access to the lesion site.

Key points

- **In frog, but probably not in mammals, cell numbers are maintained in the inner nuclear layer after ganglion cell death and ganglion cells probably undergo dendritic hypertrophy.**
- **Dendritic trees are of normal or reduced dimensions after axotomy in mammals.**
- **Some ganglion cells form secondary axons during optic nerve regeneration in fish; in mammals, intra-retinal axonal sprouting occurs, especially if axons withdraw from the lesion site and if neurotrophins are applied to the eye.**

Continued retinal neurogenesis and axonal regeneration

An hypothesis that has received considerable support from studies of both the central and peripheral nervous systems is that directed axonal regeneration occurs spontaneously only in systems with continued neurogenesis (see Holder & Clark, 1998 and Table 1). The primary visual system has been much cited in support of this possibility. It is well known that both fish (Meyer, 1978) and amphibians (Straznicky & Gaye, 1971) add ganglion cells to their retinae throughout life and can regenerate their optic nerves to appropriate targets; by contrast birds and mammals complete retinal cell generation early in life and axonal regeneration is abortive (Kiernan, 1979).

The hypothesis can be further tested by studies in adult lizard and aged *X. laevis*. In both, optic nerve regeneration is extensive, but labelling of dividing cells indicates that retinal ganglion cell generation is no longer taking place (Taylor *et al.*, 1989; Beazly *et al.*, 1998). As we describe below (see "Retinotopicity of Regenerated Projections"), lizards at early stages of optic nerve regeneration form at least a low-fidelity map although it is subsequently completely degraded. The result argues that a system no longer undergoing neurogenesis nevertheless retains at least some potential for directed axonal growth but may lack the signals to consolidate appropriate connections. We are currently conducting an experiment to determine whether a retinotopic projection is established after optic nerve regeneration in aged *X. laevis*, the study by Taylor and colleagues being concerned with optic axon regeneration *per se* rather than with map restoration. Moreover, in mammals with peripheral nerve grafts, there is as yet no compelling evidence to suggest re-establishment of retinotopically appropriate maps (Thanos *et al.*, 1997). However, tissue culture studies suggest that the mature superior colliculus contains signals that can be distinguished by nasal and temporal optic axons (Wizenmann *et al.*, 1993).

Table 1 Summary of the presence of retinal neurogenesis in adult life and the responses of retinal ganglion cells to axotomy in fish, frog, lizard and rat

	Cell survival	Neuro-genesis	Axonal outgrowth	Specificity for visual centres	Map	Function
Fish	Yes (90%)	Yes	Yes	Yes	Yes	Yes
Frog	Yes (20–70%)	Yes	Yes	Yes	Yes	Partial
Lizard	Yes (65%)	No	Yes	Partial	No	No
Mammals						
no intervention	No (0%)	No	No	?	No	No
neurotrophins	Yes (up to 17%)	No	Partial	?	No	Partial
peripheral nerve graft	Yes (up to 2–3%)	No	Partial	?	No	Partial

Key points

- The continued generation of retinal ganglion cells is not a pre-requisite for optic axon regeneration; however, it may correlate with the restoration/consolidation of retinotopic maps.

The responses of retinal ganglion cell axons and of non neuronal cells at the site of injury

In the previous sections, we reviewed the responses of the retinal ganglion cell body, its dendrites and the intra-retinal portion of the optic axon to lesioning the optic nerve. Here, we consider the reactions of the retinal ganglion cell axons within the nerve as well as the behaviour of non-neuronal cells which are either resident within the visual pathway or which migrate into it after injury (Figure 1). The term proximal is used to denote that portion of the optic nerve between the eye and the lesion site whereas the term distal refers to the nerve between the lesion site and the brain.

Axonal sprouting

As first described by Ramón y Cajal (1928), in silver impregnated sections of the lesioned rabbit optic nerve, severed axons proximal to the lesion form growth cones within a few days. In addition, the severed axons produce a number of branches or sprouts which themselves are tipped by growth cones. Such observations have since been confirmed by many workers. For example, transecting the intraorbital portion of the optic nerve in adult rat results in the formation of large numbers of putative axonal sprouts proximal to the lesion (Zeng *et al.*, 1994). The retinal origin of these axonal sprouts was confirmed by injection of horseradish peroxidase (HRP) into the vitreous humour and ultrastructural examination. Labelling small numbers of regenerating axons with the carbocyanine dye DiI in the living animal and examination in longitudinal sections revealed that the sprouts have a complex appearance with many side-branches, varicosities and loops.

Ultrastructurally, the sprouts are small unmyelinated axonal profiles which are surrounded and in contact with a variety of cellular types. These include swollen, degenerating axons, myelin debris, oligodendrocytes, myelinated axons with a normal appearance which are disconnected but have yet to show signs of degeneration as well as astrocytes and macrophages. Although no quantitative data are available, sprouts were formed within 1 day, appeared to be maximal between 5 and 7 days, had declined to low levels by 2 weeks and were rare at 8 weeks.

A companion paper (Zeng *et al.*, 1995) examined the response of optic axons to intracranial axotomy. In contrast to intraorbital surgery (Zeng *et al.*, 1994), a large central degenerative core formed which extended as a finger back towards the eye. The degenerative core probably resulted from ischaemia after damage to the branches of the circle of Willis that supply the intracranial portion of the optic nerve; intraorbital lesion spared the ophthalmic supply with little evidence of ischaemia. Bordering the degenerative core was an intermediate zone with fewer necrotic profiles while the periphery of the nerve appeared essentially normal. Nevertheless, regenerative axonal sprouts were seen in all three regions and, as after intraorbital lesion (Zeng *et al.*, 1994), were found to be in close association with a wide variety of cellular profiles including astrocytes, intact and yet to degenerate myelinated axons as well as myelin debris.

The time course of axonal sprouting was more protracted than that after intraorbital lesion with high numbers of sprouts being seen at 2 and 4 weeks; the number declined only gradually by between 6 and 8 weeks post-operatively. The protracted axonal sprouting may relate to the extensive degenerative core that forms after intracranial lesion. Infiltration of macrophages and fibroblasts has been implicated in the success of optic nerve regeneration in goldfish (Battisti *et al.*, 1995; Hirsch *et al.*, 1995) and has been used to promote a regenerative response in rat (Lazarov-Spiegler *et al.*, 1996; see "Macrophages" below). The presence of such cells within the degenerative core after intracranial lesion may produce factors conducive to prolonging regenerative sprouting.

Prolonged axonal sprouting after intracranial lesion appears to relate to the survival of retinal ganglion cells, as described above (see "Reasons for the death of axotomised ganglion cells"). Thus at 6 months post-lesion, more ganglion cells (30%) survive intracranial than intraorbital (5–10%) lesion (Villegas-Perez *et al.*, 1993). However, axon regeneration into peripheral nerve grafts is more successful after intraorbital (Vidal-Sanz *et al.*, 1987) that after intracranial (Richardson *et al.*, 1982) anastomosis. Possibly, the success of axon growth into a peripheral nerve graft is related less to cell survival *per se* than to the sprouting response. Thus, the rapid onset of sprouting after intraorbital lesion may allow a better penetration of the graft than the delayed sprouting that occurs after intracranial lesion.

There is also evidence for axonal sprouting after optic nerve lesion in cold-blooded vertebrates. However, the extent of sprouting does not appear to be related to whether successful optic nerve regeneration can be achieved. Thus, rather than occurring at early stages after lesion as in rat, sprouting appears to be a later event that occurs after axons have successfully crossed the lesion site and reached their target tissue in the brain. Electron microscopic counts of transverse sections of the distal portion of the optic nerve have shown that the number of axonal profiles is 4-fold above normal values at 6–12 weeks in goldfish (Murray, 1982). A similar elevation is seen in the distal nerve at 6 weeks in the frog *R. pipiens* and in addition, axon numbers were twice normal values in the proximal optic nerve (Stelzner & Strauss, 1986). Recent data from our laboratory, however, have shown that axonal sprouting is, at the most, minimal in the frog *L. moorei* (Dunlop *et al.*, unpublished). Anterograde labelling of individual axons did not reveal evidence for extensive axonal branching or sprouting; rather axons approached the crush site directly, took highly aberrant trajectories within the lesion site but resumed direct pathways in the distal nerve. In addition, electron microscopic counts in the distal optic nerve revealed axon numbers that, throughout regeneration, were consistently between 20 and 40% lower than normal values. We interpret the data in *L. moorei* to indicate that axonal sprouting is minimal and that ganglion cell death is associated with a failure of axons to cross the lesion site. A possible explanation for axonal sprouting in goldfish may relate to the presence of Schwann cells during optic nerve regeneration (Nona *et al.*, 1992). Such cells have been widely recognised to induce regenerative sprouting in peripheral nerves and may have a similar effect in goldfish. A comparative immunohistochemical study in *R. pipiens* and *L. moorei* would reveal whether axonal sprouting is related to the presence of Schwann cells.

Key points

- Axonal sprouting is an early response in mammals and may contribute to the successful penetration of peripheral nerve grafts.
- By contrast, sprouting in goldfish and the frog *R. pipiens* appears to occur after axons have reached the brain
- Axonal sprouting is not prominent in the frog, *L. moorei.*

Changes in non-neuronal cells: blood-brain barrier

A correlation has been found between successful axonal regeneration and a breakdown of the blood-brain barrier. Thus, for example, after optic nerve lesion, using rhodamine B-isocyanate labelled bovine serum albumin a breakdown of the blood-brain barrier can be demonstrated in goldfish but not in rat (Kiernan & Contestabile, 1980; Kiernan, 1985). We have recently exploited these observations to test in rat whether an experimentally induced breakdown of the blood-optic nerve barrier would induce optic axon regeneration (Beazley *et al.*, 1996). After intraorbital optic nerve crush, the blood-optic nerve barrier at the lesion site was breached by chronic infusion of 16% mannitol using a mini-osmotic pump; the axons appeared ultrastructurally normal indicating that the mannitol did not cause cellular damage. The breaching of the barrier was demonstrated by intracardial perfusion with rhodamine B-isocyanate conjugated to bovine serum albumin. Axonal regeneration was assessed by anterograde transport after intraocular injections of HRP. Although axons failed to penetrate the lesion site, regenerating axons tipped by growth cones in the proximal nerve were seen to grow consistently along arrays of blood vessels that were oriented parallel to the long axis of the nerve. The preference for the vasculature as a substrate, compared to other cellular elements, was observed also by Zeng *et al.*, (1994) after intraorbital crush in rat (see "Axonal sprouting" above); the findings suggest an association between blood vessels and the presence of physical and/or molecular factors that promoted axon growth.

In the frog *L. moorei*, optic nerve regeneration is closely associated with a breakdown of the blood-brain barrier (Tennant & Beazley, 1992). Thus, as the front of regenerating axons progresses along the visual pathway, there is a coincident regional breakdown of the barrier. The axons were labelled with HRP and the integrity of the blood-brain barrier assessed using rhodamine B-isocyanate conjugated to bovine serum albumin. Furthermore, in *L. moorei*, prevention of axonal regeneration by optic nerve ligation resulted in the blood-brain barrier remaining intact, suggesting that regenerating growth cones themselves act as a trigger to breach the blood-brain barrier. The breach presumably allows macromolecules and cells (see "Macrophages" below) to enter the optic nerve and induce changes within the cellular environment which are conducive to axonal regeneration.

Key points

- **A breakdown of the blood-brain barrier is associated with optic nerve regeneration in goldfish and frog but does not occur in rat.**
- **Attempts to stimulate optic nerve regeneration in rat by osmotically breaking down the blood-brain barrier stimulated axons to regenerate along blood vessels; however, axons did not penetrate the crush site.**

Macrophages

A breached blood-brain barrier will result in neural tissue being exposed to recruited inflammatory cells, in particular macrophages, the consequences of which may be both beneficial and detrimental. A classic study suggested that the success or otherwise of nerve regeneration in mammals may be linked to the invasion of macrophages (Perry *et al.*, 1987). One role of such cells was thought to be the removal of distal axon segments disconnected by axotomy, thus permitting the advance of regenerating axons. In mouse sciatic nerves, which are able to regenerate, cells identified as macrophages using the antibody F4/80$^+$ invaded the distal portion. Confirmation of the phagocytic role of these cells came from partial crushes in which macrophages were confined to regions undergoing degeneration but were absent from zones containing intact myelinated axons.

By contrast, after optic nerve lesion, macrophages were fewer and restricted to the lesion site and, by implication, could therefore not be involved in removing axonal debris (Perry *et al.*, 1987). Using the ED-1 antibody, large numbers of macrophages have also been shown to be restricted to the lesion site in the rat optic nerve at 2 days after injury (Blaugrund *et al.*, 1992). By 1 week, ED-1 labelled cells had increased further at the lesion site but only small numbers were seen outside this region. The limited recruitment of macrophages throughout the extent of the damaged mammalian optic nerve has recently been linked to a resident microglial inhibitory factor (Hirschberg *et al.*, 1995). Using *in vitro* techniques, a soluble factor was shown to be present in optic, but not sciatic, nerves. This factor not only inhibited the migration of macrophages but also prevented their differentiation into process-bearing phagocytic cells; conversely, macrophages exposed to sciatic nerves were able to migrate as well as differentiate.

In addition to their phagocytic role, macrophages have been shown to have other supportive effects for nerve regeneration. For example, in the sciatic nerve, peritoneal macrophages have been implicated in the upregulation of mRNA for nerve growth factor (NGF) and its receptor (Heumann *et al.*, 1987). Furthermore, macrophages appear to be involved in modulating the inhibitory properties of oligodendrocytes as well as astrocytes (Eitan *et al.*, 1992; David *et al.*, 1990: see "Astrocytes" below). The positive effects of macrophages have recently been exploited by Lazarov-Spiegler *et al.*, (1996) to promote regeneration of optic axons in rat. Peritoneal blood monocytes were isolated from adult rat and cultured in the presence of sciatic or optic nerve segments. Activated monocytes were then transplanted to transected optic nerves and the success of axonal regeneration assessed at 8 weeks by retrograde labelling with Di-IO-Asp applied 2 mm distal to the original transection site, that is 7-8mm from the eye. Although the effect was modest in absolute terms, treatment with sciatic nerve-conditioned macrophages resulted in the regeneration of fifteen times as many retinal ganglion cell axons (mean = 0.15% of normal values) compared to non-conditioned medium treated controls (0.01%); treatment with optic nerve-conditioned macrophages resulted in the regeneration of 5 times as many axons as in controls.

In contrast to the above positive effects, macrophages are also implicated in the acceleration of secondary degeneration by, for example, the production of superoxide anions (Colton & Gilbert, 1987). Cells were assayed for superoxide using a microcytochrome C reductase assay. Furthermore, macrophages are also involved in the secondary effects of ischaemic damage (Guilan & Robertson, 1990). Thus, transient ischaemia in the rabbit spinal cord results in impairment of neurological function which worsens 1–2 days after injury. The ischaemia is associated, not necessarily causally, with the appearance of macrophages at the site of primary damage. Administration within 6 hours of injury of chlorquinine and colchicine, which inhibit phagocytic and secretory functions in macrophages, decreased the number of macrophages and resulted in improved bladder and motor function. The therapeutic use of macrophages and their secretory products must therefore harness, directly or indirectly, their growth promoting properties while tempering their involvement in triggering an inflammatory response.

Compared to mammals, in goldfish the macrophage response encompasses the length of the optic nerve and its timing coincides with the removal of axonal debris and successful axon regeneration (Battisti *et al.*, 1995). The more widespread response in goldfish compared to mammals appears to reflect the existence of a substantial population of resident cells of monocytic origin along the length of the nerve, identified using OX-42. A crush lesion results in a dramatic increase in these cells along the length of the nerve; the response is rapid, occurring within 1 hour, peaks at 4 weeks after lesion and is still above normal values at 14 weeks.

In addition to the macrophage response, the success of axonal regeneration in goldfish may also relate to the pronounced fascicular organisation of the optic nerve, which is broadly similar to that of the mammalian PNS. In normal goldfish, all mature optic axons are myelinated and grouped into a number of large fascicles, termed the intrafascicular region, which also contains astrocytes, oligodendrocytes and OX-42 positive microglia (Battisti *et al.*, 1995). The surrounding interfascicular region is composed of fibroblasts and abundant collagen (Hirsch *et al.*, 1995) as well as OX-42 positive granular macrophages (Battisti *et al.*, 1995). A crush lesion in goldfish appears to break only the optic axons and maintains the integrity of the fibrous interfascicular portion thus providing channels for axon regeneration. Interestingly, the interfascicular structure prevents granular macrophages from entering the axon fascicles and thus being involved in the removal of axonal debris (Battisti *et al.*, 1995). Electron microscopy confirmed that only the intrafascular microglia were involved in the phagocytic response and the removal of the distal portions of axons disconnected by axotomy.

A similar widespread increase in the number of macrophages has been observed in the optic nerves of *X. laevis* tadpoles, peaking just before the onset of axonal regeneration (Wilson *et al.*, 1992). In the amphibian optic nerve, the fascicular structure is less pronounced than in goldfish, axons being grouped into fascicles solely by slender astrocytic processes; a collagenous interfascicular portion is absent. As a consequence, optic nerve lesion in amphibia probably results in more extensive damage than in goldfish with a breaching of the glia limitans by invading macrophages (Wilson *et al.*, 1992). In normal animals, staining with the 5F4 antibody revealed a small number of resident microglia with long slender processes typical of a resting state; these cells were present along the length of the optic nerve. After lesion, the number of 5F4 positive cells increases but many now had an amoeboid morphology and were considered to be macrophages. The increased number of cells probably resulted from both invasion by blood-borne cells and activation of resident microglia. Between 1 and 6 hours of lesion, there was a rapid accumulation of 5F4 amoeboid macrophages at the lesion site; by 12-24 hours these cells were also found in the distal nerve and the response peaked by 5 days. The timing of the macrophage response in *X. laevis* tadpoles suggests that these cells are involved in the rapid removal of axonal debris in the distal nerve which then allows regenerating axons to grow towards the brain.

In adult frogs, recent light and electron microscopic data from our laboratory (Dunlop *et al.*, unpublished) has revealed a somewhat different role for macrophages in optic axon regeneration compared to *X. laevis* tadpoles and adult goldfish. The adult optic nerve in the frog *L. moorei* has a similar delicate fascicular structure to that of *X. laevis* tadpoles. A crush lesion in *L. moorei* results in a gap, lacking cellular elements, of approximately 1mm between the proximal and distal portions of the optic nerve; nevertheless, the meningeal sheath remains intact as a peripheral conduit. Optic axons in *L. moorei* thus do not have the benefit of spared interfascicular channels; such channels are prominent in normal goldfish and may provided structural continuity across the lesion site. Our results in *L. moorei* show that the inner surface of the meningeal sheath is rapidly coated with a layer of macrophage-like cells, between 1 and several cells thick. We suggest that these cells enter the optic nerve via the breached blood-optic nerve barrier (Tennant & Beazley, 1992). The first regenerating axons appear to cross the gap between 0.5 and 1 week by growing along the collar of macrophages and by two weeks the gap is filled by regenerating axons. Interestingly, up to two weeks, there appears to be a virtual absence of glial cells (see "Astrocytes" below). In goldfish, it has also recently been shown that glial cells do not act as the primary substrate for regenerating axons within the lesion site (Hirsh *et al.*, 1995). Rather, fibroblasts, identified by staining with fibronectin (FN), are prominent within the lesion site at early stages and newly regenerating axons grow in close association with this connective tissue.

The macrophages in *L. moorei* thus appear to act as a physical, and possibly growth-promoting, substrate to assist axons in crossing the gap created by the lesion. Light and electron microscopic evidence also suggests a minimal occurrence of macrophages in regions of the optic nerve outside the lesion site (Dunlop *et al.*, unpublished). Indeed, in contrast to *X. laevis* tadpoles (Wilson *et al.*, 1992), the removal of the disconnected axons in adult *L. moorei* is slow with substantial numbers persisting in the optic nerve at 3 weeks post-lesion, that is after many of regenerating axons have reached the optic tectum. Indeed, we have also shown previously that the terminals of disconnected axons persist in the optic tectum for up to 12 weeks after nerve crush and that they co-exist with incoming regenerating arbors (Humphrey *et al.*, 1992). A pronounced macrophage response therefore seems not to be involved in the removal of axonal debris in *L. moorei*; nevertheless, the persistence of debris does not appear to be an impediment to successful regeneration.

Key points

- In goldfish and frog, there is a marked increase in the macrophage population at the lesion site and along the nerve.
- Macrophages and fibroblasts act as a physical and possibly growth-promoting substrate for axon growth across the lesion site.
- By contrast, in rat, the macrophage response is minimal and limited to the lesion site response.
- Macrophages are able to enhance axon outgrowth by secreting factors that modify surrounding glial cells but also have detrimental effects via their involvement in the inflammatory response.

Astrocytes

Traditionally, astrocytes have been regarded, at least in part, as responsible for the failure of CNS regeneration in warm-blooded vertebrates. In response to cortical injury, astrocytes within the lesion site show a rapid decline, possibly as a result of perivascular degeneration. By contrast, around the wound, mitotic astrocytes are seen and they form thin sheet-like processes connected by gap junctions. This reaction culminates in the long term formation of scar tissue that encapsulates the injury site and thus presents a physically impenetrable barrier to regenerating axons (Maxwell *et al.*, 1990).

The short term response of astrocytes within the lesion site itself, however, is rather different and is characterised by rapid and substantial apoptotic cell death (Maxwell *et al.*, 1990). In the rat optic nerve, glial fibrillary acidic protein (GFAP) staining has revealed that astrocytic removal is also swift with a disappearance of such cells from the lesion site being reported by 2 days after lesion; their death appears to result from the phagocytic activity of invading macrophages (Blaugrund *et al.*, 1992). At longer intervals, at least until 3 weeks, the lesion site remains conspicuously devoid of astrocytes whilst, on either side, GFAP staining is prominent (Blaugrund *et al.*, 1993). Staining with GAP-43 revealed that, as expected, axons reached only the proximal tip of the severed nerve.

By contrast, in goldfish optic nerve, the loss of astrocytes is transient (Levine, 1991; Stafford *et al.*, 1990) and vimentin staining has shown that astrocytes reappear within the crush site by two weeks (Blaugrund *et al.*, 1993). Interestingly, labelling with GAP-43 showed that regenerating axons penetrate the crush site slightly in advance of the vimentin positive cells. These findings are in agreement with Hirsch *et al.*, (1995; see "Macrophages" above) who showed that, in the absence of astrocytes, fibroblasts act as a substrate for regenerating axons. Thus, in goldfish, successful axonal regeneration through the lesion site appears to occur in the absence of astrocytes but nevertheless the lesion site is invaded by these cells shortly thereafter.

There is considerable evidence that, under certain conditions, astrocytes can act as a favourable substrate for axon outgrowth. For example, the regenerative axonal sprouts that form in the proximal part of the rat optic nerve and up to the lesion site (Zeng *et al.*, 1995) in response to either intraorbital or intracranial optic nerve lesion grow in contact with surrounding astrocytes, as well as with myelin and oligodendrocytes (see "Myelin and oligodendrocytes" below). In addition, within the distal optic nerve of both goldfish and amphibians, there is ample ultrastructural evidence for a close association between regenerating axons and astrocytes (Murray, 1982; Strobel & Steurmer, 1994). Indeed, in *X. laevis*, grafted astrocytic scars created by enucleation do not prevent axonal regeneration (Reier, 1979).

An *in vitro* study has also highlighted the permissive rather than obstructive nature of the mammalian astrocytic "scar" (David *et al.*, 1990). Cryostat sections of adult rat optic nerve lesioned 5 days previously were able to support cell attachment and neurite outgrowth only at sites close to the lesion; distal segments of the nerve were non-permissive. The permissive nature of the lesion site was found to be associated with the infiltration of macrophages. Thus, treatment of normal optic nerve with macrophages or macrophage-conditioned medium resulted in excellent neurite outgrowth. In other words, the non-permissive optic nerve had been transformed to a permissive state. A final test was to remove optic nerves 4–6 months post-lesion, a time at which myelin debris is minimal and the tissue is composed almost entirely of astrocytes. Neurite outgrowth occurred equally at the lesion site and distal to it. Thus the localised change in permissiveness at the lesion site by 5 days had become more widespread so as to include the length of the optic nerve.

Recent evidence in mammals provides evidence that damaged rather than intact astrocytes are associated with regenerative failure (Davies *et al.*, 1997; chapter 4 by Fitch and Silver, this volume). In this study, microtransplantation was used to introduce dorsal root ganglion cells into the corpus callosum of adult rats, a technique that in most cases caused minimal damage to surrounding host tissue. An absence of axonal regeneration was correlated with damaged astrocytes and an upregulation of chondroitin sulphate proteoglycan. Conversely, extensive axonal outgrowth occurred when damage was minimal and only low levels of this proteoglycan were seen. In goldfish optic nerve, however, chondroitin sulphate proteoglycan is normally present, is upregulated during optic nerve regeneration and is thought to be expressed by both interfascicular cells and axons (Battisti *et al.*, 1992). Although an absence of chondroitin sulphate proteoglycan appears obligatory for axonal regeneration in mammals, its presence in goldfish clearly does not prevent regeneration.

The other side of the coin is the ability of axons to recognise astrocytes. *In vitro* studies have shown that neurons which express Thy-1 are unable to extend neurites on astrocytes (Tiveron *et al.*, 1992). Thy-1 is widely expressed in the nervous system and characterises mature neurons after their axons have reached target tissue; by contrast, cells which continue to grow throughout life, for example, olfactory neurons, do not express Thy-1. Since retinal ganglion cells of cold-blooded vertebrates continue to grow into adulthood, we predict that these neurons would lack Thy-1 thus allowing them to grow in contact with astrocytes.

Key points

- **Astrocytes are removed from the lesion site in goldfish, frog and rat; the absence is transient in goldfish and frog but permanent in rat.**
- **In goldfish, axons penetrate the lesion site just ahead of astrocytes suggesting that astrocytes do not guide regenerating axons. In mammals, astrocytes are able to support extensive axon outgrowth but, under certain conditions, react to trauma by becoming non-permissive.**

Myelin and Oligodendrocytes

There has been considerable recent interest in the possibility that mammalian oligodendrocytes and the central myelin that they produce are inhibitory to regenerating axons whereas Schwann cells and their peripheral myelin are not. Schwab & Thoenen (1985) used an *in vitro* assay to provide dissociated superior cervical or dorsal root ganglion cells isolated from rat with the choice of growing into optic or sciatic nerve grafts. Light and electron microscopy showed that axons grew prolifically within the sciatic, but not the optic, nerves. Subsequently, two protein fractions were isolated from rat central myelin which *in vitro* were nonpermissive for cell attachment and neurite outgrowth (Caroni & Schwab, 1988a). As predicted, material derived from rat sciatic nerves was permissive. Interestingly, myelin fractions isolated from trout and frog spinal cord were also permissive. This is a surprising finding since spinal cord regeneration is only partially successful in the goldfish with 11 out of 17 brain nuclei being shown to participate in regeneration (Sharma et al., 1993). Furthermore, in frog, spinal cord regeneration is successful up to metamorphosis but is abortive from 1–2 days of tail resorption (Beattie *et al.*, 1990). Antibodies (IN-1 and IN-2) to the protein fractions were then shown to neutralize *in vitro* the nonpermissive properties of CNS white matter (Caroni & Schwab, 1988b).

The hypothesis that mammalian central myelin is inhibitory predicts that central myelin derived from regenerating systems, for example the goldfish optic nerve, would lack inhibitory markers and would thus be permissive for axon growth. Indeed, *in vitro* assays showed that growth of explanted fish retinal axons on fish myelin is substantial and is not improved by treatment with IN-1 (Bastmeyer *et al.*, 1992; 1993). Furthermore, seeding the retinal explant cultures with liposomes containing fish myelin caused only minimal growth cone collapse. The conclusion was that fish retinal explants lack mammalian-like myelin associated growth inhibitors (Wanner *et al.*, 1995). However, liposomes containing bovine CNS myelin caused fish growth cones to collapse, an effect that was avoided by prior addition of IN-1 antibodies. The result suggests that fish growth cones are able at least to recognise mammalian myelin-associated growth inhibitors. Similar *in vitro* techniques showed that, as predicted, myelin and oligodendrocytes from *X. laevis* optic nerve were also permissive for neurite growth whereas spinal cord preparations were not (Lang *et al.*, 1995).

In vitro studies by other workers, however, showed that goldfish optic nerve does contain myelin associated growth inhibitory molecules but that the inhibitory effects are expressed only by non-injured optic nerves (Sivron *et al.*, 1994); optic nerves that had been injured 1–2 weeks prior to explantation lacked inhibitory properties. Indeed, the studies which revealed an absence of inhibitory molecules consistently used goldfish optic nerves that had been lesioned 2–3 weeks prior to explantation (Wanner *et al.*, 1995; Lang *et al.*, 1995).

The down-regulation of myelin-associated inhibitory properties after injury of the goldfish optic nerve has been shown to result from cytotoxic factors secreted by macrophages (Cohen *et al.*, 1990), cells which increase in number during axonal regeneration in this species (see "Macrophages" above). The cytotoxic factor has been identified as an interleukin 2-like substance that reduces the number of process-bearing mature oligodendrocytes in culture (Cohen *et al.*, 1990). Wholesale removal of oligodendrocytes during optic nerve regeneration may therefore be one factor associated with a successful outcome in fish. *In vivo* studies in *X. laevis* have shown a similar phenomenon. Thus, oligodendrocytes with advanced myelin markers, such as myelin basic protein, proteolipid protein and anti-galactocerebroside, are absent from the nerve at 4 weeks after lesion and are replaced by cells that are thought to represent immature oligodendrocytes (Lang & Stuermer, 1996). The re-appearance of mature oligodendrocytes by 8 weeks implies an adaptive plasticity in these cells that may be related to successful axonal regeneration; indeed similar changes were not seen within oligodendrocytes of postmetamorphic *X. laevis* spinal cord.

In rat, it has recently been shown that platelet derived growth factor (PDGF), produced by astrocytes, modulates the inhibitory properties of oligodendrocytes *in vitro* (Lang *et al.*, 1996). Thus the growth cone collapsing properties of oligodendrocytes can be abolished by exposure to PDGF or to medium conditioned by astrocytes. Furthermore, PDGF was shown to reduce IN-1 immunoreactivity. Thus, although axonal regeneration does not occur in rat CNS, an interesting association appears to exist between the two classes of macroglia whereby astrocytes may have the potential *in vivo* to alter the expression of inhibitory molecules associated with oligodendrocytes.

Although there is ample *in vitro* evidence for an inhibitory role of oligodendrocytes, *in vivo* there is also substantial evidence that argues against these cells contributing substantially towards abortive regeneration within the mammalian CNS. Thus, for example, the administration of IN-1 antibodies using hybridoma cells implanted adjacent to the rat optic nerve lesioned at 16–18 days postnatal resulted only in a modest improvement in axonal regeneration at 3–6 weeks (Weibel *et al.*, 1994). The mean distance of regeneration beyond the lesion site was 1.5mm compared to 0.5mm for control antibody secreting hybridomas. More importantly, however, the number of regenerating axons was extremely small.

There is also a large body of data in which axons have been shown to regenerate for considerable distances in adult white matter tracts. For example, embryonic mouse hippocampal, neocortical and superior colliculus neurons are able to grow for at least 10mm when transplanted to the corpus callosum or cingulum of adult rats; these cells also extend axons in close association with host tract astrocytes (Davies *et al.*, 1994). Microtransplantation of postnatal day 8 dorsal root ganglion cells to the corpus callosum or fimbria also resulted in axon growth of 6–7mm by 6 days (Davies *et al.*, 1997; see also chapter 4 by Fitch & Silver, this volume). Further evidence for a minimal involvement of myelin-associated growth inhibitors in optic nerve regeneration comes from the Browman-Wyse rat model (Berry *et al.*, 1992). One variant of this mutant lacks oligodendrocytes and CNS myelin in the proximal optic nerve. Axons consistently failed to regenerate in the myelin-free and astrocyte dominant segment. When axonal regeneration was observed, it was always associated with the presence of Schwann cells and in these instances axons crossed the lesion site and penetrated the peripheral myelin debris of the distal segment.

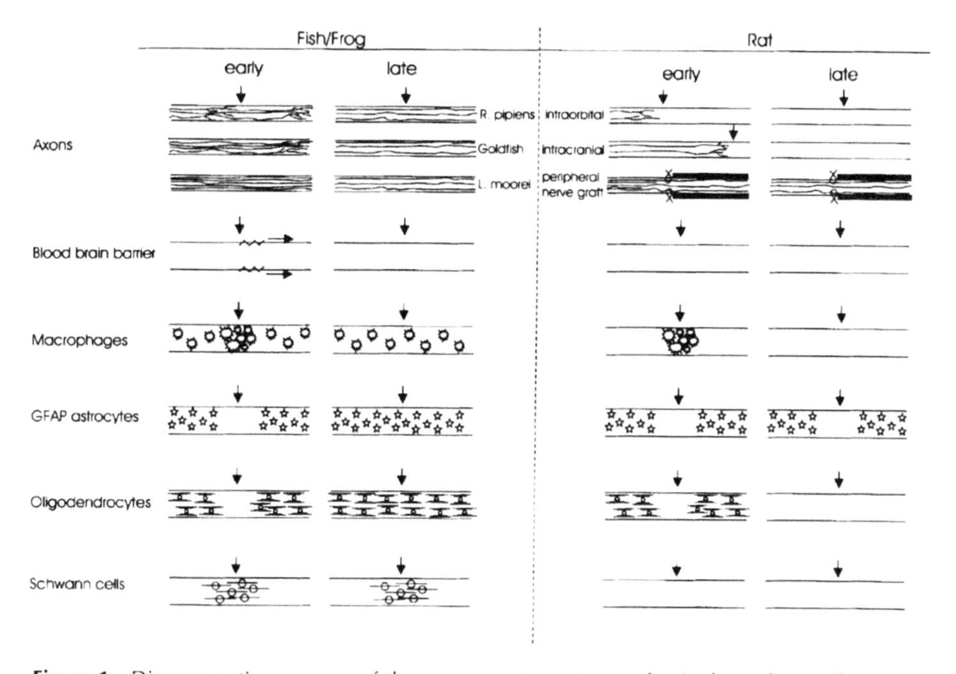

Figure 1 Diagrammatic summary of the responses to axotomy of retinal ganglion cell axons and of resident and recruited non-neuronal cells at early and late stages after optic nerve regeneration in fish/frog and rat. The optic nerve is depicted with the eye to the left and the brain to the right. The lesion site is indicated by a vertical arrow. Axonal sprouting is shown by branched structures within the nerve. A peripheral nerve graft is indicated by a thick sheath and suture threads. The breakdown in the blood-brain barrier is indicated by zig-zags and its progression towards the brain by horizontal arrows. Macrophages, astrocytes, oligodendrocytes and Schwann cells are indicated by symbols.

Key points

- Myelin-associated growth inhibitory molecules are present in non-injured optic nerves of goldfish and rat, but in goldfish become down regulated thus providing a permissive environment for axon growth after injury.

- Substantial *in vivo* evidence suggests that white matter tracts can be permissive of axon growth.

Projections to visual centres

Specificity of regenerated projections

Successful optic nerve regeneration requires axons to seek out primary visual centres. This task seems to be carried out most efficiently by fish and amphibians (Beazley, 1984). Presumably, barrier molecules set up during development remain in place throughout life and are recognised by regenerating optic axons. Thus axons remain confined to the visual pathway, even though their side-specificity is somewhat relaxed at the chiasm; in frog the small percentage of optic axons that invade the opposite optic nerve are later withdrawn (Bohn & Stelzner, 1981; Tennant *et al.*, 1993a). Even if the optic nerve is redirected to other brain regions such as the forebrain, many optic axons appear to seek out the optic tectum, usually via the optic tract (Scalia, 1987). Moreover, in frog, different ganglion cell classes normally terminate in separate layers of the optic tectum, an organisation demonstrated by electrophysiological recording to be re-established after optic nerve regeneration (Keating & Gaze, 1970).

In lizard, a recent study tracing axons labelled with carbocyanine dyes revealed that regenerating optic axons largely follow visual pathways, with their terminations favouring primary visual centres. However, other regions are innervated, especially secondary visual centres, suggesting a less stringent specificity than that in displayed by either fish or amphibians (Dunlop *et al.*, in press).

The specificity of regenerated mammalian optic axons is hard to access due to the absence of spontaneous axonal regeneration within the brain and thus the requirement for a peripheral nerve graft to be inserted directly into a visual brain centre, usually the superior colliculus or pretectum. However, both morphological and electrophysiological evidence indicates that axons regenerating into the superior colliculus via a peripheral neve graft are at all times confined to the superficial retinorecipient layers (Keirstead *et al.*, 1989). A degree of specificity is also suggested by a study in which optic axon terminals were labelled with HRP and examined ultrastructurally after the optic nerve regenerated via a peripheral nerve graft implanted, not into a visual centre, but into the cerebellum. Optic axons assumed a differential distribution with more terminals in the granule cell layer than in the molecular layer (Zwimpfer *et al.*, 1992).

Key points

- **Optic axons regenerate exclusively to primary visual centres in fish and amphibians and in lizard favour these centres but also invade secondary visual ones.**
- **It is unclear whether mammalian optic axons that regenerate via a peripheral nerve graft display a preference for visual over other brain centres.**

Retinotopicity of regenerated projections

Key features for the restoration of vision are that the regenerated projections form a retinotopic array within visual centres and that they are appropriately connected to post-synaptic cells; in this way, a spatially appropriate representation of visual space is relayed to the brain (Figure 2). To confirm these events, a combination of electrophysiological, anatomical and behavioural techniques is required.

Until recently it has been largely assumed that if the optic nerve can be triggered to regenerate, then a retinotopic projection will be restored. This is indeed the case in fish and amphibians. Electrophysiological evidence suggests that the initial projections to the optic tectum are diffuse but assume an approximate topographical alignment before undergoing refinement (Humphrey & Beazley, 1982, 1983, 1985; Schmidt, 1993). Correlated with this sequence, axons appear to enter the tectum in an essentially random array, search widely and over weeks become localised at the retinotopically appropriate site (Stuermer, 1988a,b). However, in lizard, the outcome of optic nerve regeneration differs markedly from that of fish or frog. Labelling with carbocyanine dyes has shown that extensive optic nerve regeneration takes place with axons regaining the optic tectum within one to two months (Dunlop *et al.*, unpublished). Moreover, preliminary evidence, both physiological and anatomical, suggests that connections form to post-synaptic cells. However, one essential feature is missing in that retinotopy is not restored.

In 1997, we reported anatomical evidence that, one year after optic nerve crush, the regenerated projection lacks retinotopic order (Beazley *et al.*, 1997). Moreover, we now have gained anatomical evidence for a lack of retinotopicity throughout optic nerve regeneration. We have also used electrophysiological techniques to confirm the absence of retinotopy from the long term regenerated projection (Stirling *et al.*, 1999). However, seemingly in contradiction to our anatomical studies, our electrophysiological experiments indicate a low-fidelity map is transiently present at approximately 5–7 months, the stage when projections are most pronounced anatomically.

The order is imprecise and most apparent along the temporo-nasal axis of the retina, projecting along the rostro-caudal tectal axis. We do not yet know whether all animals exhibit this transient weak organisation, but it is clear that by 7 months and beyond any semblance of order has been lost. The electrophysiological and anatomical results can probably best be reconciled by the explanation that some terminal regions are electrically active whilst others are silent. However, unlike fish or frog, the consolidation process does not take place in lizard and even the low-fidelity map is lost. The topographically random nature of the long-term connections to post-synaptic cells explains the permanent loss of visually mediated behaviour, including a failure to detect and capture prey.

Data for mammals are very limited in terms of the ability of their regenerated optic axons to form topographically ordered arrays of terminal arbors. Although increasing levels of pan-retinal ganglion cell survival can now be achieved and axonal regrowth induced into a peripheral nerve graft, regenerated projections extend at most over only one third of the superior colliculus (Keirstead *et al.*, 1989; Thanos *et al.*, 1997). The restriction is found despite ultrastructural evidence that regenerated axonal arbors continue to grow and form new terminals for extended periods (Carter *et al.*, 1994).

A recent report describes an ordering of axons within a peripheral nerve graft with axons terminating within a restricted region of rostro-lateral tectum in a pattern matching that present within the graft (Thanos *et al.*, 1997). The retinotopy in the zone of innervation, however, was not related to the normal retinotectal topography; indeed, it was completed reversed. Temporal retina projected more caudally than nasal retina, rather than more rostrally as in the normal projection. Similarly, dorsal retina projected more medially than ventral retina, rather than normal lateral location. The parallel course of regenerated axons within the graft suggests an absence of cues which might have reordered the axons, as occurs along the normal visual pathway during development (Chelvanayagam *et al.*, 1998). As a result, axons exit the graft in an array that reflects their retinotopically incorrect position within it; furthermore, being unable to search within the superior colliculus for appropriate target tissue, the axons seem able to form only a terminal array matching their order in the graft.

Key points

- **In fish and amphibians, retinotopic maps are restored after optic nerve regeneration. In lizard, a low fidelity map is formed transiently but then is lost. To date, there is no evidence that normal retinotopy is restored in mammals with a peripheral nerve graft; one problem is that only a restricted region of the superior colliculus becomes reinnervated.**

Mechanisms for map restoration and consolidation

Mechanisms for map restoration are best understood for fish and amphibians. Evidence points to a two-stage process of map restoration during optic nerve regeneration in fish and frog. Although optic axons are substantially mis-directed when they re-enter the tectum, their search strategy ensures they 'home in' on approximately the retinotopically appropriate region. Interestingly, in fish but not in frog, the axons appear to be 'locked' into a linear growth mode whilst they remain within the most superficial part of the tectum. They start to search, often widely for appropriate regions only once they have descended from the superficial into the deeper layer (Stuermer, 1988a,b).

Once the regenerating atoms have started to search, they appear initially to rely on cytochemical cues to guide them. The nature of these cues remains obscure. They may be remnants of map-making molecules, such as the ephrin family, that are distributed as gradients, during initial map formation (discussed in Beazley *et al.*, 1997). Alternatively, the remains of the disconnected distal segments of severed axons may provide adequate signals (Sharma & Romeskie, 1977; Humphrey *et al.*, 1992). Whatever the nature of the initial cues, their role during optic nerve regeneration may be merely to align the gross axes of the map appropriately.

Once the search strategy sets up the approximate axes of the map, a second mechanism can come into play. This involves an activity-dependent process whereby axons which 'fire together, wire together'. Thus adjacent regions of retina will produce similar volleys of electrical activity causing those axon terminals that are adjacent in the tectum to be reinforced; conversely, terminals receiving input from more distant retinal regions will be down-graded. The activity-dependent mechanism has been shown to involve NMDA receptors (Schmidt, 1993). The mechanism offers an explanation for the compensatory re-organisations in which the entire retina projects retinotopically to the remaining part of a partially ablated tectum or the projection from surgically produced hemi-retina expands across the entire tectum (Schmidt, 1993; Strasnicky *et al.*, 1978).

To date we can only speculate as to the nature of the breakdown in map restoration in lizard. We have shown recently using carbocyanine labelling that, as in fish and frog, regenerating axons lack retinotopicity as they enter the optic tectum. Moreover, the axons appear to be largely restricted to rostro-caudally aligned trajectories across the superficial tectum. The behaviour is reminiscent of that seen in fish whilst their regenerating optic axons are located with the most superficial tectal layer. There are at least three theoretical possibilities for the absence from lizard of retinotopy after optic nerve regeneration and these are not mutually exclusive. It may be that axons do not search for cues or that they lack receptors capable of detecting them. Secondly, cues may no longer be available in the target. Thirdly, despite searching and recognising target cues, mechanisms may not allow a nascent map to be consolidated and, as a result, it is then lost.

Our preliminary results of labelling regenerating axons in the tectum of lizard suggest that, as a result of their initial mis-direction and subsequent restriction largely to rostro-caudally aligned trajectories, they search preferentially along this tectal axis. This is the axis along which the temporal to nasal retinal axis is normally represented. By contrast, little searching seems likely in the latero-medial tectal axis, which normally receives representation from dorsal to ventral retinal axis. Thus, the regenerating terminals might be capable of searching for tectal signals required to re-establish the appropriate alignment only along the temporo-nasal retinal to the rostro-caudal tectal axis. The possibility accords with the limited order that is seen within the low-fidelity maps recorded electrophysiologically at 5–7 months of regeneration. However, presumably, the tectal cues are insufficient to maintain even this transient arrangement and do not allow it to be translated into a stable map by an activity-dependent mechanism.

In mammals, the failure of regenerated projections to extend across the entire superior colliculus hinders definitive statements being made about the potential for the restoration of appropriate maps and their consolidation. However, one experiment provides some clues. In rat, one optic nerve was severed 2 weeks before the superior colliculus was removed (Wizenmann *et al.*, 1993). Extracts were prepared and challenged *in vitro* with growing optic axons from nasal or temporal retina of chicks. The axons selectively adhered to the retinotopically appropriate tectal extracts, indicating that map-making cues are up-regulated in the superior colliculus of adult rat optic nerve injury. The finding offers hope that appropriate maps might be re-established in mammals. However, it may be that the situation would mimic that in lizard with the nascent map being be lost at a later stage, in the absence of further interventions.

Key points

- **In fish and amphibians, signals (as yet unknown) in visual centres allow an approximate map to reform; this nascent map is secondarily consolidated by an activity-dependent mechanism.**
- **In lizard, the trajectories of regenerating axons may be restricted, allowing map restoration limited to the temporo-nasal retinal to the rostro-caudal tectal axis; no secondary consolidation of nascent maps takes place and order is lost.**
- **In mammals, signals re-appear in the superior colliculus of adult rat that *in vitro* allow growing chick axons from nasal and temporal retina to distinguish between rostral and caudal tectum.**

Connections to post-synaptic cells and return of visual function

In all vetebrate classes it seems likely that, when regenerating optic axons reach visual centres, they eventually form post-synaptic connections. The usefulness of this connectivity depends on factors such as whether the regenerated visual projections are retinotopically ordered. Thus in fish and amphibians, since retinotopy is restored after optic nerve regeneration, the connections to post-synaptic cells restore visually mediated behaviour. Some impairment of binocular vision has been reported in frog, presumably reflecting the loss of a proportion of ganglion cells (Dunlop *et al.*, 1997) but, nevertheless, functions such as prey capture are re-established. In 1989, Aguayo and his colleagues reported electrophysiological data that in rat optic axons regenerating to the superior colliculus via a peripheral nerve graft form functional synaptic connections to post-synaptic cells (Keirstead *et al.*, 1989). Moreover, Thanos and his colleagues reported that, despite a restriction of the regenerated projection to a small portion of superior colliculus and abnormalities in the alignment of the retinotopic map, information was sufficient to allow animals to distinguish between vertical and horizontal stripes in a Y-maze. Furthermore, visual evoked potentials confirmed that regenerated projections were relayed to the cortex (Thanos *et al.*, 1997).

Key points

- **Post-synaptic connections are reformed after optic nerve regeneration and useful vision is restored in fish and amphibians. In mammals, optic axons regenerating via a peripheral nerve graft form post-synaptic connections and some visual discrimination returns. In lizard, useful vision is not restored.**

Applications of the findings in animals to man

To initiate successful optic nerve regeneration in man, we need to ensure that the system behaves more like that of a fish or an amphibian than a mammal. Studies in frog indicate that even limited ganglion cell survival is sufficient for useful vision. Strategies in mammals now allow levels of ganglion cell survival after axotomy approaching those present in frog. However, only limited axonal regeneration has yet been triggered along the mammalian optic pathway. Moreover, despite proving an invaluable model providing insights into the conditions necessary for successful axonal regeneration, a peripheral nerve graft is unlikely to be clinically applicable. There are functional implications of removing a length of autologous peripheral nerve. Probably of greater import is the deep-seated location of primary visual centres in man, in marked contrast to their accessibility in rodents, which would preclude the implantation of a graft between the eye and a primary visual centre. Therefore strategies to modify the characteristics of resident and recruited non-neuronal cells within the optic pathway to ensure axonal regrowth are particularly relevant. Equally pressing is the issue of encouraging axons to search widely within visual centres to allow appropriate retinotopic maps to be restored and consolidated. Nevertheless, insights from the non-mammalian species are likely to continue to provide strategies towards overcoming the multiple obstacles before successful optic nerve regeneration can restore sight in man.

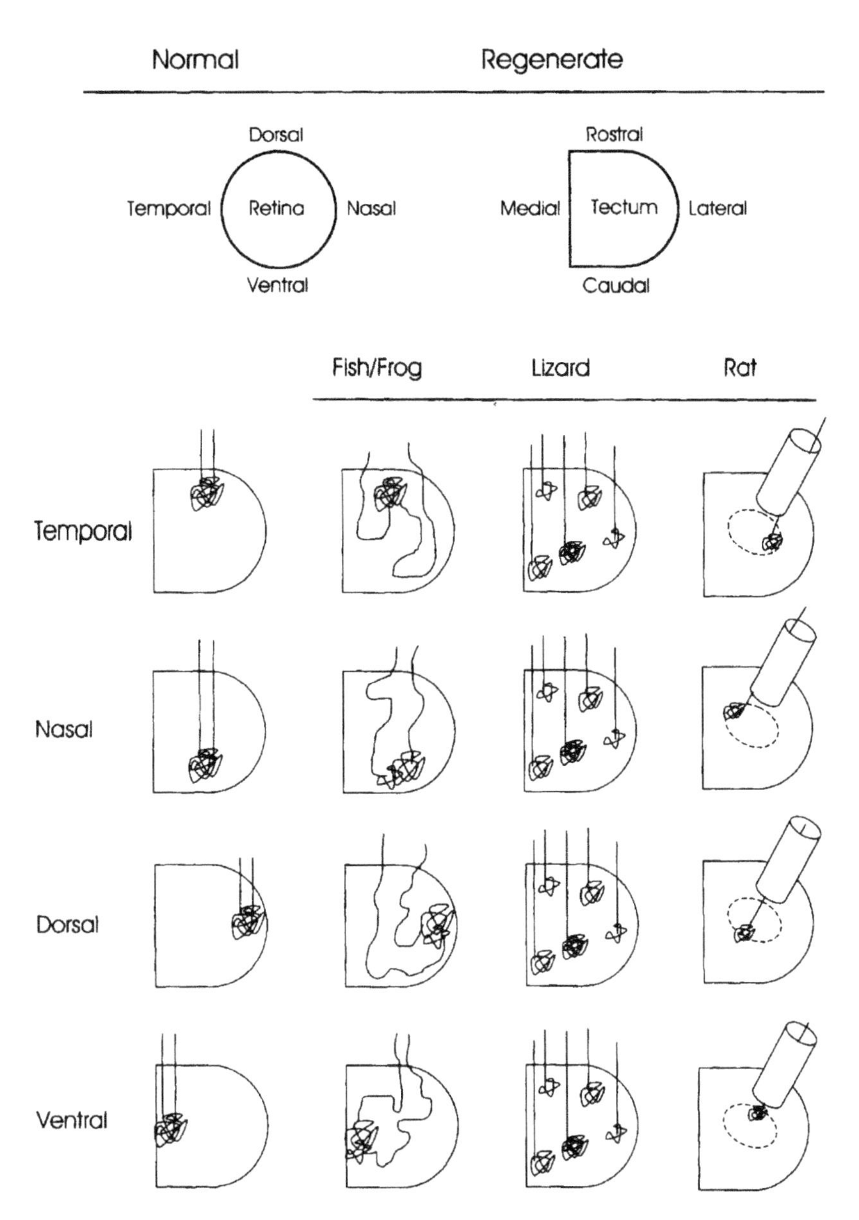

Figure 2 Diagrammatic summary of retinal projections in normal (left-hand column) animals and after optic nerve regeneration in fish/frog, lizard and in rat after axonal regeneration via a peripheral nerve graft (right-hand columns). The left retina is drawn as viewed from behind the animal and the right tectum is viewed dorsally. In normal animals, axons from temporal, nasal, dorsal and ventral retina terminate respectively in rostral, caudal, lateral and medial tectum. In fish/frog, axons reform a retinotopic map in the tectum by terminating in their appropriate locations after searching widely (Stuermer, 1988a,b). In lizard, axons from each retinal quadrant terminate across the tectal extent and thus fail to re-form a retinotopic map (Beazley *et al.*, 1997). In rat, axons are arranged in a parallel array within the peripheral nerve graft and terminate in regions which reflect their locations within the graft; however, the location of the terminals is retinotopically inappropriate (Thanos *et al.*, 1997).

Acknowledgement
Grant support: National Health and Medical Research Council Program Grant 952304.

References

Barron K D McGuinness C M Misantone L J Zanakis M F Grafstein B and Murray M (1985) RNA content of normal and axotomized retinal ganglion cells of rat and goldfish. Journal of Comparative Neurology 236, 265–273

Bastmeyer M Beckman M Schwab M E and Stuermer C A O (1991) Growth of regenerating goldfish axons is inhibited by rat oligodendrocytes and CNS myelin but not by goldfish optic nerve tract oligodendrocyte like cells and fish CNS myelin. Journal of Neuroscience 11, 626–640

Bastmeyer M Bähr M and Stuermer C A O (1993) Fish optic nerve oligodendrocytes support axonal regeneration of fish and mammalian retinal ganglion cells. Glia 8, 1–11

Battisti W P Shinar Y M Schwartz M Levitt P and Murray M (1992) Temporal and spatial patterns of expression of laminin, chondroitin sulphate proteoglycans and HNK-1 immunoreactivity during regeneration in the goldfish optic nerve: A light and electron microscopic analysis. Journal of Neurocytology 21, 557–573

Battisti W P Wang J Bozek K and Murray M (1995) Macrophages, microglia and astrocytes are rapidly activated after crush injury of the goldfish optic nerve: A light and electron microscopic analysis. Journal of Comparative Neurology 354, 306–320

Beattie M S Bresnahan J C and Lopate G (1990) Metamorphosis alters the response to spinal cord transection in *Xenopus laevis* frogs. Journal of Neurobiology 21, 1108–1122

Beazley L D (1984) Formation of specific nerve connections in the visual system of lower vertebrates. In Current topics in research on synapses 1, 55–117

Beazley L D Darby J E and Perry V H (1986) Cell death in the retinal ganglion cell layer during optic nerve regeneration in the frog *Rana pipiens*. Vision Research 26, 543–556

Beazley L D Tennant M and Dunlop S A (1996) The effect of a chronic breakdown of the blood-optic nerve barrier on the severed optic nerve in adult rat. Restorative Neurology and Neuroscience 10, 95–101

Beazley L D Sheard P W Tennant M Starac D and Dunlop S A (1997) Optic nerve regenerates but does not restore topographic projections in the lizard *Ctenophorus ornatus*. Journal of Comparative Neurology 377, 105–120

Beazley L D Tennant M Stewart T M and Anstee S D (1998) Optic nerve regeneration in the adult lizard *Ctenophorus ornatus* shows that neurogenesis is not obligatorily linked to axonal regeneration but may be a pre-requisite for map restoration. Vision Research 316, 789–793

Becker D L and Cook J E (1990) Changes in goldfish retinal ganglion cells during axonal regeneration. Proceedings of the Royal Society of London B 241, 73–77

Berkelaar M Clarke D B Wang Y C Bray G M and Aguayo A J (1994) Axotomy results in delayed death and apoptosis of retinal ganglion cells in adult rats. Journal of Neuroscience 14, 4368–4343

Berry J C Carlile J and Hunter A (1996) Peripheral nerve explants grafted into the vitreous body of the eye promote the regeneration of retinal ganglion cell axons severed in the optic nerve. Journal of Neurocytology 25, 147–170

Berry M Hall S Rees L Carlile J and Wyse J P H (1992) Regeneration of axons in the optic nerve of the adult Browman-Wyse (BW) mutant rat. Journal of Neurocytology 21, 426–448

Blaugrund E Duvdevani R Lavie V Solomon A and Schwartz M(1992) Disappearance of astrocytes and invasion of macrophages following crush injury of adult rodent optic nerves: implications for regeneration. Experimental Neurology 118, 105–115

Blaugrund E Lavie V Cohen I Solomon A Schreyer J and Schwartz M (1993) Axonal regeneration is associated with glial migration: Comparison between the injured optic nerves of fish and rats. Journal of Comparative Neurology 330, 105–112

Bohn R C and Reier P J (1982) Anomalous axonal outgrowth at the retina caused by injury to the optic nerve or tectal ablation in adult *Xenopus*. Journal of Neurocytology 11, 211–234

Bohn R C and Stelzner D J (1981) The aberrant retino-retinal projection during optic nerve regeneration in the frog. I. Time course of formation and cells of origin. Journal of Comparative Neurology 196, 605–620

Carmignoto G Maffe L Candeo P Canella R and Comelli C (1989) Effect of NGF on the survival of rat retinal ganglion cells following optic nerve section. Journal of Neuroscience 9, 1263–1272

Caroni P and Schwab M E (1988a) Two membrane protein fractions from rat central myelin with inhibitory properties for neurite growth and fibroblast spreading. Journal of Cell Biology 106, 1281–1288

Caroni P and Schwab M E (1988b) Antibody against myelin-associated inhibitor of neurite growth neutralizes nonpermissive substrate properties of CNS white matter. Neuron 1, 85–96

Carter D A Bray G M and Aguayo A J (1994) Long-term growth and remodeling of regenerated retino-collicular connections in adult hamsters. Journal of Neuroscience 14, 590–598

Cenni M C Bonfanti L Martinou J C Ratto G M Strettoi E and Maffei L (1996) Long-term survival of retinal ganglion cells following optic nerve section in adult *bcl*-2 transgenic mice. European Journal of Neuroscience 8, 1735–1745

Chelvanayagam D K Dunlop S A and Beazley L D (1998) Axon order in the visual pathway of the quokka wallaby. Journal of Comparative Neurology 390, 333–341

Cohen A Sivron T Duvdevani R and Schwartz M (1990) Oligodendrocyte cytotoxic factor associated with fish optic nerve regeneration: implication for mammalian CNS regeneration. Brain Research 537, 24–32

Colton C A and Gilbert D L (1987) Production of superoxide anions by a CNS macrophage, the microglia. FEBS Letters 223, 284–288

Darby J E Carr R A and Beazley L D (1990) Retinal ganglion cell death during regeneration of the frog optic nerve is not accompanied by appreciable cell loss from the inner nuclear layer. Anatomy and Embryology 182, 487–492

David S Bouchard C Tsatas O and Giftochristos N (1990) Macrophages can modify the nonpermissive nature of the adult mammalian central nervous system. Neuron 5, 463–469

Davies S J A Field P M and Raisman G (1994) Long interfascicular axon growth from embryonic neurons transplanted into adult myelinated tract. Journal of Neuroscience 14, 1596–1612

Davies S J A Fitch M T Membert S P Hall A K Raisman G and Silver J (1997) Regeneration of adult axons in white matter tracts of the central nervous system. Nature 390, 680–683

Doster S K Lozano A M Aguayo A J and Willard M B (1991) Expression of the Growth-associated protein GAP-43 in adult rat retinal ganglion cells following axon injury. Neuron 6, 635–647

Dunlop S A Humphrey M F and Beazley L D (1992) Displaced retinal ganglion cells in normal frogs and those with regenerated optic nerves. Anatomy and Embryology 185, 431–438

Dunlop S A Roberts J D Armstrong K N Edwards S J Reynolds S J Thom M D and Beazley L D (1997) Impaired vision for binocular tasks after unilateral optic nerve regeneration in the frog *Litoria moorei*. Behavioural Brain Research 84, 195–201

Dunlop S A Tran N G Tee L B G Papadimitriou J and Beazley L D (in press) Retinal projections throughout optic nerve regeneration in the ornate dragon lizard, *Ctenophorous ornatus*. Journal of Comparative Neurology

Eitan S Zisling R Cohen A Belkin M Hirschberg D L Lotan M and Schwartz M (1992) Identification of an interleukin 2–like substance as a factor cytotoxic to oligodendrocytes and associated with central nervous system regeneration. Proceedings of National Academy of Science USA 89, 5442–5446

Fukuda Y and Watanabe M (1996) Regeneration of cat's optic nerve and its functional recovery. Proceedings of the International Symposium of Neuroscience, Brazil 56, 69–78

Garcia I Martinou I Tsujimoto Y and Martinou J-C (1992) Prevention of programmed cell death of sympathetic neurons by *bcl-2* proto-oncogene. Science 258, 302–304

Gills J-P and Wadsworth J A C (1967) Retrograde trans-synaptic degeneration in the inner nuclear layer of the retina. Investigative Ophthalmology & Visual Science 6, 437–448

Glasgow E Druger R K Fuchs C Lane W S and Schechter N (1994) Molecular cloning of gefiltin (ON_1): serial expression of two neurofilament mRNAs during optic nerve regeneration. European Molecular Biology Organisation Journal 13, 297–305

Guilian D and Robertson C (1990) Inhibition of mononuclear phagocytes reduces ischemic injury in the spinal cord. Annals of Neurology 27, 33–42

Hahn J S Aizenman E and Lipton S A (1988) Central mammalian neurons normally resistant to glutamate toxicity are made sensitive by elevated extracellular Ca^{2+}; Toxicity is blocked by the N-methyl-D-asparate antagonist MK-801 Proceedings of the National Academy of Science USA 85, 6556–6560

Heumann R Lindholm D Bandtlow C Meyer M Radeke M J Misko T P Shooter E and Thoenen H (1987) Differential regulation of mRNA encoding nerve growth factor and its receptor in rat sciatic nerve during development, degeneration and regeneration: Role of macrophages. Proceedings of the National Academy of Science USA 84, 8735–8739

Hirsch S Cahill M A and Stuermer C A O (1995) Fibroblasts at the transection site of the injured goldfish optic nerve and their potential role during retinal axonal regeneration. Journal of Comparative Neurology 360, 599–611

Hirschberg D L and Schwartz M (1995) Macrophage recruitment to acutely injured central nervous system is inhibited by a resident factor: a basis for an immune-brain barrier. Journal of Neuroimmunology 61, 89–96

Holder N and Clarke J W D (1988) Is there a correlation between continuous neurogenesis and directed axon regeneration in the vertebrate nervous system? Trends in Neuroscience 11, 94–99

Humphrey M F (1987) Effect of different optic nerve lesions on retinal ganglion cell death in the frog *Rana pipiens*. Journal of Comparative Neurology 266, 209–219

Humphrey M F (1988) A morphometric study of the retinal ganglion cell response to optic nerve severance in the frog *Rana pipiens*. Journal of Neurocytology 17, 293–304

Humphrey M F and Beazley L D (1982) An electrophysiological study of early retinotectal projection patterns during optic nerve regeneration in *Hyla moorei*. Brain Research 239, 595–602

Humphrey M F and Beazley L D (1983) An electrophysiological study of early retinotectal projection patterns during regeneration following optic nerve crush inside the cranium in *Hyla moorei*. Brain Research 269, 153–158

Humphrey M F and Beazley L D (1985) Retinal ganglion cell death during optic nerve regeneration in the frog *Litoria moorei*. Journal of Comparative Neurology 236, 382–402

Humphrey M F Renshaw G M C Kitchener P D and Beazley L D (1995) Substance P, Bombesin and Leucine-Enkephalin immunoreactivities are restored in the frog tectum after optic nerve regeneration. Journal of Comparative Neurology 354, 295–305

Humphrey M F Dunlop S A Shimada A and Beazley L D (1992) Disconnected optic axons persist in the visual pathway during regeneration of the retino-tectal projection in the frog. Experimental Brain Research 90, 630–634

Jacobson M (1991) Developmental Neurobiology. Plenum Press, New York, London

Jelsma T N Friedman H H Berkelaar M Bray G M and Aguayo A J (1993) Differential forms of the neurotrophin receptor *trk*-B mRNA predominate in rat retina and optic nerve. Journal of Neurobiology 24, 1207–1214

Keating M J and Gaze R M (1970) The depth distribution of visual units in the contralateral optic tectum following regeneration of the optic nerve in the frog. Brain Research 21, 197–206

Keirstead S A Rasminsky M Fukuda Y Carter D A Aguayo A J and M. Vidal-Sanz M (1989) Electrophysiologic responses in hamster superior colliculus evoked by regenerating retinal axons. Science 246, 255–257

Kiernan J A (1979) Hypotheses concerned with axonal regeneration in the mammalian nervous system. Biological Reviews 54, 155–197

Kiernan J A (1985) Axonal and vascular changes following injury to the rat's optic nerve, Journal of Anatomy 141, 139–154

Kiernan J A and Contestabile A (1980) Vascular permeability associated with axonal regeneration in the optic system of the goldfish. Acta Neuropathologica (Berlin) 51, 39–45

Kuljis R O and Karten H J (1985) Regeneration of peptide-containing retinofugal axons into the optic tectum with reappearance of a Substance P-containing lamina. Journal of Comparative Neurology 240, 1–15

Lang D M Rubin B P Schwab M E and Stuermer C A O (1995) CNS myelin and oligodendrocytes of the *Xenopus* spinal cord — but not optic nerve — are nonpermissive for axon growth. Journal of Neuroscience 15, 99–109

Lang D M and Stuermer C A O (1996) Adaptive plasticity of *Xenopus* glial cells *in vitro* and after CNS fiber tract lesions *in vivo*. Glia 18, 92–106

Lang D M Hille M G Schwab M E and Stuermer C A O (1996) Modulation of the inhibitory substrate properties of oligodendrocytes by platelet-derived growth factor. Journal of Neuroscience 16, 5741–5748

Lazarov-Spiegler O Solomon A S Zeev-Brann A B Hirschberg D L Lavie V and Schwartz M (1996) Transplantation of activated macrophages overcomes central nervous system regrowth failure. The FASEB Journal 10, 1296–1302

Levin A L Clark J A and Johns L K (1996) Effect of lipid peroxidation inhibition on retinal ganglion cell death. Investigative Ophthalmology & Visual Science 37, 2744–2749

Levin L A Schlamp S C Spieldoch R L Geszvain K M and Nickells R W (1997) Identification of the *bcl*-2 Family of genes in the rat retina Investigative Ophthalmology & Visual Science 38, 2545–2553

Levin L A and Geszvain K M (1998) Expression of ceruloplasmin in the retina: Induction after optic nerve crush. Investigative Ophthalmology & Visual Science 39, 157–163

Levine R L (1991) Gliosis during optic fibre regeneration in the goldfish : an immunohistochemical study. Journal of Comparative Neurology 312, 549–560

Maffei L Carmignoto G V Perry V H Candeo P and Ferrari G (1990) Schwann cells promote the survival of rat retinal ganglion cells after optic nerve section. Proceedings of the National Academy of Science USA 87, 1855–1859

Maturana H R Lettvin J Y and McCulloch W S (1960) Anatomy and physiology of vision in the frog *Rana pipiens*. Journal of General Physiology 43, 129–175

Maxwell W L Follows R Ashhurst D E and Berry M (1990) The response of the cerebral hemisphere of the rat to injury. 1 The mature rat. Philosophical Transactions of the Royal Society of London B 328, 479–500

McKerracher L Essagian C and Aguayo A J (1993) Marked increase in ß-Tubulin mRNA expression during regeneration of axotomized retinal ganglion cells in adult mammals. Journal of Neuroscience 13, 5294–5300

Mey J and Thanos S (1993) Intravitreal injections of neurotrophic factors support the survival of axotomised retinal ganglion cells in adult rats *in vivo*. Brain Research 602, 304–317

Meyer R L (1978) Evidence from thymidine labelling for continuing growth of the retina and tectum in juvenile goldfish. Experimental Neurology 59, 99–111

Murray M (1982) A quantitative study of regenerative sprouting by optic axons in goldfish. Journal of Comparative Neurology 209, 352–362

Murray M Sharma S C and Edwards M E (1982) Target regulation of synaptic number in the compressed retinotectal projection of the goldfish. Journal of Comparative Neurology 20, 374–385

Ng T F So K-F and Chung S K (1995) Influence of peripheral nerve grafts on the expression of GAP-43 in regenerating retinal ganglion cell axons in adult hamsters. Journal of Comparative Neurology 24, 487–496

Peinado-Ramon P Salvador M Villegas-Perez M P and Vidal-Sanz M (1996) Effects of axotomy and intraocular administration of NT-4, NT-3, and brain-derived neurotrophic factor on the survival of adult rat retinal ganglion cells. A quantitative *in vivo* study. Investigative Ophthalmology & Visual Science 37, 489–500

Perry V H Brown M C and Gordon S C (1987) The macrophage response to central and peripheral nerve injury — a possible role for macrophages in regeneration. Journal of Experimental Medicine 165, 1218–1223

Quigley H A Nickells R W Kerigan L A Pease M E Thibault D J and Zack D J (1995) Retinal ganglion cell death in experimental glaucoma and after axotomy occurs by apoptosis. Investigative Ophthalmology & Visual Science 36, 774–786

Ramón Y Cajal S (1928) Degeneration and Regeneration of the Nervous System. Translated and edited by R A May Oxford University press, London, Humphrey Milford

Reier P J (1979) Penetration of grafted astrocytic scars by regenerating optic nerve axons in *Xenopus* tadpoles. Brain Research 164, 61–68

Richardson P M Issa V M K and Shemie S (1982) Regeneration and retrograde degeneration of axons in the rat optic nerve. Journal of Neurocytology 11, 949–966

Saito H-A (1983) Morphology of physiologically identified X-, Y- and W-type retinal ganglion cells of the cat. Journal of Comparative Neurology 221, 279–288

Sawai H Clarke D B Kittlerova P Bray G M and Aguayo A J (1996) Brain- derived neurotrophic factor and neurotrophin-4/5 stimulate growth of axonal branches from regenerating retinal ganglion cells. Journal of Neuroscience 16, 3887–3894

Scalia F (1987) Synapse formation in the olfactory cortex by regenerating optic axons: Ultrastructural evidence for polyspecific chemoaffinity. Journal of Comparative Neurology 263, 497–513

Scalia F V Arango V and Singman E L (1985) Loss and displacement of ganglion cells after optic nerve regeneration in adult *Rana pipiens*. Brain Research 344, 267–280

Schmidt J T (1978) Retinal fibers alter tectal markers during expansion of the half retinal projection in goldfish. Journal of Comparative Neurology 177, 279–300

Schmidt J T (1993) Activity-driven mechanisms for sharpening retinotopic projections: Correlated activity, NMDA receptors, calcium entry, and beyond. In: Formation and Regeneration of Nerve Connections, ed. Sharma S C & Fawcett J W pp. 185–204 Boston, Birkhäuser

Sharma S C and Romeskie M (1977) Immediate compression of the goldfish retinal projection to a tectum devoid of degenerating debris. Brain Research 133, 367–370

Sharma S C Jadhao A G and Prasada Rao P D (1993) Regeneration of supraspinal neuron projection in the adult goldfish. Brain Research 620, 221–228

Sheard P W and Beazley L D (1988) Retinal ganglion cell death is not prevented by application of tetrodotoxin during optic nerve regeneration for the frog *Hyla moorei*. Vision Research 28, 461–470

Silveira L C L Russelakis-Carneiro M and Perry V H (1994) The ganglion cell response to optic nerve injury in the cat: differential responses revealed by neurofibrillar staining. Journal of Neurocytology 23, 75–86

Sivron T Schwab M E and Schwartz M (1994) Presence of growth inhibitors in fish optic nerve myelin: postinjury changes. Journal of Comparative Neurology 343, 237–246

So K-F and Aguayo A J (1985) Lengthy regrowth of cut axons from retinal ganglion cells after peripheral transplantation into the retina of adult rats. Brain Research 328, 349–354

Stafford C A Shehab S A S Nona S N and Cronly-Dillon J R (1990) Expression of glial fibrillary acidic protein (GFAP) in goldfish optic nerve following injury. Glia 3, 33–42

Stelzner D J and Strauss J A (1986) A quantitative analysis of frog optic nerve regeneration: Is retrograde ganglion cell death or collateral axonal loss related to selective reinnervation? Journal of Comparative Neurology 245, 83–106

Straznicky K (1988) On the dendritic arbors of retinal ganglion cells. Neuroscience. Letter (Supplement) 30, 539

Straznicky K and Gaze R M (1971) The growth of the retina in *Xenopus laevis*: An autoradiographic study. Journal of Embryology and Experimental Morphology 26, 67–79

Straznicky K Tay D and Lunam C (1978) Changes in retinotectal projection in adult *Xenopus* following partial retinal ablation. Neuroscience Letters 8, 105–111

Stirling R V Dunlop S A and Beazley L D (1998) An *in vitro* technique to map retinotectal projections in reptiles. Journal of Neuroscience Methods 81, 85–89

Strobel G and Stuermer C A O (1994) Growth cones of regenerating retinal axons contact a variety of cellular profiles in the transected goldfish optic nerve. Journal of Comparative Neurology 346, 435–448

Stuermer C A O (1988a) Trajectories of regenerating retinal axons in thegoldfish tectum. 1 A comparison of normal and regenerated axons at later regeneration stages. Journal of Comparative Neurology 267, 55–68

Stuermer C A O (1988b) Trajectories of regenerating retinal axons in the goldfish tectum. 2 Exploratory branches and growth cones on axons at early regeneration stages. Journal of Comparative Neurology 287, 69–91

Taylor J S H Lack J L and Easter S S (1989) Is the capacity for optic nerve regeneration related to continous retinal ganglion cell production in the frog? A test of the hypothesis that neurogenesis and axon regeneration are obligatorily linked. European Journal of Neuroscience 1, 626–638

Tennant M and Beazley L D (1992) A breakdown of the blood-brain barrier is associated with optic nerve regeneration in the frog. Visual Neuroscience 9, 149–155

Tennant M Bruce S R and Beazley L D (1993a) Survival of ganglion cells which give rise to a retino-retinal projection during optic nerve regeneration in the frog. Visual Neuroscience 10, 681–612

Tennant M Moore S R and Beazley L D (1993b) Transient neovascularisation of the frog retina during optic nerve regeneration. Journal of Comparative Neurology 336, 605–612

Thanos S (1988) Alterations in the morphology of ganglion cell dendrites in the adult retina after optic nerve transection and grafting of peripheral nerve segments. Cell and Tissue Research 254, 599–609

Thanos S Mey J and Wild M (1993) Treatment of the adult retina with microglia-suppressing factors retards axotomy-induced neuronal degradation and enhances axonal regeneration *in vivo* and *in vitro*. Journal of Neuroscience 13, 455–466

Thanos S and Mey J (1995) Type-specific stabilization and target-dependent survival of regenerating ganglion cells in the retina of adult rats. Journal of Neuroscience 15, 1057–1079

Thanos S Naskar R and Heiduschka P (1997) Regenerating ganglion cell axons in the adult rat establish retinofugal topography and restore visual function. Experimental Brain Research 114, 483–491

Tiveron M Barboni C E Bernardo Pliego Rivero F Gormley A M Seeley P J Grosveld F and Morris R (1992) Selective inhibition of neurite outgrowth on mature astrocytes by Thy-1 glycoprotein. Nature 355 745–748

Vidal-Sanz M Bray G M Villegas-Perez M P Thanos S and Aguayo A J (1987) Axonal regeneration and synapse formation in the superior colliculus by retinal ganglion cells in the adult rat. Journal of Neuroscience 7, 2894–2909

Villegas-Perez M P Vidal-Sanz M Raminsky M Bray G M Aguayo A J (1993) Rapid and protracted phases of retinal ganglion cell loss following axotomy in the optic nerve of adult rats. Journal of Neurobiology 24, 23–36

Wanner M Lang D M Bandtlow C E Schwab M E Bastmeyer M and Stuermer C A O (1995) Re-evaluation of the growth-permissive substrate properties of goldfish optic nerve myelin and myelin proteins. Journal of Neuroscience 15, 7500–7508

Weibel D Cadelli D and Schwab M E (1994) Regeneration of lesioned rat optic nerve fibers is improved after neutralization of myelin-associated neurite growth inhibitors. Brain Research 642, 259–266

Wilson M A Gaze R M Goodbrand I A and Taylor J S H (1992) Regeneration in the *Xenopus* tadpole optic nerve is preceded by a massive macrophage/microglial response. Anatomy and Embryology 186, 75–89

Wizenmann A Thies E Klosterman S Bonhoeffer F and Bähr M (1993) Appearance of target-dependent guidance information for regenerating axons after CNS lesions. Neuron 11, 685–696

Zeng B Y Anderson P N Campbell G and Lieberman A R (1994) Regenerative and other responses to injury in the retinal stump of the optic nerve in adult albino rats: transection of the intraorbital optic nerve. Journal of Anatomy 185, 643–661

Zeng B Y Anderson P N Campbell G and Lieberman A R (1995) Regenerative and other responses to injury in the retinal stump of the optic nerve in adult albino rats: transection of the intracranial optic nerve. Journal of Anatomy 186, 495–508

Zwimpfer T Z Aguayo A J and Bray G M (1992) Synapse formation and preferential distribution in the granule cell layer by regenerating retinal ganglion cell axons guided to the cerebellum of adult hamsters. Journal of Neuroscience 12, 1144–1159

7

The role of cytoskeleton in regeneration of central nervous system axons

Alana Conti and Michael E Selzer

Introduction

During embryonic development, axons grow at high speeds by a process that appears to involve a pulling action exerted by a specialized structure at the leading edge, the growth cone. The present chapter will review briefly the mechanism of axon growth during development and compare this with what is known about the mechanism of axon elongation during regeneration in the central nervous system. Evidence in the lamprey suggests that these two processes may be very different and that regeneration may involve a protrusive force generated by neurofilament transport into the growing axon tip. If true, this could have important consequences for the use of axonal regeneration as a therapeutic strategy in treating injuries of the CNS.

Key points

- Axon growth in embryos appears to involve a pulling action exerted by a specialized structure at the leading edge, the growth cone.
- Axon growth in regeneration may involve a protrusive force generated by neurofilament transport into the growing axon tip.

Mechanisms of growth cone elongation in the developing central nervous system

The purpose of the growth cone during CNS development is to orient and extend the growing axon towards its eventual targets. However, most of what we know about the growth cone and its role in axon elongation has been derived from the study of embryonic vertebrate and *Aplysia* bag cell neurons in dissociated cell culture and developing invertebrate neurons growing in two dimensional dissected preparations (Bentley & O'Connor, 1994; Letourneau *et al.*, 1994; Challacombe *et al.*, 1996; Suter & Forscher, 1998). The results are impressively convergent, but it is possible that some of the mechanisms of axon growth *in vivo* are different from the generally accepted view that has been generated from *in vitro* studies. Several intracellular components involved in growth cone elongation have been identified but the cytomechanical events of axonal growth are still not completely understood. The growth cone consists of two parts, the lamellipodium, a flat sheath-like structure immediately in front of the axon, and filopodia, finger-like protrusions from the leading edge of the lamellipodium. Current theories have identified three major intracellular cytoskeletal components responsible for the cytomechanical forces in the axon's leading edge: actin microfilaments, myosin and microtubules (Sabry *et al.*, 1991; Challacombe *et al.*, 1996; Lin *et al.*, 1996). Intermediate filaments (primarily neurofilaments) are not found in the embryonic growth cone (Bridgman & Dailey, 1989) but are believed to be the primary skeletal elements that maintain the shape and diameter of the extending axon (Nixon *et al.*, 1994; Lin & Szaro, 1995). The degree to which they may be important in the overall elongation of the axon beyond the stage of growth cone advance is not clear and will be discussed later.

Current concepts of the mechanisms involved in growth cone migration and axon elongation are illustrated in Figure 1A, adapted from the work of Forscher and colleagues (Lin *et al.*, 1996). The growth cone has a proximal flat portion, the lamellipodium, which consists of a central domain rich in microtubules and a peripheral domain rich in actin. Filopodia, finger-like extensions of the peripheral domain are especially rich in actin, a polar molecule that can assemble into filamentous-actin (f-actin) from its component subunits, globular actin (g-actin). f-actin polymerization occurs through a process termed "treadmilling." g-actin is added to the distal (plus) end of the f-actin microfilament, causing it to elongate. At the proximal (minus) end, actin is constantly depolymerizing and being translocated retrogradely via an ATP-dependent mechanism. It has been postulated that actin polymerization is responsible for the primary forward driving force and elongation (3–6 μm/min) of the growth cone (Marsh & Letourneau, 1984; Forscher & Smith, 1988; Smith, 1988; Forscher *et al.*, 1992; O'Connor & Bentley, 1993). Through the use of surface beads and high resolution video microscopy, Forscher and colleagues have demonstrated retrograde translocation of actin molecules and filament disassembly, which counteract the anterograde flow of actin monomers (Forscher *et al.*, 1992). Inhibition of actin polymerization by cytochalasin does not inhibit this observed retrograde actin flow. Therefore, the retrograde flow is driven by a mechanism independent of that which controls actin polymerization. Recent evidence suggests that myosin is responsible for shuffling actin molecules retrogradely in order for recycling to occur at the minus end of the developing filament (Lin *et al.*, 1996). Inhibition of the myosin molecular motor and subsequent slowing of retrograde actin flow results in protrusive axonal growth that is proportional to the extent of myosin inhibition. Thus growth cone elongation may occur in two stages. The tensile stress associated with actin assembly at the leading edge is counteracted by the resistive forces of myosin-actin bridges. At this time, actin filaments are shuffled by myosin from the leading edge as quickly as or more quickly than they are assembled and retrograde actin flow is observed. No forward growth, or even filopodial retraction, is observed at this time. As the leading edge adheres to the substratum and newly formed actin filaments become stabilized, myosin-driven retrograde actin flow slows and the growth cone exhibits protrusive forward elongation. Evidence in *Aplysia* suggests that the stabilization of actin may be mediated by transmembrane cell adhesion molecules, which aggregate at points of tension on the surface of filopodia, link the extracellular matrix to the actin network and thus prevent the retrograde flow of actin

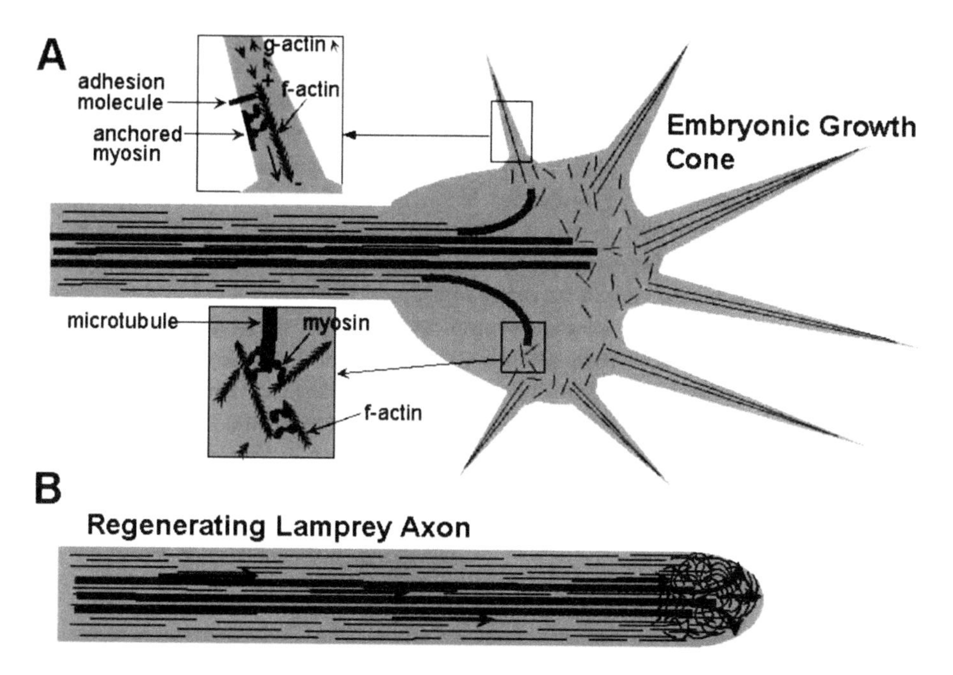

Figure 1 Comparison of the proposed mechanism of axonal regeneration in the lamprey spinal cord with that believed to underlie the movement of growth cones during embryonic development. A, Scheme of the embryonic growth cone simplified from the model of Forscher and colleagues (Lin *et al.*, 1996) Filopodial elongation is achieved by addition of g-actin subunits to the distal end of f-actin microfilaments. Filopodial position is stabilized by linkage between actin microfilaments the extracellular matrix via transmembrane adhesion molecules. A myosin molecular motor exerts a retrograde force on the actin microfilaments, and links the microfilamentous network to the microtubules in the central region of the lamellipodium. When the actin filaments are stabilized distally, the force generated by actin-myosin interaction applies traction to the microtubular network, pulling the microtubules in the direction of the stabilized filopodia, thereby pulling the axon forward. B, Hypothesis for the role of neurofilaments in the mechanism of regeneration of reticulospinal axons in the lamprey. Neurofilaments are transported along microtubules into the growth cone. The pressure generated by this process results in an increase in the packing density of neurofilaments and a protrusive force that contributes to axon elongation (Jacobs *et al.*, 1997).

(Suter *et al.*, 1998). More proximally, in the central portion of the growth cone, myosin motors produce tension between the actin network and microtubules. When retrograde actin flow has been shut down in a filopodium, the tension developed between the actin network and the microtubules pulls the microtubules toward that filopodium and channels elongation of the central core of the lamellipodium in its direction.

Much less is known about the mechanisms of axon growth *in vivo*. However, the appearance of growth cones in the developing insect nervous system and their disoriented pathfinding when actin polymerization is inhibited (Bentley & Toroian-Raymond, 1986) suggest that they are using mechanisms similar to those described in tissue culture. Growth cones with filopodia and lamellipodia have also been described many times in the developing nervous systems of mammalian and submammalian vertebrates, both in the PNS (Tosney & Landmesser, 1985) and CNS (Bovalenta & Mason, 1987; Nordlander & Singer, 1987; Bovalenta & Dodd, 1990; Kuwada & Bernhardt, 1990; Norris & Kalil, 1990; Yaginuma *et al.*, 1991). The result is a very rapid growth. Estimates for the movement of growth cones on appropriate substrates such as laminin are in the range of 1–3mm/day (Marsh & Letourneau, 1984; Fan *et al.*, 1993). Estimates for regeneration of peripheral nerve are approximately 1–5mm per day (Selzer, 1980), although the shape and cytoskeletal contents of regenerating peripheral nerve growth cones may be different from those of developing axons (see below). Cytochalasin, which blocks actin polymerization, causes filopodia and lamellipodia to collapse and growth cone movement to slow greatly. Poisoning microtubules with colchicine eliminates the remaining movement (Marsh & Letourneau, 1984). Thus it is very likely that the mechanisms that are being elucidated by the study of embryonic neurons in culture apply to the growth of axons during early embryonic development.

Key points

- The growth cone consists of a lamellipodium, a flat sheath-like structure immediately in front of the axon, and filopodia, finger-like protrusions from the leading edge of the lamellipodium.
- 3 major intracellular cytoskeletal components are responsible for the cytomechanical forces in the axon's leading edge: actin microfilaments, myosin and microtubules.
- Intermediate filaments (primarily neurofilaments) are not found in embryonic growth cones, but are believed to be primary skeletal elements that maintain shape and diameter of extending axons.
- Lamellipodium, consists of central domain rich in microtubules and peripheral domain rich in actin.
- Filopodia are especially rich in actin.
- f-actin polymerization occurs through a process termed "treadmilling." g-actin is added to the distal (plus) end of the f-actin microfilament, causing it to elongate.
- At the proximal (minus) end, actin is constantly depolymerizing and being translocated retrogradely via an ATP-dependent mechanism.
- Actin polymerization postulated to be responsible for primary forward driving force and elongation (3–6 mm/min) of the growth cone.

In summary, axon elongation of embryonic neurons in dissociated cell culture and *in vivo* involves a complexly organized apparatus, the growth cone, in which actin polymerization in filopodia that have found an appropriate adhesive substrate is thought to be responsible for the orientation of growth. Tension between the actin network and microtubules generated by a myosin molecular motor pulls the central core of the growth cone in the direction of stabilized actin microfilaments. The result is a high speed (up to 6µm/min) growth that is exquisitely sensitive to extracellular directional cues, although the intracellular signaling mechanisms that translate the various fixed and diffusible guidance cues into the required cytoskeletal rearrangements have not been determined. Neurofilaments fill the newly generated axonal extension and are responsible for establishing the mature calibre and shape of the axon, but are not generally believed to participate in the forces that move the growing tip forward. Nevertheless, there is growing evidence that neurofilaments are important to the ability of vertebrates to grow and maintain long axons. Thus the possible role of neurofilaments in axon elongation will be considered.

Development of axons in the absence of neurofilaments

Neurofilaments of higher vertebrates are heteropolymers of three peptides, low, middle and high molecular weight subunits (NF-L, NF-M and NF-H). Neurofilament assembly *in vivo* requires at least NF-L plus one other subunit (Lee *et al.*, 1993). NF-L deficient mutant quail (quiverer; (Ohara *et al.*, 1993) and transgenic mice that are deficient in NF-L (Zhu *et al.*, 1997) lack neurofilaments. These animals develop axons that, although small in calibre, are otherwise functional. Such observations have suggested that neurofilaments are not an essential part of the mechanism of axon outgrowth, at least during embryonic development. This conclusion may need to be re-examined. Lin & Szaro (1995) co-injected antibodies specific for *Xenopus* NF-M together with an intracellular lineage tracer into one blastomere of *Xenopus laevis* embryos at the 2 cell stage. They then studied the descendent embryonic spinal cord neurons in dissociated cell culture. For the first two hours, neurons descended from the antibody-injected blastomere showed normal neurite outgrowth, but by 21 hours these neurites were stunted in length despite normal distributions of f-actin and α-tubulin. Similarly, injection of RNA encoding a truncated form of NF-M into one blastomere resulted in stunted axonal development in dissociated neural tube cultures and in peripheral nerves *in vivo* (Lin & Szaro, 1996). These findings suggest that even in the embryo, neurofilaments may contribute to the effectiveness of axon elongation beyond the stage of growth cone advance.

Growth cones in late development and invasion of grey matter

While most of the heretofore studied axon elongation occurs by the described actin/myosin-dependent mechanism acting in morphologically complex growth cones, some axonal elongation occurs with much simplified growth cones, and even without filopodia or lamellipodia. Several authors have pointed out that growth cones are most complex when they are making pathfinding decisions and become simplified once they are following predefined pathways (Bray, 1982; Bovalenta & Mason, 1987; Nordlander & Singer, 1987). In general, growth cones become progressively simplified as development progresses (Nordlander & Singer, 1987), but because it is difficult to determine

Key points

- Neurofilaments of higher vertebrates are heteropolymers of three peptides, NF-L, NF-M and NF-H.
- NF-L deficient quail and mice are able to develop axons in embryo.
- *Xenopus* NF-M antibodies injected into *Xenopus* 2-celled blastomeres showed deficient axon outgrowth.
- In the embryo, neurofilaments may contribute to the effectiveness of axon elongation beyond the stage of growth cone advance.

fine growth cone structure in histological sections (as opposed to whole mounted and *in vitro* preparations), it is often impossible to determine from published images whether a growing tip truly lacks filopodia. When Marsh & Letourneau (1984) grew chick embryo dorsal root ganglion cells in the presence of cytochalasin B, an inhibitor of actin assembly, filopodia and lamellipodia collapsed but axons continued to grow at a much slower rate. These results suggested that secondary mechanisms of neurite outgrowth exist that are not dependent on filopodial or lamellipodial activity.

A naturally occurring growth in the absence of filopodia and lamellipodia was described by O'Leary and colleagues (O'Leary & Terashima, 1988). DiI was injected into the frontal or posterior occipital cortex of developing rats in order to visualize the corticopontine connections. These developed not by primary axon growth to the pons, but by delayed "interstitial" sprouting of buds from corticospinal axons that had already entered the spinal cord. These pontine collaterals grew relatively slowly without typical growth cones, i.e., without lamellipodia and filopodia. The cytoskeletal contents of the simple growth cones is unknown, but their growth is directed in a highly specific manner. Experiments with cortical and pontine co-cultures suggested that these buds are guided by diffusible chemoattractants (Heffner *et al.*, 1990). Thus the absence of filopodia and lamellipodia does not prevent interstitial buds from following environmental guidance cues. Whether growth without filopodia and lamellipodia is characteristic for collateral sprouting in mature CNS is not known.

In similar experiments, Norris & Kalil (1990) studied growth cones in the developing corpus callosum of the hamster. Nearly all of the growth cones exhibited filopodia regardless of their location in the pathway or their temporal emergence, except for the cones that had entered the cortex via delayed interstitial sprouting. Growth cones invading the grey matter were bullet-shaped with compact processes and were consistently smaller than those growing within the white matter tract. Moreover, they grew at half the rates of growth cones navigating within the corpus callosum. Thus they were similar to the interstitial sprouts invading the pons as described by O'Leary during the development of the corticospinal pathway. Videomicroscopic observations on living brain slices through the corpus callosum confirmed these observations (Halloran & Kalil, 1994), although in video demonstrations by Dr. Kalil, the presence of some lamellipodial and filopodial structures was suggested, even for interstitial sprouts within the target grey matter (K. Kalil, personal communication). Whether the differences relate to methodologic or brain region variances, or whether there is heterogeneity in growth cone structure even within a given region is not clear.

The role of cytoskeletal proteins in peripheral nerve regeneration

Much of the generally accepted view of the mechanisms of peripheral nerve regeneration is based on correlations between the time course of changes in expression and transport of cytoskeletal proteins and that of regeneration (Bisby & Tetzlaff, 1992). Following nerve section, synthesis and transport of actin and tubulin increase, while that for neurofilament decreases. These changes are reversed when regeneration is complete and the target (muscle) is reinnervated (Oblinger & Lasek, 1988; Muma *et al.*, 1990). The changes do not occur if regeneration is prevented artificially (Tetzlaff *et al.*, 1988). A conditioning lesion of the nerve enhances the rate of regeneration seen following a subsequent transection. While the increases in synthesis of actin and tubulin are not affected by the conditioning lesion, the reduction in neurofilament synthesis is exaggerated (Tetzlaff *et al.*, 1996). In addition, the rates of regeneration of the proximal and distal axons of dorsal root ganglion cells correlated more closely with the transport rates of tubulin than neurofilaments (Wujek & Lasek, 1983). These and other observations have led to the view that regeneration of peripheral nerve is dependent on the transport and assembly of actin and microtubules, but not neurofilaments, which participate in the subsequent morphological maturation and enlargement of the regenerated fibers. Recently a mouse deficient in light weight neurofilament subunit (NF-L) has been produced by Julien and coworkers. Because the middle (NF-M) and high (NF-H) molecular weight subunits are unable to assemble into filaments without NF-L, the axons of these mice lack neurofilaments. Following peripheral nerve crush, regeneration and maturation of myelinated axons were slowed (Zhu *et al.*, 1997), although both regeneration and myelination eventually did occur. Since the regeneration of unmyelinated axons were not evaluated, it may be that all of the observations related to maturation and that the leading edge of axonal regeneration was not slowed. In any case, it seems likely that the growth cone serves primarily to guide the growing axon, while the final length of the axon requires the ingrowth of more rigid cytoskeletal elements to consolidate the structure. Whether these must be microtubules or whether in some circumstances, neurofilaments play a role in peripheral nerve regeneration is not clear. What is surprising is the relative lack of information about the structure of growth cones in regenerating peripheral nerve. Ramón y Cajal drew them as simple bulbous enlargements, but he did not have the benefit of intracellular labeling or electron microscopy. More recent studies, particularly by Ide and colleagues, have demonstrated numerous membrane bound organelles, including several types of vesicles, multivesicular bodies and mitochondria, suggestive of high metabolic rates and active membrane turnover (Ide, 1996). These organelles are present in an amorphous or finely filamentous matrix, but organized cytoskeletal elements such as microfilaments, microtubules and neurofilaments are rare. To our knowledge, typical filopodia have not been described. Although this could be due to fixation artifact, fixation has not prevented the observation of filopodia elsewhere. It seems more likely that the absence of filopodia and lamellipodia reflects the maturity of the neuron and the specific biochemical and structural features of the endoneurial tube through which the regeneration occurs.

Axotomy-induced changes in expression of cytoskeletal proteins

Axotomy of CNS neurons results in many changes in gene expression, including expression of cytoskeletal proteins. These have been studied by Tetzlaff and his colleagues, who compared the changes with those in motoneurons following peripheral nerve lesions (Tetzlaff *et al.*, 1994). By contrast with facial motoneurons, in which axotomy produced increases in mRNA and protein synthesis for actin and total tubulin, and decreases for neurofilament, the responses of axotomized rubrospinal neurons were attenuated and transient, and corresponding mRNA levels in corticospinal neurons and retinal ganglion cells were decreased for all these cytoskeletal proteins (Tetzlaff *et al.*, 1991; Bisby & Tetzlaff, 1992). (How regeneration into peripheral nerve grafts alters the expression of mRNA for these proteins in retinal ganglion cells is described below). Because rubrospinal neurons showed sustained increases in mRNA for GAP-43 and the developmentally regulated T-α-1-tubulin, the authors concluded that these neurons were capable of mounting the gene expression changes typical of a regenerative response, but that failure of regeneration and subsequent neuronal atrophy was due to extraneuronal factors (Tetzlaff *et al.*, 1991). The shape and contents of growth cones during regeneration in CNS has been very poorly studied. Because regeneration does not occur spontaneously in mammalian CNS, no observations are possible except in artificial circumstances, e.g., during the growth of CNS axons within peripheral nerve bridges or fetal CNS transplants. Even in these circumstances, morphological observations of the leading edge of the regenerating axon have been so scant that nothing can be concluded about the cytoskeletal contents or even the shapes of the growth cones. Thus virtually all that is known comes from lower vertebrate models of CNS regeneration.

Key points

- Following peripheral nerve section, synthesis and transport of actin and tubulin increase, but decrease for neurofilament.
- These changes reverse following successful regeneration and reinnervation of targets, but not if regeneration is prevented.
- Regeneration of peripheral nerve has been thought to be dependent on the transport and assembly of actin and microtubules, but not neurofilaments.
- However, in mice lacking neurofilaments, regeneration and maturation of myelinated axons were slowed, although they did eventually occur.
- Likely that growth cones serve primarily to guide the growing axons, while final length of axons requires ingrowth of more rigid cytoskeletal elements. Not clear if only microtubules are involved or if neurofilaments play a role in peripheral nerve regeneration.

Cytoskeletal proteins in regeneration of axons in CNS of lower vertebrates
Growth cones lacking filopodia and lamellipodia have been observed in several models of axonal regeneration in the CNS. The molecular composition of these simple growth cones has not been studied extensively. Thus far, only two cytoskeletal proteins have been identified as major components; tubulin and neurofilament. Alterations of mRNA and protein levels have been observed for both components during neuronal regeneration in the CNS. Few reports combine morphological descriptions of growth cones with identification of their molecular constituents. Furthermore, the changes in molecular composition vary greatly across animal models and injury/regeneration paradigms. For this reason, it is difficult to generalize about changes in expression of these cytoskeletal proteins during regeneration. What little is known about the cytoskeletal components of such simple-shaped growth cones has come from models of regeneration in the goldfish and rat optic nerve and the lamprey spinal cord.

Lanners & Grafstein (1980) provided an early report of simple-shaped growth cones and their molecular composition following transection of the goldfish optic tract. As early as 2 hours post-injury, bundles of neurofilaments were observed in the bulbous swellings of cut axons with no evidence for microtubule invasion. This pattern continued for 1–2 days in the absence of filopodia and lamellipodia. Newly formed sprouts with large axon diameter contained neurofilament bundles as well as microtubules by 3–4 days. When correlated with previous reports, the axonal outgrowth preceded changes in total RNA and protein synthesis, but subsequent studies by Schechter and his colleagues showed that neurofilament protein expression was greatly upregulated during the course of regeneration. In an attempt to quantify changes in retinal RNA levels, Tesser *et al.* (1986) showed that synthesis rates of goldfish optic nerve neurofilament proteins (ON1 and ON2) increase 10- and 30-fold, respectively, 3 weeks following optic nerve crush, as measured by *in vitro* translation from total retinal RNA (Tesser *et al.*, 1986). Synthesis rates returned to normal by 35 days post-crush. Thus, unlike the situation in peripheral nerve, an early upregulation of neurofilament mRNA and protein synthesis occurs during regeneration in the goldfish optic nerve.

The presence of neurofilaments in the growth cones of regenerating optic nerve axons in the goldfish and the upregulation of neurofilament expression in their retinal ganglion cells, suggest a possible role for neurofilament in regeneration, since in the first reports, there did not appear to be evidence for actin-filled filopodia and lamellipodia. A more recent study indicated that some growth cones have filopodia, although the majority of examples illustrated had simple profiles and their cytoskeletal contents were not described (Strobel & Stuermer, 1994). In frog optic nerve, regenerating axons in the distal segments displayed both bulbous tips and more elaborate growth cones suggestive of lamellipodia and filopodia (Scalia & Matsumoto, 1985) but again, their cytoskeletal contents were not described.

By contrast with goldfish optic nerve, adult rat optic nerve does not regenerate after transection and axotomized retinal ganglion cells undergo different changes in mRNA levels for tubulin and the middle molecular weight subunit of neurofilament (NF-M, McKerracher *et al.*, 1993a). Although β-tubulin mRNA increased by 140% one day after injury, these levels had dropped to below control values by one week. NF-M mRNA levels decreased to 80% of controls and remained suppressed for the entire 6 month period of the experiment. Attaching a peripheral nerve graft to the proximal stump of the severed optic nerve rescues a small proportion of the retinal ganglion cells from death and approximately 20% of the surviving retinal ganglion cells regenerate axons into the graft (Aguayo *et al.*, 1991). In this circumstance, the axotomized retinal ganglion cells showed a 50% decrease in NF-M mRNA shortly after injury (McKerracher *et al.*, 1993b). β-tubulin mRNA also decreased to an average of 63% of controls, but some more intensely labeled cells (up to 300% above control levels) were observed. Retrograde labeling showed that these corresponded to the cells whose axons regenerated into the peripheral nerve graft. By contrast with mRNA levels, transport rates for both tubulin and neurofilament doubled during regeneration. Thus the response of retinal ganglion cells to axotomy was very different from that of peripheral nerve, in which transport of neurofilaments was decreased. How transport of neurofilament could double in the face of reduced message levels will be discussed below.

Key points

- Goldfish optic nerve regenerates following axotomy, but adult rat optic nerve does not.
- Axotomized retinal ganglion cells in goldfish and rat optic nerve undergo different changes in mRNA levels for tubulin and the middle molecular weight subunit of neurofilament (NF-M).
- Neurofilament synthesis increased in axotomized goldfish optic nerve but decreased in rat. β-tubulin mRNA increased but only transiently in rat.
- Peripheral nerve graft in rat rescues small proportion of retinal ganglion cells from death; approximately 20% of surviving retinal ganglion cells regenerate axons into the graft.

Axon regeneration in the lamprey spinal cord — evidence for a role for neurofilaments

One of the difficulties in interpreting the morphological observations on growth cones described above is that the findings were usually incidental and not a main point of the study. Authors may have assumed that the growth cones of regenerating neurons were similar to those of developing axons, and features distinguishing them were not looked for. Instead they had to be deduced by the interested reader. However, in the sea lamprey, the shape of the growth cone and its cytoskeletal contents have been examined in detail. In this animal, spinal cord transection is followed by recovery of locomotor function and this is accompanied by regeneration of a large proportion of the descending axons from the brainstem (Yin & Selzer, 1983; Cohen *et al.*, 1988; Davis & McClellan, 1994). Intracellular injection of tracer showed that growth cones of these regenerating axons lacked filopodia and lamellipodia, and were filled with neurofilaments (Lurie *et al.*, 1994). In their distal-most extremes the growth cones contained numerous vacuolar membrane bound organelles, but no microfilaments were seen. The packing density of neurofilaments in these growth cones was more than double that of the axon proximally (Pijak *et al.*, 1996). Immunohistochemical observations tended to support the possibility that the neurofilaments in the growth cones were under some pressure. It has been postulated that the phosphorylation of sidearms of the higher molecular weight neurofilament subunits increases their spacing by providing electrorepulsive forces that cause enhanced sidearm extension (Nixon *et al.*, 1994). The neurofilament of lamprey is a homopolymer of a single 180 kD subunit (NF-180) that also exists in several states of phosphorylation (Pleasure *et al.*, 1989). As in higher vertebrates, neurofilament phosphorylation is high and packing density low in the largest axons of the lamprey spinal cord (Pijak *et al.*, 1996). This correlation broke down during regeneration. The slender regenerating neurites of the giant reticulospinal neurons had highly phosphorylated neurofilaments, even though the packing densities of the neurofilaments in these neurites were greater than that of control axons of comparable size (Pijak *et al.*, 1996). Packing densities were greatest in the distal growth cones, even though their neurofilaments were also highly phosphorylated (Pijak *et al.*, 1996; Hall *et al.*, 1997). Thus it is possible that in the lamprey growth cones, an additional force is acting to overcome the electrostatic forces that would ordinarily provide for wide neurofilament spacing.

The appearance of the growth cones of lamprey reticulospinal neurons is similar to the description of "retraction bulbs" or "sterile clubs" in mammalian CNS systems suggestive of abortive regeneration. However, this cannot explain the appearance of the lamprey growth cones. First, the simple shaped terminals were seen not only during the period of axon die-back, but also in axons that had regenerated into and beyond the spinal transection (Lurie *et al.*, 1994). Second, similar growth cones characterized the ectopic regeneration of long axon-like neurites from the dendritic tree following axotomy of the same giant reticulospinal neurons very close to the cell body (Hall & Cohen, 1988). The distal tips of these neurites could not have been dying back, since they were never cut and did not exist before the close axotomy. Yet they contained abundant phosphorylated neurofilament, which filled the growth cones distal to the domain of the microtubules, and had very little actin (Hall *et al.*, 1997). Electron microscopic and immunohistochemical observations suggested that at early stages in the outgrowth of these ectopic axons, the dendrites lost many of their microtubules and increased their content of neurofilaments. The cell body and proximal dendrites showed a reduction in staining for acetylated tubulin, while the distal dendritic tips stained for phosphorylated neurofilament (Hall *et al.*, 1997).

In the above studies, the simple bulbous shape of the growth cones, their lack of f-actin, and their early filling with densely packed phosphorylated neurofilaments suggested the possibility that axonal regeneration in the CNS of lamprey may be based in part on an internal protrusive force generated by the transport of neurofilaments into the growing axon tip, rather than an actin-based pulling by filopodia. The proposed mechanism is diagrammed in Figure 1B. The neurofilament hypothesis received additional support by the observations of Jacobs *et al.*, (Jacobs *et al.*, 1997), who compared NF-180 mRNA levels in axotomized and unaxotomized brain neurons following hemisection of the lamprey spinal cord. Message levels decreased dramatically in all axotomized neurons by 4 weeks post-hemisection. The reductions tended to be more severe in neurons whose axons do not regenerate well than in the good regenerators. Thereafter, NF-180 expression returned toward normal levels in proportion to the probability of regeneration for the various identified spinal projecting neurons. This selective re-expression of neurofilament message in neurons whose axons normally regenerate well was not a consequence of the regeneration because it occurred even if regeneration was blocked by excision of a length of spinal cord distal to a transection. Thus in the lamprey, the ability to upregulate NF message following an initial reduction appears to be an intrinsic property of neurons whose axons have a high probability to regenerate.

Key points

- **Unlike adult mammals, spinal cord transection in lampreys is followed by recovery of locomotor function and is accompanied by regeneration of a large proportion of descending axons from brainstem.**
- **Regenerating lamprey cord axons lacked filopodia and lamellipodia, but were filled with neurofilaments.**
- **Axonal regeneration in the CNS of lamprey may be based partly on an internal protrusive force generated by transport of neurofilaments into growing axon tips, rather than an actin-based pulling by filopodia (Figure 1B).**

The initial downregulation of NF-180, like that of NF-M in regenerating rodent retinal ganglion cells, does not prove that neurofilaments are unimportant in regeneration. When axotomy greatly reduces the volume of axoplasm that must be supported, an absolute decrease in transcript levels could even represent a volume specific increase, i.e., an increase in the amount of mRNA per volume of axoplasm to be supplied with neurofilaments. In the lamprey, this was reflected by an increase in brain NF-180 protein levels at the same time that NF-180 mRNA levels were reduced in most axotomized neurons (Jacobs *et al.*, 1997). In rat, it was reflected by an increase in NF transport by axons whose neurons had downregulated NF-M (McKerracher *et al.*, 1993b).

Conclusions and implications for the enhancement of regeneration following CNS injury

The growth of axons in the developing nervous system is heavily regulated by the behavior of the growth cone, a complex structure whose main function appears to be to respond rapidly to positional cues and guide the axon toward its target. This axonal elongation generally occurs at stages of development when distances covered are relatively small and the distinction between growth cone advance and axon elongation may be mute. However, once the animal is mature, additional factors come into play. First, the length of the axon far outstrips the dimensions of the growth cone. Second, there may be a need for additional tensile strength to prevent rupture of the long axon by limb and body movements. In addition, at least in the CNS, the adhesive molecular matrix along which the axon grew during development may have become so modified or lost that the growth cone no longer has a mechanism for actin filament stabilization. Finally, it is now clear that, in addition to environmental changes, intraneuronal changes occur with maturation that restrain growth cone motility. Evidence from some models of regeneration, particularly the transected lamprey spinal cord, suggests that in the absence of a vigorous actin/myosin based filopodial pulling mechanism, axons may regenerate by a somewhat slower, protrusive mechanism that takes advantage of the invasion of the growth cone by cytoskeletal elements. Microtubules are probably critical in this process because they are involved in the mechanism of axonal transport that would be needed to supply the protrusive force, but in some cases, the elements being transported that provide the protrusive force may be neurofilaments. If this is so, then it might be possible to enhance regeneration in the CNS by increasing the synthesis neurofilaments and their subsequent transport.

It has been suggested that overexpression of neurofilaments might be toxic to neurons, since overexpression of NF-H in mice produces a model of neuronal degeneration similar to human motor neuron disease (MND) (Julien *et al.*, 1995), while other models of neuronal degeneration are accompanied by similar axon swellings or perikaryal inclusions that label for neurofilaments (Cleveland *et al.*, 1996). However, recent evidence suggests that the neurofilament accumulations are not themselves toxic, but may represent byproducts of another underlying pathological process such as the interference with the axonal transport system. For example, transgenic mice expressing the mutant Cu, Zn superoxide dysmutase (SOD 1) gene found in approximately 20% of cases with familial MND develop proximal axonal swellings that stain for neurofilaments, and experience axonal degeneration and massive motor neuronal death. A similar pathology is seen in the "dystonia musculorum" mouse (dt), which is deficient in the BPAG-1 linker protein. When these mutants were transfected with the truncated NF-H-Lac Z gene, which prevents the assembly of neurofilaments, the neurofilament-filled proximal axonal swellings were not formed, but the axonopathy and neuronal degeneration developed just as in the original SOD 1 and dt mutant mouse strains (Eyer *et al.*, 1998). In addition, overexpressing NF-L rescued the motor neurons from the effects of overexpressing the human NF-H in the transgenic mouse model of MND (J.-P. Julien, presented at the 1998 Gordon Conference on Intermediate Filaments). Thus accumulation of neurofilaments does not account for the axonal and neuronal degeneration of at least some of the transgenic mouse models of neurodegenerative disease in which neurofilamentous inclusions are found as pathological concomitants. Moreover, overexpression of neurofilament subunits does not produce pathological effects by producing too many neurofilaments. Rather it appears that the problem is one of stoichiometric imbalance. Therefore, if it turns out that the transport of neurofilaments into the growth cone provides a protrusive force that contributes to regeneration, it may be possible to overexpress the neurofilmentous subunits in a favourable stoichiometric combination that will enhance regeneration without causing neuronal death.

Key points

- In absence of connective tissue sheaths, regeneration of long axons in CNS may be especially dependent on tensile strength of intermediate filaments to avoid rupture.
- CNS environment in adult animals may lack molecular elements for optimal actin filament stabilization and growth cone elongation.
- Regeneration may require protrusive forces generated by transport of cytoskeletal elements, e.g., neurofilaments, into the growing tip.
- Regeneration might be enhanced by increasing neurofilament synthesis and transport into growing tip. However, there may be need for proper stoichiometric balance in the overexpression of three neurofilament subunits, since overexpression of NF-H alone produces neuronal degeneration in mutant mice.

References

Aguayo A J Rasminsky M Bray G M Carbonetto S McKerracher L Villegas-Perez M P Vidal-Sanz M and Carter D A (1991) Degenerative and regenerative responses of injured neurons in the central nervous system of adult mammals. Philosophical Transactions of the Royal Society of London B331, 337–343

Bentley D and O'Connor T P (1994) Cytoskeletal events in growth cone steering. Current Opinion in Neurobiology 4, 43–48

Bentley D and Toroian-Raymond A (1986) Disoriented pathfinding by pioneer neurone growth cones deprived of filopodia by cytochalasin treatment. Nature 323, 712–715

Bisby M A and Tetzlaff W (1992) Changes in cytoskeletal protein synthesis following axon injury and during axon regeneration. Molecular Neurobiology 6, 107–123

Bovalenta P and Dodd J (1990) Guidance of commissural growth cones at the floor plate in embryonic rat spinal cord. Development 109, 435–447

Bovalenta P and Mason C (1987) Growth cone morphology varies with position in the developing mouse visual pathway from retina to first targets. Journal of Neuroscience 7, 1447–1460

Bray D (1982) The mechanism of growth cone movements. Neuroscience Research Program Bulletin 20, 821–829

Bridgman P C and Dailey M E (1989) The organization of myosin and actin in rapid frozen nerve growth cones. Journal of Cell Biology 108, 95–109

Challacombe J F Snow D M and Letourneau P C (1996) Actin filament bundles are required for microtubule reorientation during growth cone turning to avoid an inhibitory guidance cue. Journal of Cell Science 109, 2031–2040

Cleveland D W Bruijn L I Wong P C Marszalek J R Vechio J D Lee M K Xu X S Borchelt D R Sisodia S S and Price D L (1996) Mechanisms of selective motor neuron death in transgenic mouse models of motor neuron disease. Neurology 47, S54–61

Cohen A H Mackler S A and Selzer M E (1988) Behavioral recovery following spinal transection: functional regeneration in the lamprey CNS. Trends in Neurosciences 11, 227–231

Davis G R and McClellan A D (1994) Extent and time course of restoration of descending brainstem projections in spinal cord-transected lamprey. Journal of Comparative Neurology 344, 65–82

Eyer J Cleveland D W Wong P C and Peterson A C (1998) Pathogenesis of two axonopathies does not require axonal neurofilaments. Nature 391, 584–587

Fan J Mansfield S G Redmond T R G-W P and Raper J A (1993) The organization of F-actin and microtubules in growth cones exposed to a brain-derived collapsing factor. Journal of Cell Biology 121, 867–878

Forscher P Lin C H and Thompson C (1992) Novel form of growth cone motility involving site-directed actin filament assembly. Nature 357, 515–518

Forscher P and Smith S J (1988) Actions of cytochalasins on the organization of actin filaments and microtubules in a neuronal growth cone. Journal of Cell Biology 107, 1505–1516

Hall G F and Cohen M J (1988) The pattern of dendritic sprouting and retraction induced by axotomy of lamprey central neurons. Journal of Neuroscience 8, 3584–3597

Hall G F Yao J Selzer M E and Kosik K S (1997) Cytoskeletal changes correlated with the loss of neuronal polarity in axotomized lamprey central neurons. Journal of Neurocytology 26, 733–753

Halloran M C and Kalil K (1994) Dynamic behaviors of growth cones extending in the corpus callosum of living cortical brain slices observed with video microscopy. Journal of Neuroscience 14, 2161–2177

Heffner C D Lumsden A G and O'Leary D D (1990) Target control of collateral extension and directional axon growth in the mammalian brain. Science 247, 217–220

Ide C (1996) Peripheral nerve regeneration. Neuroscience Research 25, 101–121

Jacobs A J Swain G P Snedeker J A Pijak D S Gladstone L J and Selzer M E (1997) Recovery of neurofilament expression selectively in regenerating reticulospinal neurons. Journal of Neuroscience 17, 5206–5220

Julien J P Cote F and Collard J F (1995) Mice overexpressing the human neurofilament heavy gene as a model of ALS. Neurobiology of Aging 16, 487–490

Kuwada J Y and Bernhardt R R (1990) Axonal outgrowth by identified neurons in the spinal cord of zebrafish embryos. Experimental Neurology 109, 29–34

Lanners H N and Grafstein B (1980) Early stages of axonal regeneration in the goldfish optic tract: an electron microscopic study. Journal of Neurocytology 9, 733–751

Lee M K Xu Z Wong P C and Cleveland D W (1993) Neurofilaments are obligate heteropolymers *in vivo*. Journal of Cell Biology 122, 1337–1350

Letourneau P C Condic M L and Snow D M (1994) Interactions of developing neurons with the extracellular matrix. Journal of Neuroscience 14, 915–928

Lin C H Espreafico E M Mooseker M S and Forscher P (1996) Myosin Drives Retrograde F-Actin Flow in Neuronal Growth Cones. Neuron 16, 769–782

Lin W and Szaro B G (1995) Neurofilaments help maintain normal morphologies and support elongation of neurites in *Xenopus* laevis cultured embryonic spinal cord neurons. Journal of Neuroscience 15, 8331–8344

Lin W and Szaro B G (1996) Effects of intermediate filament disruption on the early development of the peripheral nervous system of *Xenopus* laevis. Developmental Biology 179, 197–211

Lurie D I Pijak D S and Selzer M E (1994) Structure of reticulospinal axon growth cones and their cellular environment during regeneration in the lamprey spinal cord. Journal of Comparative Neurology 344, 559–580

Marsh L and Letourneau P C (1984) Growth of neurites without filopodial or lamellipodial activity in the presence of cytochalasin B. Journal of Cell Biology 99, 2041–2047

McKerracher L Essagian C and Aguayo A J (1993a) Temporal changes in ß-tubulin and neurofilament mRNA levels after transection of adult rat retinal ganglion cell axons in the optic nerve. Journal of Neuroscience 13, 2617–2626

McKerracher L Essagian C and Aguayo A J (1993b) Marked increase in ß-tubulin messenger-RNA expression during regeneration of axotomized retinal ganglion-cells in adult mammals. Journal of Neuroscience 13, 5294–5300

Muma N A Hoffman P N Slunt H H Applegate M D Lieberburg I and Price D L (1990) Alterations in levels of mRNAs coding for neurofilament protein subunits during regeneration. Experimental Neurology 107, 230–235

Nixon R A Paskevich P A Sihag R K and Thayer C Y (1994) Phosphorylation on carboxyl terminus domains of neurofilament proteins in retinal ganglion cell neurons *in vivo*: influences on regional neurofilament accumulation, interneurofilament spacing, and axon caliber. Journal of Cell Biology 126, 1031–1046

Nordlander R H and Singer M (1987) Axonal growth cones in the developing amphibian spinal cord. Journal of Comparative Neurology 263, 485–496

Norris C R and Kalil K (1990) Morphology and cellular Interactions of growth cones in the developing corpus callosum. Journal of Comparative Neurology 293, 268–281

Oblinger M M and Lasek R J (1988) Axotomy-induced alterations in the synthesis and transport of neurofilaments and microtubules in dorsal root ganglion cells. Journal of Neuroscience 8, 1747–1758

O'Connor T P and Bentley D (1993) Accumulation of actin in subsets of pioneer growth cone filopodia in response to neural and epithelial guidance cues in situ. Journal of Cell Biology 123, 935–948

Ohara O Gahara Y Miyake T Teraoka H and Kitamura T (1993) Neurofilament deficiency in quail caused by nonsense mutation in neurofilament-L gene. Journal of Cell Biology 121, 387–395

O'Leary D D M and Terashima T (1988) Cortical axons branch to multiple subcortical targets by interstitial axon budding: implications for target recognition and "wating periods". Neuron 1, 901–910

Pijak D S Hall G F Tenicki P J Boulos A S Lurie D I and Selzer M E (1996) Neurofilament spacing, phosphorylation, and axon diameter in regenerating and uninjured lamprey axons. Journal of Comparative Neurology 368, 569–581

Pleasure S J Selzer M E and Lee V M (1989) Lamprey neurofilaments combine in one subunit the features of each mammalian NF triplet protein but are highly phosphorylated only in large axons. Journal of Neuroscience 9, 698–709

Sabry J H O'Connor T P Evans L Toroian-Raymond A Kirschner M and Bentley D (1991) Microtubule behavior during guidance of pioneer neuron growth cones in situ. Journal of Cell Biology 115, 381–395

Scalia F and Matsumoto D E (1985) The morphology of growth cones of regenerating optic nerve axons. Journal of Comparative Neurology 231, 323–338

Selzer M E (1980) Regeneration of peripheral nerve. In: The Physiology of Peripheral Nerve Disease, edited by A J Sumner, pp 358–431. Philadelphia, Saunders

Smith S J (1988) Neuronal cytomechanics: the actin-based motility of growth cones. Science 242, 708–715

Strobel G and Stuermer C A (1994) Growth cones of regenerating retinal axons contact a variety of cellular profiles in the transected goldfish optic nerve. Journal of Comparative Neurology 346, 435–448

Suter D M Errante L D Belotserkovsky V and Forscher P (1998) The Ig superfamily cell adhesion molecule, apCAM, mediates growth cone steering by substrate-cytoskeletal coupling. Journal of Cell Biology 141, 227–240

Suter D M and Forscher P (1998) An emerging link between cytoskeletal dynamics and cell adhesion molecules in growth cone guidance. Current Opinion in Neurobiology 8, 106–116

Tesser P Jones P S and Schechter N (1986) Elevated levels of retinal neurofilament mRNA accompany optic nerve regeneration. Journal of Neurochemistry 47, 1235–1243

Tetzlaff W Alexander S W Miller F D and Bisby M A (1991) Response of facial and rubrospinal neurons to axotomy: changes in mRNA expression for cytoskeletal proteins and GAP-43. Journal of Neuroscience 11, 2528–2544

Tetzlaff W Bisby M A and Kreutzberg G W (1988) Changes in cytoskeletal proteins in the rat facial nucleus following axotomy. Journal of Neuroscience 8, 3181–3189

Tetzlaff W Kobayashi N R Giehl K M Tsui B J Cassar S L and Bedard A M (1994) Response of rubrospinal and corticospinal neurons to injury and neurotrophins. Progress in Brain Research 103, 271–286

Tetzlaff W Leonard C Krekoski C A Parhad I M and Bisby M A (1996) Reductions in motoneuronal neurofilament synthesis by successive axotomies: a possible explanation for the conditioning lesion effect on axon regeneration. Experimental Neurology 139, 95–106

Tosney K and Landmesser L (1985) Growth cone morphology and trajectory in the lumbosacral region of the chick embryo. Journal of Neuroscience 5, 2345–2358

Wujek J R and Lasek R J (1983) Correlation of axonal regeneration and slow component B in two branches of a single axon. Journal of Neuroscience 3, 243–251

Yaginuma H Homma S Kunzi R and Oppenheim R W (1991) Pathfinding by growth cones of commisural interneurons in the chick embryo spinal cord: A light and electron microscopic study. Journal of Comparative Neurology 304, 78–102

Yin H S and Selzer M E (1983) Axonal regeneration in lamprey spinal cord. Journal of Neuroscience 3, 1135–1144

Zhu Q Couillard-Despres S and Julien J P (1997) Delayed maturation of regenerating myelinated axons in mice lacking neurofilaments. Experimental Neurology 148, 299–316

8

What types of bridges will best promote axonal regeneration across an area of injury in the adult mammalian spinal cord?

Mary Bartlett Bunge

Introduction

Each year in the United States there are 10,000 new spinal cord injuries, which occur most frequently in teenaged persons and young adults, and add to the nearly quarter of a million cases. Due to recent advances in medical care, spinal cord injured persons now have a greatly enhanced life expectancy. The most common injury site is the 5th cervical (C5) level (15.7%) followed by C4 (12.7%), C6 (12.6%), T12 (7.6%), C7 (6.3%), and L1 (4.8%). Detailed pathological studies of human cervical spinal cord injury have revealed that lesions are of four types: contusion evolving to cavity, massive compression, solid cord syndrome, and laceration (Bunge *et al.*, 1997). Every type of lesion involves axotomy; in most cases, the neuronal somata and target neurons of these axons remain viable. Bunge and colleagues (1997) found that in 35% of 46 cases studied there was complete anatomical discontinuity across the cord.

Fifty-two percent of the injuries lead to tetraplegia; 18% and 29% of the injuries are complete and incomplete, respectively. A trend noted from the '70s to the '90s is that the percentage of complete tetraplegia declined by nearly 10%. This change has resulted largely from prevention efforts and improvements in emergency medicine techniques rather than improved surgical therapies or drugs. The one treatment for spinal cord injury in widespread use in the United States is the administration of large doses of a glucocorticoid, methylprednisolone, within the first eight hours of injury. This leads to a modest improvement, equivalent to one spinal level of function on average (see also chapter 1 by Brown, this volume). Rehabilitation strategies are steadily improving but are not yet widely available.

Clearly, to make a profound change in reversing the effects of spinal cord injury, new treatments are needed. Developing effective neuroprotective and neuroregenerative regimens is compelling, both from the point of view of personal loss and the enormous cost of treatment and maintenance of a spinal cord injured person for very long periods of time. We need to develop methods to preserve as much tissue as possible following the injury. Current studies of applying neuroprotective compounds in experimental spinal cord injury models are underway but will not be discussed here. This chapter will consider strategies to promote axonal regeneration across the area of spinal cord injury to restore function.

Spinal cord injury is devastating, not so much because a large number of neurons is lost but because the interruption of fibres across the cord diameter leads to functional loss of those fibres along the remainder of the cord. Whereas circuitry may remain in place within the cord, instructions from the brain are lost, thus eliminating the possibility of volitional control. Therapeutic advances must be devised to reconnect the cord and the brain. One strategy for reconnection is to construct bridges that will enable the severed axons to regrow across the area of injury and into the spinal cord below (for descending axons) or above (for ascending axons) where they will encounter their target neurons. What types of bridges have been tested? Which ones appear to be most effective so far? This review will concentrate on the use of peripheral nerve or Schwann cells as bridging material, primarily in completely transected adult rat spinal cord but see also "other cell types", below and chapter 9 by Harvey, this volume. Due to the suggested limitation on the number of references, reviews will constitute a major type of citation.

Key Points

- Improved medical care of spinal injury patients has improved life expectancy.
- Cervical spinal lesions are commonest (especially C4-6).
- Four lesion types: contusion evolving to cavity, massive compression, solid cord syndrome, and laceration.
- 52% injuries lead to quadraplegia.
- Modest improvement with early methylprednisolone treatment.
- Reconnection of brain and spinal cord by constructing bridges of material at site of injury to promote axon growth.

Peripheral nerve

The rationale to try peripheral nerve grafting in the cord was to place an environment in which axonal regeneration is known to occur into a milieu in which regeneration is known not to take place spontaneously. Landmark work published in the early 1980s by Aguayo and co-workers (reviewed in 1985) included the demonstration that a piece of peripheral nerve placed between the completely transected stumps of adult rat thoracic spinal cord elicited regeneration of axons from both stumps into the graft (Richardson *et al.*, 1980). This work had taken a cue from investigations near the turn of the century by workers in Spain, among them Tello. Ramón y Cajal (1928) speculated that the peripheral nerve environment provides nutritive (trophic) and orienting (tropic) substances that are lacking in the adult central nervous system (CNS). Whatever the substances provided by peripheral nerve, the work by the Aguayo team clearly demonstrated that, if adult CNS neurons are presented an appropriate environment, they are capable of regenerating their extensions. At least some if not all adult neurons are not limited by an innate deficiency to regenerate.

In the intraspinal peripheral nerve bridging study by Richardson *et al.* (1980) just mentioned, 5,850 myelinated axons were found in the graft when nearby dorsal root ganglia had been avulsed. The central axons that had regenerated into the nerve graft were myelinated by the endogenous Schwann cells in the nerve. Using a newly developed tracing method, the authors demonstrated that fibres regenerated across the graft. The sites of tracer injection did not permit detection of re-entry of such axons into the spinal cord. The regenerative growth was not from brainstem neurons which are so important for locomotion. It is only when peripheral nerve grafts are positioned close to the brainstem, at the cervical level, that there is a response from brainstem neurons (reviewed in Aguayo, 1985). Functional improvement was not observed.

In a variation of this paradigm, one end of a 35 mm long piece of peripheral nerve was inserted rostrally into the lower medulla and the other end, caudally into either the lower cervical or upper thoracic spinal cord without performing a cord transection, although there was local damage at the sites of insertion. David & Aguayo (1981) found that when retrograde tracer was applied to the graft 30 mm from the medulla, 261 medullary neurons (4 animals) were labelled; when the tracer was presented to the graft 30 mm from the cord insertion site, 189 grey matter cord neurons (3 animals) contained the tracer. These results demonstrated that, provided a suitable environment, axons from CNS neurons could regenerate as much as 30 mm. The fibres did not extend more than 2 mm into the CNS tissue, however. The lack of apparent neurologic deficits in this paradigm precluded assessing functional repair, and connectivity was not examined. This extraspinal bridging paradigm appears promising because the site of a lesion could be bypassed and regenerated axons could be brought to desired cord levels.

By combining additional strategies with thoracic spinal cord peripheral nerve grafting, some improvement in hindlimb movement has been observed (Cheng *et al.*, 1996). In completely transected adult rat spinal cord, 18 pieces of thin peripheral nerves were used to span the 3–5 mm gap. They were carefully positioned to extend from white matter to grey matter, to mirror the pathways of certain tracts. In addition, the spinal column was stabilized and the nerve segments were held in place by fibrin glue which contained acidic fibroblast growth factor (FGF-1). If one of these steps was omitted, the improvement in locomotion and neuroanatomical tracing evidence of growth of axons from the brain and brainstem into the grafts and beyond were not found. By using anterograde and retrograde tracing, axonal regeneration was detected from corticospinal tract and a variety of brainstem neurons into the distal cord. By utilizing a number of behavioural tests, the authors reported movements in the three joints of the hindlimbs, partial support of body weight, and contact placing of the hindlimbs, suggesting corticospinal tract involvement. Coordinated fore and hindlimb movements were not seen. More work will be needed to determine the importance of each component of the multifaceted strategy (Young, 1996), and it will be important to transect the grafts to assess the dependence of the improved functional outcome upon the regeneration of fibres through the grafts.

The targeting of the grafts from white matter to grey matter was undoubtedly an important step. There are known myelin-related inhibitory molecules in white matter, such as the one whose neurite growth inhibitory ability is neutralized by an antibody termed IN-1 (reviewed in Schwab & Bartholdi, 1996). Developmental studies also point to the inhibitory nature of myelin to axonal growth. During development of chick spinal cord, axonal regrowth ceases as myelin appears; if myelination is interrupted, the permissive period for axonal regeneration is prolonged (Kierstead *et al.*, 1995). Thus, it is generally considered that grey matter is more permissive for axonal growth. There are, however, examples of nerve fibre regeneration into myelinated tracts; one is mentioned below in relation to proteoglycan inhibitory barriers (Davies *et al.*, 1997).

The report by Cheng *et al.* (1996) is the first to present evidence of corticospinal fibre regeneration into peripheral nerve grafts. Neither the Aguayo team (reviewed in Aguayo, 1985) nor Houle *et al.* (1994) observed evidence of corticospinal tract fibres in peripheral nerve grafts. When Schnell & Schwab (1993) dorsally hemisected the thoracic cord (to sever the corticospinal tracts bilaterally), administered IN-1, and transplanted embryonic cord, newborn pons, collagen, amnion extracellular matrix, gelfoam, laminin-coated nitrocellulose filters, carbon filaments, or glass fibres into the lesion, the corticospinal fibres consistently skirted the transplant material and extended only through the remaining ventral cord.

Acidic fibroblast growth factor, a normal spinal cord constituent, is upregulated following spinal cord injury. It is considered to be involved in repair because the lack of a signal sequence implies that it is released from cells only after damage. Also, acidic fibroblast growth factor is known to decrease gliosis and improve neuronal survival and nerve fibre growth (Eckenstein, 1994). For these reasons, Cheng *et al.* (1996) decided to add it to their experimental paradigm. This factor may have been important in promoting growth of the corticospinal fibres into peripheral nerve grafts. Acidic fibroblast growth factor (in fibrin glue) presented to Schwann cell grafts positioned between completely transected stumps of spinal cord lessened dieback of the corticospinal tract and enhanced sprouting and modest entry of the fibres into the Schwann cell graft, a peripheral nerve environment, of course (Guest *et al.*, 1997b). The combination strategy devised by Cheng *et al.* (1996) is potentially an important advance in bridging techniques. As of this writing, confirmation of these results from another laboratory has not yet been published.

Combining neurotrophic factors with peripheral nerve transplants improves regeneration. Ye & Houle (1997) observed that combining neurotrophic factors with peripheral nerve transplants apposed to hemisected cervical cord increases the regenerative response of specific subchronically and chronically injured supraspinal neurons. Also, Fernandez and colleagues (1990) had shown earlier that the infusion of nerve growth factor (NGF) into autologous peripheral nerve grafts inserted into adult rat spinal cord dorsal columns caused a greater number of spinal cord neurons to regenerate axons into the grafts and some of these axons grew from neurons farther from the graft site than when NGF was not used. Neurotrophins may not only improve axonal growth but also may help to overcome the graft-host barrier to fiber growth from the graft into the cord. In a recent study by Oudega & Hagg (1996), the infusion of NGF into spinal cord parenchyma beyond a peripheral nerve graft led to increased numbers of regenerating sensory fibres that entered the cord beyond the graft. Thus, peripheral nerve grafts remain promising as a component of an eventual strategy for spinal cord repair. But their effectiveness will depend upon combining them with additional treatment strategies.

Schwann cells

Potential problems with peripheral nerve grafts are their availability and the possibility of graft rejection if allografts are used. An attractive alternative would be Schwann cell grafts. Purified populations of cultured rodent Schwann cells became available by the late seventies. Schwann cells are largely responsible for the regeneration-promoting environment of peripheral nerve; when the peripheral nerve is frozen and thawed before transplantation to kill the cellular content, then the peripheral nerve is not effective as a transplant. Schwann cell regeneration-promoting capabilities are related to the secretion of numerous trophic factors, expression of cell adhesion molecules and integrins on their surfaces, and production of a number of extracellular matrix components, such as laminin, which is known to be permissive for neurite growth (reviewed in Guénard *et al.*, 1993). Schwann cells are able to function in the CNS environment; there are many studies indicating that Schwann cells are able to myelinate or ensheathe axons from CNS neurons. A substantial body of literature using *in vitro* preparations has demonstrated that Schwann cells promote the growth of axons from central neurons (reviewed in Bunge & Hopkins, 1990).

Another advantage of the use of Schwann cells as bridging transplants is that large numbers of Schwann cells may now be generated *in vitro* in a short period of time. In fact, human as well as rat Schwann cells may now be rapidly expanded in culture (reviewed in Bunge & Kleitman, 1998). Recent modifications have enabled a nearly 100,000-fold increase in human Schwann cell number within about one month's time after harvest of the nerve, with a final purity of 90–95%. A 1 cm piece of nerve can yield enough Schwann cells to prepare a graft 30 meters long and 2.6 mm in diameter. This is an important consideration because Schwann cells may be obtained from a peripheral nerve biopsy from a spinal cord injured person and expanded in culture for autotransplantation into the area of injury. Human Schwann cells have been found to be as effective as rat Schwann cells in the bridging paradigm to be described below. Lastly, Schwann cells may be genetically modified during the culture period. Also, Tuszynski *et al.* (1998) observed that the host inflammatory response to Schwann cell grafts is less than that to fibroblasts, with fewer polymorphonuclear leukocytes and macrophages present at shorter time intervals in Schwann cell grafts (see Fitch & Silver, chapter 4 this volume).

Bridges of cultured Schwann cells have been found to be effective in uniting transected stumps of adult rat thoracic spinal cord and promoting growth of fibres from both stumps into the graft (Xu *et al.*, 1997; see also Xu *et al.*, 1995a). Purified populations of Schwann cells in an extracellular matrix (Matrigel) are drawn into polymer guidance channels, whereupon the graft contracts inside the channel, leading to the formation of a cable of Schwann cells. This construct is then placed into the gap between the severed stumps of the spinal cord and the stumps are inserted 1 mm into each end of the channel with the enclosed Schwann cell graft in apposition to the cord stumps. After a month, the cellular bridge appears to be well connected to both stumps of the cord and is well-vascularized.

When a tracer, such as Fast Blue, was injected into the middle of the Schwann cell cable at one month post-transplantation, a mean of 1064 ± 145 (SEM) spinal neurons was labelled as far away as cervical level 3 (C3) and sacral level 4 (S4) (Xu *et al.*, 1997). Inside the graft, there was a mean of 1990 ± 594 myelinated axons and 8 times as many unmyelinated but ensheathed axons at the graft midpoint, as determined by electron microscopy. Few myelinated or unmyelinated axons were present in control grafts containing only Matrigel. Many experiments have shown that Schwann cells must be in the graft in order for axons to be present weeks later. When cables of Schwann cells received tracer, brainstem neurons were not retrogradely labelled from the graft, although immunoreactive serotonergic and noradrenergic axons grew for a short distance into the rostral end of the graft. The introduction of anterograde tracer above the graft demonstrated that axons may grow the length of the graft but they did not exit the graft to enter the distal cord. Thus, the Schwann cell cable grafts supported regeneration of both ascending and descending axons, regeneration of spinal and sensory neurons, and limited regrowth of serotonergic and noradrenergic fibres from the rostral stump, but little growth from brainstem neurons, no corticospinal fibre regeneration into the graft, and no fibre growth from the graft into the distal cord (cf chapter 3 by Anderson & Lieberman, this volume). When Schwann cell suspensions were transplanted into a thoracic cord compression injury, most regenerating axons in the transplant originated from dorsal root ganglia; no corticospinal or brainstem axons were found in the graft (Martin *et al.*, 1996).

The use of a similar type of channel enclosing rat Schwann cells but only the size of half the cord diameter led to better regeneration than with a channel into which the entire cord stumps were inserted (Xu *et al.*, 1999). The model consisted of a right spinal cord hemisection at the 8th thoracic segment, implantation of the Schwann cell-containing mini-channel (1.25mm in outer diameter), and restoration of the cerebrospinal fluid circulation by suturing the dura. The intact left hemicord was transected before performing anterograde tracing to prevent labelling of axons coursing around rather than through the graft. A mean of 1,000 myelinated axons and nine times more unmyelinated axons were found at the graft midpoint. In addition to propriospinal and sensory axons, fibres from as many as 19 brainstem regions (mean of traced neurons, 125) grew into the graft without supplementary treatments. Also, significantly, some regenerating axons in the Schwann cell grafts penetrated the distal graft-host barrier to re-enter the host environment as far as 3.5mm. These axons grew towards the grey matter where they formed terminal bouton-like structures. Possible reasons for the improvement in the regenerative response using the hemi-channel may be the restoration of cerebrospinal fluid circulation and relatively more stable cord-graft interfaces due to the limited mechanical damage to the local vertebrae compared with the larger channel paradigm.

Cultured human Schwann cell cable grafts inside cord diameter-size polymer channels or transplanted without the channels have also been tested in the transection model (Figure 1, Guest *et al.*, 1997a). The human Schwann cells, grafted into nude rats which accept cells from other species, have been found to be as effective as rat Schwann cells in this paradigm. In fact, the regenerative response is better in some ways, but this reflects use of the nude rat, in contrast to the Fischer rat which is used for the rat Schwann cell transplant studies. When the grafting was done in a comparable way

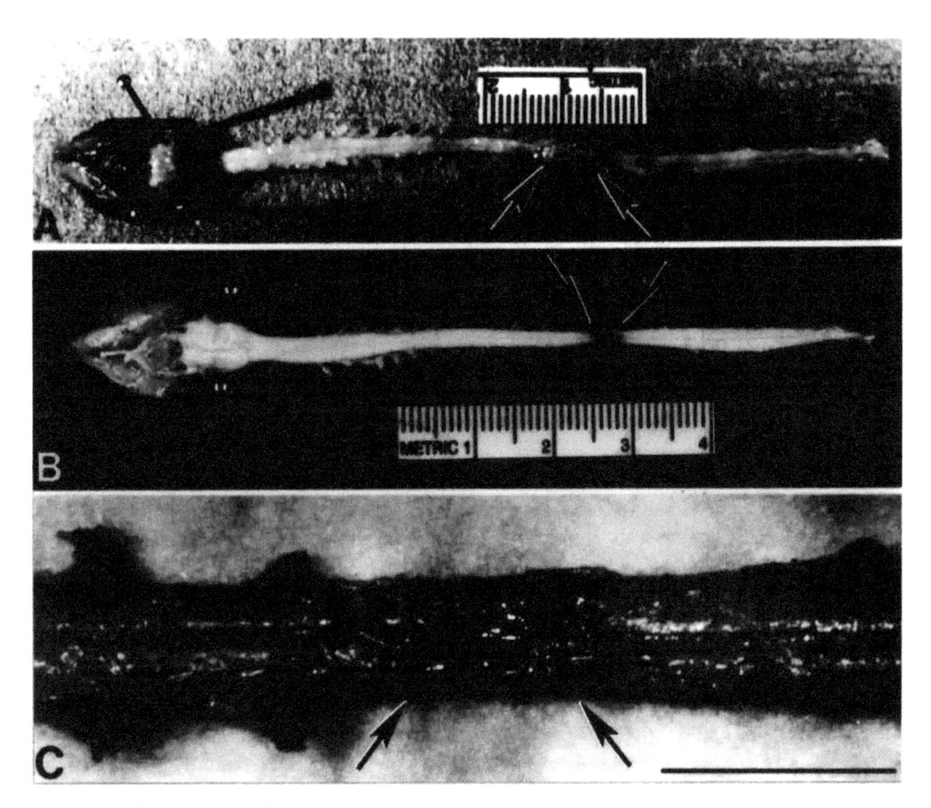

Figure 1 Illustrations of human Schwann cell grafting, either inside a polymer channel **(A)** or without an enclosing channel **(B, C)**, into completely transected adult nude rat thoracic cord. These preparations were dissected from the animal after preservation. **A.** A 6–7 mm channel enclosing a Schwann cell graft, designated by arrows, spans the T8-T10 spinal cord levels. **B.** The arrows point to a 3–4 mm Schwann cell graft transplanted without a channel at the T8 spinal cord level. **C.** Ventral view with dura intact showing one of the best examples of the union of a human Schwann cell graft with the adjacent cord stumps (arrows). Little scarring or cavitation in the cord stumps is visible. Bar in C, 5 mm. From Guest *et al.* (1997a) with permission.

to the rat Schwann cell transplantation studies, a mean of 1442 ± 514 myelinated axons was found in the graft (Guest *et al.*, 1997a). The orientation of the regenerating fibres within the graft was strictly longitudinal, as has been found in rat Schwann cell cables. By prelabelling the human Schwann cells, they were observed to survive for the 40–45 days following grafting and remain largely confined to the graft area. Serotonergic and noradrenergic fibres were detected within the grafts but essentially not beyond in the distal spinal cord. Unlike the rat Schwann cell grafts, a small number of fibres (propriospinal and sensory), labelled following injection of an anterograde tracer rostral to the graft, extended throughout the graft and into the distal cord for a maximum of 2.6mm (Figure 2).

Because fibres regenerated not only into the graft but also into the distal cord, two types of behavioural analysis were done, the inclined plane test and the open field

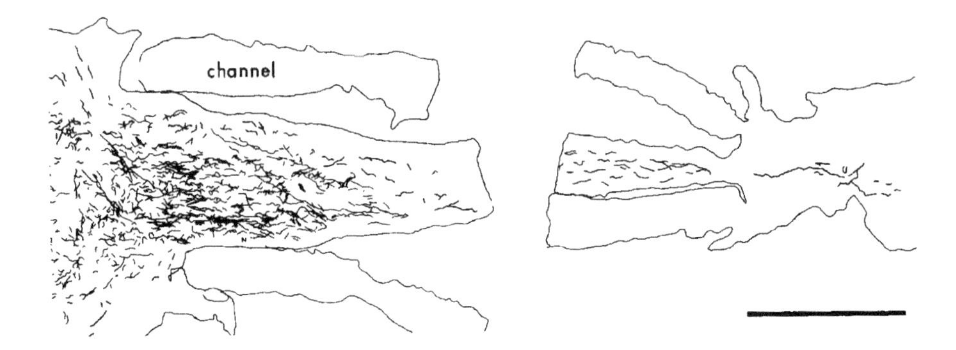

Figure 2 Camera lucida drawing of regenerated fibres following anterograde tracing from above the graft, as observed in six serial sections. Many more fibres are seen in the rostral portion of the transplant at the left but, nonetheless, some fibres have grown into the distal cord at the right of the figure. The channel indicates the position of the Schwann cell graft. Bar, 2.5 mm. From Guest *et al.* (1997a) with permission.

locomotion test developed by Drs. Basso, Beattie and Bresnahan (Basso *et al.*, 1995). In both cases, there was a modest but statistically significant improvement in the behavioural scores when human Schwann cell-cable implanted animals were compared with animals transplanted with similar cables but capped at the distal end to prevent outgrowth into the distal cord (Guest *et al.*, 1997a). In a few of the animals with bridging grafts, there was evidence of some form of contact placing response but, because the animals lacked retrograde tracing evidence of corticospinal tract regeneration, Guest *et al.* (1997a) assumed that these responses represented propriospinal placing mediated by local neuronal circuits. In sum, the human Schwann cell/nude rat model was more effective than the Fischer rat model in that there was a response from some brainstem neurons (raphe and coeruleus) despite the distant thoracic location (see discussion above under "Peripheral nerve"), some axons extended from the graft into the distal cord, and modest but significant behavioural recovery was observed.

As with peripheral nerve, Schwann cell grafting will require additional strategies to improve the regenerative response, particularly from brainstem and cortical neuronal somata. One added strategy tried by Chen *et al.* (1996) was the administration of methylprednisolone at the time of transection and Schwann cell transplantation. Methylprednisolone is, as already mentioned, now used routinely in the United States within eight hours after human spinal cord injury (see Brown, chapter 1, this volume). When methylprednisolone was administered, the Schwann cell bridge contained many myelinated axons (mean, 3237) plus many more unmyelinated axons, and numerous spinal neurons (mean, 2083) extended axons into the graft. Also, brainstem neurons responded (mean, 57); serotonergic and noradrenergic axons were found 2.0–2.5 mm into the graft. Very significantly, following anterograde tracer injection 5–6 mm rostral to the graft, a modest growth of fibres from the graft into the distal cord was observed, in contrast to animals that did not receive methylprednisolone. In addition, a striking finding was that more spinal cord tissue inserted into each end of the polymer channel survived with methylprednisolone administration; the inserted tissue largely deteriorated without methylprednisolone. The improvement of axonal regeneration from both spinal cord and brainstem neurons into thoracic Schwann cell grafts may be related to this reduction in secondary spinal cord tissue loss adjacent to the graft.

Very recent work by Oudega *et al.* (1999) showed that dieback of one of the tracts, the vestibulospinal tract, was significantly diminished when methylprednisolone was administered after completely transecting the spinal cord (with no Schwann cell graft). In this simple transection model, cord tissue loss and the number of ED1-positive macrophages/monocytes were also significantly decreased for weeks in both cord stumps when methylprednisolone was administered (Oudega *et al.*, 1999). Bartholdi & Schwab (1995) reported that the invasion of injured spinal cord tissue by ED1-positive cells was markedly suppressed for 24 hours following a dorsal hemisection lesion.

The addition of neurotrophins to the Schwann cell grafting paradigm has been explored as well by Xu *et al.* (1995b). Neurotrophins are substances that are required for neuronal survival, support growth of neuronal extensions, and influence neuronal function (Lindsay *et al.*, 1994). Two neurotrophins, Brain-Derived Neurotrophic Factor (BDNF) and Neurotrophin-3 (NT-3), were delivered together into the capped caudal end of a channel containing the Schwann cell graft for the first 14 days, with the animals being maintained for an additional two weeks. One month later, a mean of 1523 ± 292 myelinated axons was present in Schwann cell/neurotrophin grafts and more retrogradely labeled neurons (967 ± 104) were present throughout the rostral cord. In the Schwann cell/neurotrophin but not Schwann cell/vehicle graft, at least 5 mm from the rostral cord-graft interface, some nerve fibres were immunoreactive for serotonin, a neurotransmitter specific for raphe neurons. Retrograde tracing from Schwann cell/neurotrophin grafts revealed labelled neurons in a number of brainstem nuclei, with 67% of these being in vestibular nuclei. The mean number of labelled brainstem neurons in the Schwann cell/neurotrophin group (92) contrasted with the mean in the Schwann cell vehicle group (6). These results clearly demonstrated that BDNF and NT-3 infusion enhanced propriospinal axon regeneration and, more significantly, promoted axonal regeneration of specific distant populations of brainstem neurons into grafts at the midthoracic level. Thus, regeneration of some neuronal populations distant from the spinal cord transection and implant can be recruited by combinations of trophic factors and a favourable cellular substrate.

In other experiments by Menei *et al.* (1998), the neurotrophin, BDNF, was presented to the lesioned spinal cord by introducing the human prepro BDNF cDNA into Schwann cells by means of infection with a retroviral vector. The experimental paradigm was somewhat different from that described above. The engineered Schwann cells were deposited in a 5 mm long trail in the distal cord beyond the transection site and into the transection site itself. No polymer channel was used in these experiments. The use of Hoechst prelabelled Schwann cells demonstrated that trails were maintained for at least a month (Figure 3A, D). When Brook *et al.* (1994) transplanted Schwann cells in vertical columns extending up to 4 mm through the thalamus and across the choroid fissure into the hippocampus in the adult rat, most cells remained in this orientation for up to three weeks. The Schwann cells acted as bridges that enabled directed axonal growth across boundary membranes of the brain, carrying substantial numbers of nerve fibres from one area to another (Brook *et al.*, 1994).

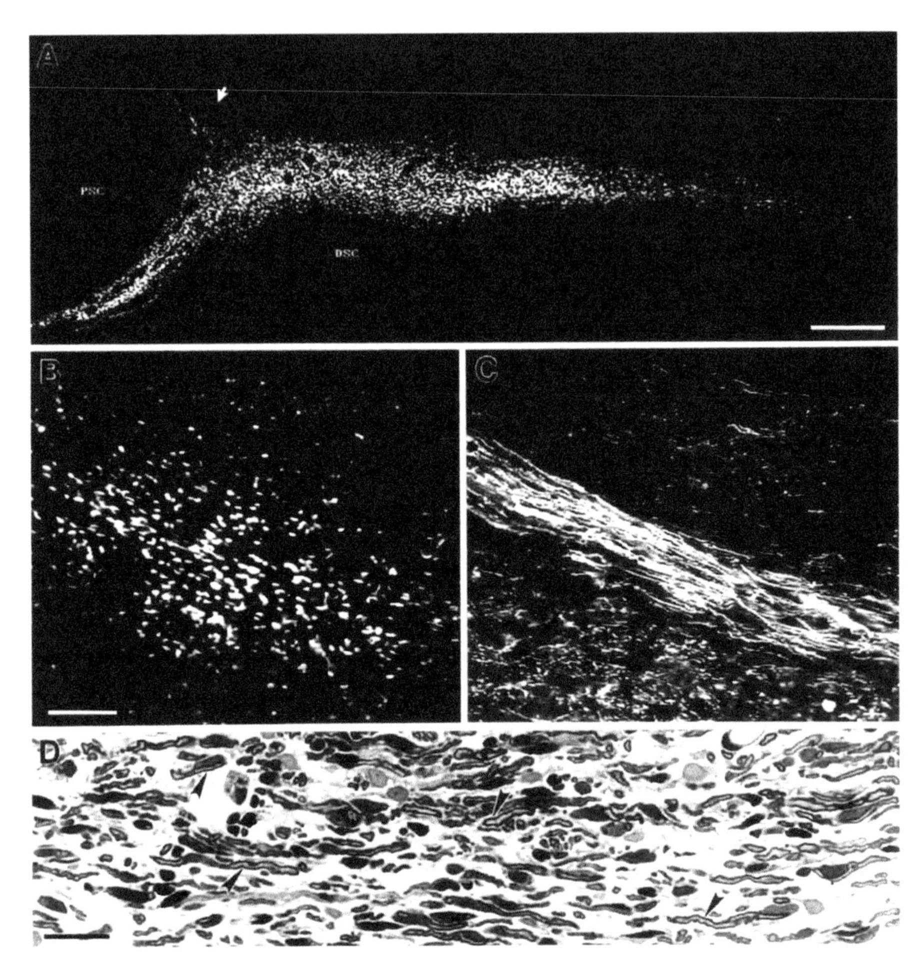

Figure 3 Illustrations of Hoechst-labelled rat Schwann cell trails one month after transplantation. **Panel A** shows the trail extending from the proximal spinal cord stump (PSC), with the transection site designated by the arrow, and extending into the distal spinal cord (DSC). **Panels B and C** are a pair of illustrations showing the Hoechst-labelled Schwann cell trail in B, along which immunostained nerve fibres are seen to extend in close and parallel array (C). **Panel D** contains a sagittal section of part of a trail beyond the transection site, as visualized in a semi-thin plastic section. Numerous Schwann cell-myelinated axons (arrowheads) are seen. Bars: A, 100 μm; B and C, 200 μm; D, 50 μm. From Menei *et al.* (1998) with permission.

After one month, in animals with transections only, serotonergic and noradrenergic axons were not seen beyond the transection (Menei *et al.*, 1998). When control (non-modified) Schwann cells were transplanted, numerous axons were seen in the trails beyond the transplanted transection site (Figure 3B, C), including some serotonergic fibres. More serotonergic and some noradrenergic axons were seen in the trail beyond the transection site when the engineered cells were transplanted. A retrograde tracer, Fast Blue, was injected at the distal end of the 5 mm long Schwann cell trail to identify neurons that had regenerated axons across the transection and along the length of the trail. When engineered Schwann cells were grafted, as many as 135 retrogradely labelled neurons were found in the brainstem, mostly in the reticular and raphe nuclei (non-modified Schwann cells up to 22, mostly in vestibular nuclei). No labelled neurons rostral to the transection were seen when Schwann cells were not transplanted. Thus, the transplantation of Schwann cells secreting human BDNF improved the regenerative response across the transection site and into the thoracic cord. Moreover, the enhanced regeneration observed with the engineered Schwann cells appeared at least partly specific, as the largest response was from neurons known to express the receptor for BDNF, *trk*B.

Schwann cells, genetically modified with a retroviral vector to produce and secrete high levels of another neurotrophin, NGF, have been grafted into previously uninjured midthoracic adult rat spinal cord and examined from two weeks to one year later (Tuszynski *et al.*, 1998). *In vivo* expression of the human NGF transgene lasted for at least six months. At three months and later, the NGF-secreting grafts slowly increased in size, in contrast to control (non-transduced) Schwann cell grafts. After two weeks, the transduced transplants contained numerous sensory nociceptive axons from the dorsolateral funiculus and, after three months, dopaminergic and noradrenergic (coeruleospinal) axons were present and were even more numerous at six months; control grafts contained far fewer of these axonal species. Neither transduced nor control grafts contained serotonin- or choline acetyltransferase-immunopositive axons or corticospinal tract axons. Again, genetically manipulating Schwann cells to increase their production of a neurotrophin improved their ability to support CNS axonal regeneration.

Clearly, the strategies used in conjunction with transplanted Schwann cells improved the regenerative response. This is true for both growth of axons into the transplant and from the transplant into the distal cord. In three types of experiments mentioned above, use of human Schwann cells in nude rats, methylprednisolone administration, and genetic modification of rat Schwann cells, some outgrowth of fibres into the distal cord was observed, but substantial improvements in function were not evident. What additional strategy could improve the outgrowth from the graft into the cord, growth that is necessary for fibres to reach their target neurons? One potential intervention is to attempt to diminish proteoglycan production at the distal graft-host interface (Plant *et al.*, 1998); proteoglycans have long been thought to be inhibitory to axonal growth by Silver and colleagues. In a recent paper by his group in collaboration with Raisman (Davies *et al.*, 1998), it was reported that regenerating axons of transplanted neurons halted at a chondroitin sulfate proteoglycan-rich boundary. If this boundary was not formed, the axons crossed the lesion and extended rapidly in myelinated white matter tracts. We decided to investigate yet another strategy to improve outgrowth from graft to distal cord, transplantation of olfactory ensheathing glia into areas next to the Schwann cell transplant.

Ensheathing glia are found only in the olfactory system where neurons are continually generated throughout adulthood. As the new axons extend to the outer layers of the olfactory bulb (CNS), they are enwrapped in bundles in the olfactory nerves by ensheathing glia (Raisman, 1985; reviewed in Ramón-Cueto & Avila, 1998). These unique glia, when transplanted beside cut and reanastomosed dorsal roots, enabled regeneration of dorsal root ganglion fibres across the usually inhibitory dorsal root entry zone into the adult spinal cord (reviewed in Ramón-Cueto & Avila, 1998). After we placed a Schwann cell-containing polymer channel into the transection gap, suspensions of pure (98%) Hoechst-labelled ensheathing glia were stereotaxically injected into the midline of both stumps at 1mm from the edges of the Schwann cell graft (Ramón-Cueto *et al.*, 1998). Supraspinal serotonergic axons were observed to cross the transection gap through connective tissue formed on the exterior of the channel rather than through the Schwann cell graft inside the channel. They entered the distal cord, elongating in both white and periaqueductal grey matter, reaching the farthest distance analyzed, 1.5cm. Tracer introduced above the graft (cervical level 7) allowed visualization of axons that traversed the Schwann cell graft and extended across the distal interface and up to 800 μm into the distal cord. Long distance regeneration (at least 2.5cm) of injured ascending axons occurred across and rostral to the Schwann cell graft. Tracing to detect corticospinal tract regeneration was not done in this study. Transplanted ensheathing glia migrated longitudinally and laterally from the injection sites, reaching the farthest distance analyzed, 1.5cm, through both white and grey matter, glial scars, the Schwann cell graft, and connective tissue on the outside of the channel. In the controls (without ensheathing glia transplantation), serotonergic fibres were not seen in the connective tissue on the channel exterior, labelled fibres were not observed to exit the graft into the distal cord, and the long distance regeneration of ascending axons was not found. Thus, ensheathing glia provided injured axons with an appropriate milieu for long distance elongation in the cord and improved axonal regeneration across the Schwann cell graft-host cord interfaces.

Despite this morphological evidence of improved axonal regeneration into the distal cord, statistically significant improvement in stepping movements of the hindlimbs, as assessed by the open field locomotion Basso-Beattie-Bresnahan test (Basso *et al.*, 1995) was not seen compared with controls (Broton, Ramón-Cueto, Plant, and Calancie, personal communication). When the Schwann cell/ensheathing glia-transplanted animals were examined on a treadmill, however, they displayed rhythmic and alternating hindlimb movements for nearly 70% of the time on average, similar to animals transplanted with Schwann cells only (Broton, Xu, & Calancie; personal communication). Transected only animals exhibited this activity for only 20% of the time on the treadmill (Broton *et al.*, 1996). No coordination between fore and hindlimbs was found. These observations may reflect increased excitability of existing spinal cord circuits after transplantation rather than voluntary movement that requires involvement of supraspinal neurons.

There have been no reports up to now of the use of ensheathing glia in bridges spanning completely transected spinal cord. But there is a new report by Li *et al.* (1997) in which ensheathing glia were transplanted as a mini-bridge, into a very local area where only the corticospinal tract was damaged on one side between the first and second cervical segments in adult rats. The ensheathing glia appeared to be effective as a bridge because corticospinal axons, labelled by anterograde tracing, were observed in parallel bundles that traversed the full rostrocaudal extent of the lesion and at least 2-3 mm beyond in the caudal cord. An important aspect of this study was the behavioural testing of forelimb function after lesioning and transplantation and the correlation with morphological assessment of the extent of tract damage. It was critical to detect if the lesioned side contained any remaining fibres because as few as 1% corticospinal axons supported normal directed forelimb reaching (Li *et al.*, 1997). The authors found that in four animals, which functioned normally following complete lesioning, transplanted cells were present along the entire extent of the lesion area. In three animals in which forelimb reaching did not return to normal, again there was complete destruction of the tract but ensheathing glia did not bridge the entire lesion rostrocaudally. Thus, normal function was correlated with the presence of a continuous bridge of ensheathing glia across the lesion. These results also are consistent with a number of earlier studies concluding that far fewer fibres than normal will be required to restore function after spinal cord injury. This study and the previous one emphasize that the ensheathing glial cell type is a very promising one for achieving regeneration in the CNS and will be worthy of considerable study to understand the basis for its favourable influence on CNS axonal growth. Moreover, the Li *et al.* (1997) study showed that function can be restored by ensheathing glia transplantation.

It was concluded in the Li *et al.* (1997) study that the transplanted cells ensheathed and myelinated the regenerated axons; ensheathment and myelination of fibres spanning the lesion were not seen without ensheathing glia transplantation. The ensheathing glia were not labelled in such a way as to identify them in the elecron microscope (Pallini, 1998). We found in Schwann cell grafts, into which we know ensheathing glia migrated, that it was impossible to distinguish them from Schwann cells electron microscopically. Whereas Schwann cells could have entered lesion areas, without ensheathing glia there were no regenerating fibres for the Schwann cells to wrap around, thus accounting for the lack of ensheathment and myelination in the control animals. The ensheathing glia could have supported axonal regeneration, whereas Schwann cells could have ensheathed and myelinated the regenerated fibres.

How might we compare the prospects of Schwann cells and ensheathing glia for transplantation therapy? Whereas Schwann cells have shown a remarkable capability to enhance regeneration in the CNS, they are not as migratory in the CNS as are ensheathing glia (Franklin & Barnett, 1997). Where Schwann cells approach an astrocyte-containing region, they do not enter and an inhibitory barrier appears to form such as at the dorsal root entry zone. Information has been presented above to show that this barrier can be overcome by ensheathing glia which are able to migrate substantial distances from the site of transplantation. The Schwann cells, of course, can form myelin around regenerated axons, but it appears that ensheathing glia also have this capability (Franklin & Barnett, 1997). We do not yet know in what way ensheathing glia appear to be so effective in promoting axonal regeneration in the CNS. Bixby *et al.* (1988) demonstrated that the *in vitro* growth of axons from chick ciliary ganglion cells on rat Schwann cells is regulated by the cell adhesion molecules, L1 and N-cadherin, and a beta-1 integrin component. The L1 molecule also has been demonstrated to be critical for growth of mammalian CNS (rat retina) neurites on Schwann cells (Kleitman *et al.*, 1988). Ensheathing glia express L1 (see review by Ramón-Cueto & Avila, 1998). Considered to be a unique glial cell type, they exhibit characteristics of both Schwann cells and astrocytes (reviewed in Ramón-Cueto & Avila, 1998). Does their enhancement of growth depend upon **i)** their ensheathment capability, thereby protecting the axons from the surrounding milieu, **ii)** their provision of growth-promoting membrane surfaces, or **iii)** their modification of the milieu by secreted molecules that favour regeneration? This promises to be a worthwhile area of research. As mentioned above, Schwann cells can be generated in very large numbers, particularly for autotransplantation. How easily large numbers of ensheathing glia can be generated remains to be determined. It will be more difficult to obtain human ensheathing glia than human Schwann cells for culture and expansion.

Key Points

Types of bridges

- Peripheral nerve.
- Addition of other material eg acidic fibroblast growth factor (FGF-1), NGF.
- Purified Schwann cells.
- Addition of methylprednisolone and/or growth factors eg BDNF, NT-3.
- Addition of olfactory ensheathing glia.
- Fibroblasts genetically engineered to produce and secrete growth factors.
- Fetal CNS tissue.
- CNS-derived neural cell lines, immortalized (but not transformed) by retroviral vectors.

Other cell types

The focus of this chapter is to discuss peripheral nerve and Schwann cell bridges, primarily in completely transected spinal cord. But not to at least mention briefly the possibility of other bridging material would be amiss. For example, the pioneering efforts of the Gage-Tuszynski team to genetically modify fibroblasts to provide neurotrophic factors to spinal cord have been successful and have introduced the promise of this approach (refs. in Grill *et al.*, 1997). In much of the work, the modified fibroblasts were introduced into the cord without an initial injury, with the result that fibres grew into the transplanted mass of cells, the type of fibres depending upon the NGF, BDNF, NT-3, or basic FGF genes introduced.

In one of the most recent studies, Grill *et al.* (1997) severed bilaterally either the corticospinal tract or the entire dorsal half of the midthoracic cord (thus removing additional tracts). Only the more extensive lesioning led to long-lasting (2 months) functional deficit as observed with a locomotion grid task that requires sensorimotor integration, thus partially reflecting the function of supraspinal motor projections to the cord. Having established this, the investigators next grafted fibroblasts genetically modified to secrete NT-3 into cavities created by the more extensive dorsal hemisectioning procedure. Up to three months later, significant sensorimotor functional improvement occurred only in the NT-3 fibroblast- grafted animals. Accompanying this was a significant increase in corticospinal axon growth at and up to 8 mm distal to the injury site. In no case did corticospinal axons enter grafts; they were only seen in the grey matter. Neurofilament immunostained fibres, however, were observed in either control or NT-3-secreting grafts; these included sensory afferents and serotonergic and noradrenergic axons, as determined by immunolabeling. Specific activation of trkC receptors on corticospinal axons likely accounts for their response to NT-3.

The transplantation of fetal CNS tissue into the adult host, studied for many years, particularly by the Reier and Anderson and Bregman groups, has been well reviewed (e.g., Reier *et al.*, 1992; Bregman, 1994). This type of tissue survives well, differentiates, becomes vascularized, fills the lesion cavity, becomes integrated into the spinal cord and, of course, provides neurons in addition to nonneuronal cells. Host fibres are observed to grow into this type of transplant, and neurons in the transplanted tissue extend axons into the host tissue. Thus, this type of transplant does not seem to serve as a bridge as is described in this chapter, i.e. a bridge carrying axons across the entire area of injury and into the cord. Nonetheless, CNS tissue transplants have led to some degree of functional recovery as assessed by behavioural and electrophysiological techniques (Bregman, 1994; see also chapter 2 by Saunders & Dziegielewska).

Additional candidates for transplantation, particularly to replace damaged neurons, are CNS-derived neural cell lines, immortalized (but not transformed) by retroviral vectors. The use of these cells has shown that they are able to respond to local environmental cues and that the adult CNS retains the capacity to direct their differentiation to resemble surrounding host nerve cells. Other possibilities for transplantation into areas of injury are stem cells to replace neuronal and non-neuronal populations. Considerable work is in progress to determine how best to increase their numbers and to steer their differentiation into astrocytes, oligodendrocytes, or neuronal populations. They also may be genetically engineered to provide key molecules for regeneration. An exciting direction of work will be investigating ways in which surviving tissue may be treated to enhance the number and differentiation of stem cells now known to be present in the CNS. All of these options may be preferable to the transplantation of human fetal tissue which is fraught with numerous difficulties, among them ethical concerns and availability of sufficient material. The reader is referred to reviews by leading investigators in these areas, Gage (Gage *et al.*, 1995), McKay (1997), and Whittemore & Snyder (1996).

Prospectus

Due to the CNS axonal regeneration-promoting activity of peripheral nerve and Schwann cell bridges, they may play an important role in future therapeutic strategies developed to treat human spinal cord injury. But these bridges will undoubtedly be only one component of a multifaceted approach that will include the administration of neuroprotective and neurotrophic factors as well, either by infusion or genetic manipulation of cells. Regeneration is substantially improved when these constituents are added to peripheral nerve or Schwann cell bridges. Ramón y Cajal's (1928) speculation in the early 1900s that nutritive (neurotrophic) and orienting (neurotropic) substances will have to be supplied for regeneration to be successful in the adult CNS hinted at the necessity of a combination therapy. More has to be learned about treatments to over

Key Points

Influence of different bridges on types of axons that grow into/across them

- All of the bridges listed in Box on p. 184 have been reported to promote axon growth.
- In many cases the axon growth was confined to the graft.
- Types of axon involved appear to depend on nature of bridge.
- Some types of bridge promoted growth of serotonergic (5HT) and noradrenergic fibres.
- Few seem to promote growth of corticospinal axons, but some did, eg peripheral nerve on its own did not but required additional treatment such as FGF-1 (Cheng *et al.*, 1996); olfactory ensheathing glia (Li *et al.* (1997); NT-3 fibroblast- grafts (Grill *et al.* (1997); CNS transplants in immature animals (chapter 2 by Saunders & Dziegielewska).

come white matter inhibitory activity and to promote guidance to appropriate target areas of the spinal cord. The mechanisms for the successful influence of the ensheathing glia of the olfactory system on CNS regeneration should be determined. The increasing incidence of reports describing advances in spinal cord repair strategies and the accelerating pace of discoveries elucidating basic mechanisms of axonal growth, including repulsion and guidance of axons in developing spinal cord, provide the basis for the current hope and optimism that the damaged mammalian spinal cord can be repaired successfully.

Key Points

Influence of bridges on function and future prospects.

- Most types of bridges do not appear to have had any detectable effect on locomotion.
- Bridges that have been reported to improve locomotor function are: peripheral nerve + FGF-1 (Cheng *et al.*, 1996); Schwann cell cable grafts (Guest *et al.* (1997a); olfactory ensheathing glia (Li *et al.* (1997); NT-3 fibroblast- grafts (Grill *et al.* (1997).
- Bridges will be only one component of a multifaceted approach to repair of spinal injury.

Acknowledgments

Experiments in the author's laboratory reviewed here were performed by Drs. X.M. Xu, A. Chen, V. Guénard, M. Oudega, P. Menei, C. Montero-Menei, J. Guest, A. Ramón-Cueto, and G. Plant; some have been accomplished in collaboration with Dr. N.

Kleitman. Expert help for these studies was provided by K. Akong, M. Bates, J.-P. Brunschwig, R. Camarena, E. Cuervo, A. Gomez, K.J. Klose, A. Rao, D. Santiago, C. Vargas, and A. Weber. Channels were provided by Dr. P. Aebischer (Université de Lausanne) and CytoTherapeutics, Inc; human nerve, by Les Olson of the Organ Procurement Team at the University of Miami School of Medicine; neurotrophins, by Regeneron, Inc.; the 1BDNF-1B producer line, by Dr. X. Breakefield (Harvard Medical School). Drs. R. Bunge and S. Whittemore offered valuable counsel; Dr. N. Kleitman made comments to improve this review. C. Rowlette was responsible for word processing. This work was supported by NIH grants NS28059 and NS09923, the American Paralysis Association, the Hollfelder, Rudin, and Heumann Foundations, and The Miami Project to Cure Paralysis. Drs. Xu and Oudega have been Werner Heumann Memorial International Scholars; Dr. Guest was a Fellow of the American Association of Neurological Surgeons; Drs. Menei and Montero-Menei were funded by IRME to work in the Bunge Laboratory; and Dr. Ramón-Cueto was supported by the Human Frontier Sciences Program.

References

Aguayo A J (1985) Axonal regeneration from injured neurons in the adult mammalian central nervous system. In Synaptic Plasticity, edited by CW Cotman, pp 457–484, New York, The Guilford Press

Bartholdi D and Schwab M E (1995) Methylprednisolone inhibits early inflammatory processes but not ischemic cell death after experimental spinal cord lesion in the rat. Brain Research 672, 177–186

Basso D M Beattie M S and Bresnahan J C (1995) A sensitive and reliable locomotor rating scale for open field testing in rats. Journal of Neurotrauma 12, 1–21

Bixby J L Lilien J and Reichardt L F (1988) Identification of the major proteins that promote neuronal process outgrowth on Schwann cells *in vitro*. Journal of Cell Biology 107, 353–362

Brook G A Lawrence J M Shah B and Raisman G (1994) Extrusion transplantation of Schwann cells into the adult rat thalamus induces directional host axon growth. Experimental Neurology 126, 31–43

Bregman B S (1994) Recovery of function after spinal cord injury: Transplantation strategies. In Functional Neural Transplantation edited by SB Dunnett and A Bjorklund, pp 489–529, New York, Raven Press, Ltd

Broton J G Xu X M Bunge M B Lutton S Cuthbert T and Calancie B (1996) Hindlimb movements of adult rats with transected spinal cords. Society for Neuroscience Abstracts 22, 1096

Bunge M B and Kleitman N (1998) Neurotrophins and neuroprotection improve axonal regeneration into Schwann cell transplants placed in transected adult rat spinal cord. In CNS Regeneration: Basic Science and Clinical Advances, edited by MH Tuszynski and JH Kordower, pp. 631–646. New York, Academic Press

Bunge R P and Hopkins J M (1990) The role of peripheral and central neuroglia in neural regeneration in vertebrates. Seminars in Neuroscience 2, 509–518

Bunge R P Puckett W R and Hiester E D (1997) Observations on the pathology of several types of human spinal cord injury, with emphasis on the astrocyte response to penetrating injuries. In Advances in Neurology Vol 72: Neuronal Regeneration, Reorganization, and Repair, edited by FL Seil, pp 305–315, Philadelphia, Lippincott-Raven

Chen A Xu X M Kleitman N and Bunge M B (1996) Methylprednisolone administration improves axonal regeneration into Schwann cell grafts in transected adult rat thoracic spinal cord. Experimental Neurology 138, 261–276

Cheng H Cao Y and Olson L (1996) Spinal cord repair in adult paraplegic rats: Partial restoration of hind limb function. Science 273, 510–513

David S and Aguayo A J (1981) Axonal elongation into peripheral nervous system "bridges" after central nervous system injury in adult rats. Science 214, 931–933

Davies S J A Fitch M T Memberg S P Hall A K Raisman G and Silver J (1997) Regeneration of adult axons in white matter tracts of the central nervous system. Nature 390, 680–683

Eckenstein F B (1994) Fibroblast growth factor in the nervous system. Journal of Neurobiology 25, 1467–1480

Fernandez E Pallini R and Mercanti D (1990) Effects of topically administered nerve growth factor on axonal regeneraton in peripheral nerve autografts implanted in the spinal cord of rats. Neurosurgery 26, 37–42

Franklin R J M and Barnett S C (1997) Do olfactory glia have advantages over Schwann cells for CNS repair? Journal of Neuroscience Research 50, 665–672

Gage F H Ray J and Fisher L J (1995) Isolation, characterization, and use of stem cells from the CNS. Annual Review of Neuroscience 18, 159–192

Grill R Murai K Blesch A Gage F H and Tuszynski M H (1997) Cellular delivery of neurotrophin-3 promotes corticospinal axonal growth and partial functional recovery after spinal cord injury. Journal of Neuroscience 17, 5560–5572

Guénard V Xu X M and Bunge M B (1993) The use of Schwann cell transplantation to foster central nervous system repair. Seminars in Neuroscience 5, 401–411

Guest J D Rao A Olson L Bunge M B and Bunge R P (1997a) The ability of human Schwann cell grafts to promote regeneration in the transected nude rat spinal cord. Experimental Neurology 148, 502–522

Guest J D Hesse D Schnell L Schwab M E Bunge M B and Bunge R P (1997b) Influence of IN-1 antibody and acidic FGF-fibrin glue on the response of injured corticospinal tract axons to human Schwann cell grafts. Journal of Neuroscience Research 50, 888–905

Houle J D Wright J W and Ziegler M K (1994) After spinal cord injury, chronically injured neurons retain the potential for axonal regeneration. In Neural Transplantation, CNS Neuronal Injury, and Regeneration. Recent Advances, edited by H Teitelbaum, and KN Prasad, pp 103–118, Boca Raton, CRC Press

Keirstead H S Dyer J K Sholomenko G N McGraw J Delaney K R and Steeves J D (1995) Axonal regeneration and physiological activity following transection and immunological disruption of myelin within the hatchling chick spinal cord. Journal of Neuroscience 15, 6963–6974

Kleitman N Simon D K Schachner M and Bunge R P (1988) Growth of embryonic retinal neurites elicited by contact with Schwann cell surfaces is blocked by antibodies to L1. Experimental Neurology 102, 298–306

Li Y Field P M and Raisman G (1997) Repair of adult rat corticospinal tract by transplants of olfactory ensheathing cells. Science 277, 2000–2002

Lindsay R M Wiegand S J Altar C A and DiStefano P S (1994) Neurotrophic factors: from molecule to man. Trends in Neuroscience 17, 182–190

Martin D Robe P Franzen R Delrée P Schoenen J Stevenaert A and Moonen G (1996) Effects of Schwann cell transplantation in a contusion model of rat spinal cord injury. Journal of Neuroscience Research 45, 588–597

McKay R (1997) Stem cells in the central nervous system. Science 276, 66–71

Menei P Montero-Menei C Whittemore S R Bunge R P and Bunge M B (1998) Schwann cells genetically modified to secrete human BDNF promote enhanced axonal regrowth across transected adult rat spinal cord. European Journal of Neuroscience 10, 607–621

Oudega M and Hagg T (1996) Nerve growth factor promotes regeneration of sensory axons into adult rat spinal cord. Experimental Neurology 140, 218–229

Oudega M Vargas C G Weber A B Kleitman N and Bunge M B (1999) Long-term effects of methylprednisolone following transection of adult rat spinal cord. European Journal of Neuroscience, 11, 2453–2464

Pallini R (1998) Anatomy of regenerating axons (letter). Science 280, 181–182

Plant G W Dimitropolou A Bates M L and Bunge M B (1998) The expression of inhibitory proteoglycans following transplantation of Schwann cell grafts into completely transected adult rat spinal cord. Society for Neuroscience Abstracts 24, 69

Raisman G (1985) Specialized neuroglial arrangement may explain the capacity of vomeronasal axons to reinnervate central neurons. Neuroscience 14, 237–254

Ramón y Cajal S (1928) Degeneration and Regeneration of the Nervous System. New York, Oxford University Press (translated by RM May)

Ramón-Cueto A and Avila J (1998) Olfactory ensheathing glia: Properties and function. Brain Research Bulletin 46, 175–187

Ramón-Cueto A Plant G Avila J and Bunge M B (1998) Long-distance axonal regeneration in the transected adult rat spinal cord is promoted by olfactory ensheathing glia transplants. Journal of Neuroscience 18, 3803–3815

Reier P J Stokes B T Thompson F J and Anderson D K (1992) Fetal cell grafts into resection and contusion/compression injuries of the rat and cat spinal cord. Experimental Neurology 115, 177–188

Richardson P M McGuinness U M and Aguayo A J (1980) Axons from CNS neurones regenerate into PNS grafts. Nature 284, 264–265

Schnell L and Schwab M E (1993) Sprouting and regeneration of lesioned corticospinal tract fibres in the adult rat spinal cord. European Journal of Neuroscience 5, 1156–1171

Schwab M E and Bartholdi D (1996) Degeneration and regeneration of axons in the lesioned spinal cord. Physiological Reviews 76, 319–370

Tuszynski M H Weidner N McCormack M Miller I Powell H and Conner J (1998) Grafts of genetically modified Schwann cells to the spinal cord: Survival, axon growth, and myelination. Cell Transplantation 7, 187–196

Whittemore S R and Snyder E Y (1996) Physiological relevance and functional potential of central nervous system-derived cell lines. Molececular Neurobiology 12, 13–38

Xu X M Guenard V Kleitman N and Bunge M B (1995a) Axonal regeneration into Schwann cell-seeded guidance channels grafted into transected adult rat spinal cord. Journal of Comparative Neurology 351, 145–160

Xu X M Guenard V Kleitman N Aebischer P and Bunge M B (1995b) A combination of BDNF and NT-3 promotes supraspinal axonal regeneration into Schwann cells grafts in adult rat thoracic spinal cord. Experimental Neurology 134, 261–272

Xu X M Chen A Guenard V Kleitman N and Bunge M B (1997) Bridging Schwann cell transplants promote axonal regeneration from both the rostral and caudal stumps of transected adult rat spinal cord. Journal of Neurocytology 26, 1–16

Xu X M Zhang S-X Li H Aebischer P and Bunge M B (1999) Regrowth of axons into the distal spinal cord through a Schwann cell-seeded mini-channel implanted into hemisected adult rat spinal cord. European Journal of Neuroscience 11, 1723–1740

Ye J-H and Houle J D (1997) Treatment of the chronically injured spinal cord with neurotrophic factors can promote axonal regeneration from supraspinal neurons. Experimental Neurology 143, 70–81

Young W (1996) Spinal cord regeneration. Science 273, 451

9

Use of cell/polymer hybrid structures as conduits for regenerative growth in the central nervous system

Alan R Harvey

Introduction

The adult mammalian central nervous system (CNS) has only a limited capacity for regenerative growth after injury, thus CNS damage usually results in significant and long-lasting functional impairments. Cerebrovascular accidents or traumatic injury to grey matter obviously result in loss of neurons and degeneration of associated nerve pathways. Injury to fibre tracts however, such as after a stroke in the internal capsule or after a concussive or compressive spinal cord injury, may initially spare most of the projecting neurons. In these instances, reconstitution of damaged circuitry may be possible if neurons with damaged processes are supported and provided with an environment that promotes regrowth.

The focus of this chapter is on the construction of prosthetic cell/polymer matrices for use as bridges across degenerative injury sites in the brain or spinal cord and to act as conduits for regenerative axonal growth. Immediately after CNS injury, it may prove possible in the acute phase to minimize any secondary degenerative changes and prevent the formation of scar tissue, thus limiting the need for extensive bridging grafts. However, in chronic CNS injury situations, when surgical intervention is attempted some time after the injury has occurred, there will almost certainly be scarring and/or extensive cavitation in white matter tracts, resulting in the physical separation of normal tissue either side of the injury site. In these circumstances there is a particular need to provide regrowing axons with bridging implants or guidance channels.

Key points

- Importance of providing local CNS environment that promotes growth of axons when injured.
- Prosthetic cell/polymer matrices as bridges across degenerative injury sites, especially in chronic injuries.

Use of polymer bridges in the peripheral nervous system

The possible use of artificial nerve guides to bridge CNS injury sites has ample precedent in reconstructive surgical procedures in the peripheral nervous system. Fields *et al.* (1989) comprehensively reviewed the history of nerve tubulization, a technique developed to bridge gaps between distal and proximal stumps of damaged peripheral nerves. Many early studies experimented with bridging materials derived from natural or biological sources. More recently, a variety of synthetic nerve guides have been tested; these include semi-permeable acrylic polymers, polyester, polyglactin, and silicon rubber and plastics. In the peripheral nervous system such inert, biocompatible tubular nerve guides have been successfully used to bridge gaps in severed nerves in experimental animals as well as in humans.

Empty polymer tubes attached between distal and proximal peripheral nerve become filled with fibrin matrix, forming a scaffold that permits the infiltration of Schwann cells, fibroblasts, endothelial cells and, eventually, regenerative growth of axons. The advantages of polymer nerve guides are that they can be pre-filled with components of the extracellular matrix and/or specific cell types, they can be used as a source of neurotrophic substances, and the synthetic materials can be modified to alter surface polar groups, adhesivity and even develop electrical fields along the length of the implant (reviewed in Fields *et al.*, 1989).

Polymer delivery systems and cell encapsulation in the CNS

There is also an extensive literature on the potential use of slow polymer release systems and polymer-encapsulated cells for the long-term release of neuroactive molecules in the CNS (reviewed in Tan & Aebischer, 1996). The emphasis here is not so much on bridging structures in tracts, but on the chronic, site-directed release of factors that might prevent or reduce neuronal death associated with neurodegenerative diseases such as Parkinson's disease, Alzheimer's disease and motor neuron disease. The encapsulation technology is designed to surround grafted cells with a selectively permeable membrane with pores large enough to allow inward diffusion of nutrients and outward diffusion of specific neuroactive molecules, but small enough to prevent recognition or contact by host immunocompetent cells (Tan & Aebischer, 1996). Thus non-neural cells, cell lines, even xenogeneic cells can be transplanted into host CNS and their encapsulation prevents immunological rejection as well as cellular overgrowth and potential tumour formation.

Key points

- Polymer delivery systems for slow release of neurotrophic factors.
- Polymer encapsulated cells for the long-term release of neuroactive molecules in the CNS.
- Cell types used: PC12, fibroblasts, myoblasts, kidney cells.
- Cells genetically engineered to release specific neurotrophins eg NGF, GDNF.

A variety of cells has been used in experimental graft studies including PC12 cells, fibroblasts, myoblasts and kidney cells. For example, Deglon *et al.* (1996) showed that encapsulated C2C12 myoblasts transfected with the human ciliary neurotrophic factor (CNTF) gene can partially rescue motoneurons after axotomy-induced death. A clever aspect of this work is that, prior to engraftment in the CNS, the myoblast cell lines can be differentiated into a non-mitotic stage by exposure to low serum. Others, using a xenogeneic baby hamster kidney (BHK) cell line transfected with the human nerve growth factor (NGF) gene, have reported continued *in vivo* release of NGF from the implanted, encapsulated cells for more than one year (Winn *et al.*, 1996). Recently it has been shown that polymer-encapsulated cells genetically modified to release glial cell line-derived neurotrophic factor (GDNF) can be co-grafted with fetal neural tissue grafts, resulting in better fetal graft survival and improved functional recovery in host animals (Sautter *et al.*, 1998).

These are exciting results, but their success depends on isolating grafted cells from the host CNS environment. While it is possible to envisage chains of microencapsulated genetically engineered cells laid down in rows within damaged white matter tracts, most studies using cell/polymer hybrid structures as conduits for axonal regrowth in the CNS require that grafted cells are free within the polymer scaffold and thus able to interact directly with the growth cones of regenerating axons. This is exactly the situation in tubular nerve repair in the peripheral nervous system.

One encapsulation study is, however, of more immediate relevance to tract regeneration. Cadelli & Schwab (1991) entrapped a hybridoma (tumorogenic) cell line within Millipore filter capsules prior to transplantation into lesion cavities. The so-called ravioli (Cadelli & Schwab, 1991) were placed above extracellular matrix bridges that had been inserted into large aspiration lesions of the rat fimbria/fornix. The cell line secreted the antibody IN-1, which is reported to block the action of growth-inhibitory membrane proteins expressed by oligodendroglia and associated myelin (see also Fitch & Silver, chapter 4, this volume). In the presence of the encapsulated cell line, there was increased regenerative growth of lesioned cholinergic septohippocampal (acetyl-cholinesterase-positive) fibres across the bridges and increased ingrowth into target hippocampal tissue.

Cell/polymer hybrid structures used as bridges in the CNS: two-dimensional structures

One of the earliest examples of the use of a synthetic polymer to bridge a lesion defect in the CNS was published by Silver & Ogawa (1983). In their study, pieces of nitro-cellulose paper (Millipore, 0.45μm pore size) were inserted between the cerebral hemispheres of fetal or early postnatal mice that had been made acallosal after intra-uterine surgery. The implants became coated by cells, probably astrocytes, and supported the growth of axons from one hemisphere to the other. Due to the difficulty of the microsurgery and the ability to correctly orientate the implants within the fetal or neonatal brains, the success rate was low — less than 20% of the cellulose bridges spanned the midline and were appropriately positioned. Nonetheless, when properly aligned, these polymer scaffolds did occasionally support commissural axons and allow the surgical recreation of a corpus callosum-like structure.

Other studies suggest that cell-free nitrocellulose papers are not always effective in supporting axonal growth. For example, no growth of corticospinal axons was seen on the surface of uncoated nitrocellulose papers inserted into thoracolumbar levels of P1–P4 (day of birth = P0) rat spinal cord (Schreyer & Jones, 1987), and there was no or only minimal growth of retinal ganglion cell axons on similar implants placed in the lesioned rat optic tract (Chen *et al.*, 1991). Schreyer & Jones (1987) did report that a few nitrocellulose papers that had been pre-coated with laminin or cultured neonatal spinal cord cells supported corticospinal axons, however papers coated with cultured cortical cells were ineffective. Note here that implants were placed into thoracolumbar spinal cord prior to the arrival of most if not all developing corticospinal axons, thus this study examined mainly *de novo* rather than regenerative growth. Furthermore, when growth was seen, axons often reversed direction and grew along the paper in a rostral direction. Some fibres appeared to grow into the polymer network itself, but no axons grew along the entire length of any implant. Growth was seen on six implants, however a total of 247 rats underwent surgery, indicating a very low success rate.

Astrocyte-nitrocellulose polymers have also been used as bridges for dorsal root fibres in the adult rat spinal cord (Kliot *et al.*, 1990). Roots were crushed close to the dorsal root reentry zone (the interface between the peripheral and central nervous system in the spinal cord) and Millipore pennants coated with astrocytes derived from fetal spinal cords were inserted between the nerve roots and the cord. In 11 of the 26 operated animals, horseradish peroxidase-labelled sensory axons regrew into the spinal cord white matter, and in 6 rats (23%) axons were visualized in grey matter. Again however, the success rate was low; in five of these six animals axon ingrowth was restricted to tissue immediately adjacent to the coated implant. This ingrowth was somewhat similar to that seen in one of four (25%) animals that received uncoated (cell-free) Millipore implants. Only in one rat with an astrocyte-coated implant was more robust regrowth seen in the dorsal horn.

In optic tract lesion studies in young (P13–P18) rats, some regrowth of retinal ganglion cell axons was seen on astrocyte-coated nitrocellulose papers, however the greatest and most consistent regrowth was seen on papers pre-coated with cultured Schwann cells (Chen *et al.*, 1991). Of the 42 implants, 30 were attached to the dorsal lateral geniculate nucleus/optic tract rostral to the lesion, and 15 of these supported retinal axon regrowth. Growth of 500μm or more was seen in nine animals, up to a maximum distance of about 1.2mm. However, even in the best cases, retinal axons were not seen to cross the whole width of the nitrocellulose bridges (cf. Schreyer & Jones, 1987). We noted, as did others (Silver & Ogawa, 1983; Schreyer and Jones, 1987), that initially uncoated two-dimensional bridges become covered with cells after implantation into CNS tissue. Do host cells also migrate on to pre-coated Millipore implants, and if so what is the fate of the donor glial cells initially attached to the polymer?

To address this issue, we implanted astrocyte-coated nitrocellulose papers into cortical lesion cavities in adult mice. Implants coated with neonatal male cortical astrocytes were placed in female hosts, and papers coated with neonatal female astrocytes were implanted into male hosts (Harvey *et al.*, 1993). Animals were perfused 16–18 days later and tissue sections through the implanted polymers were hybridized with a Y-chromosome specific probe to identify male cells. Male to female grafts allowed visualization of donor glia and their behaviour after transplantation, female to male grafts allowed an analysis of how host cells responded to the presence of the glia-coated implants. Even by 16-18 days we found substantial intermixing of cells, with many donor glia migrating away from the polymer surface into surrounding host tissue, and host cells (mostly astrocytes) migrating on to both sides of the nitrocellulose papers. These observations suggest that any growth-promoting effects of cells initially associated with the implants may be short-lived, and that both donor and host glia may influence the extent of axonal regrowth across two-dimensional polymer bridges.

Key points

- Synthetic polymers: nitrocellulose paper, alone or coated with astrocytes.
- Tubular 3-dimensional cell/polymer implants, eg polycarbonate tubes containing Schwann cells; inclusion of supporting matrix.
- Cell-seeded guidance channels; addition of IGF-1 and PDGF
- Hydrogel 3-dimensional matrices.

Tubular cell/polymer implants in the CNS

Use of tubular, three-dimensional bridging structures reduces the loss of growth-promoting cells from implants and allows the transplanted cells to be placed within conduits that can be oriented to direct and guide regrowing axons to specific destinations. Montgomery & Robson (1993) constructed 1–2mm diameter tubes from sheets of polycarbonate paper (12μm pore size) and filled them with cultured Schwann cells. One end of the tubes was implanted into the thalamus, the other end was left protruding extracranially, outside the CNS. After 1–6 months, a cord of dense tissue was found in the centre of the tubes which looked somewhat like a peripheral nerve. The tissue cord contained Schwann cells, fibroblasts, blood vessels, laminin, collagen and myelinated and unmyelinated axons. The myelin was of peripheral origin. Tissue cords were surrounded by a perineurial-like sheath. No indication of the number of axons within the polymer/Schwann cell tubes was given, however in two rats horseradish peroxidase was placed on the extracranial end of the implants. In each animal about 200 CNS neurons were retrogradely labelled, most situated near to the cranial opening of the Schwann cell-filled tubes. Similar tubes filled with astrocytes supported less fibre ingrowth.

We were interested to find out how effective these Schwann cell/polycarbonate tubes would be if implanted entirely intracranially and used as bridges within the CNS. Polycarbonate tubes (0.5mm diameter, 2–2.5mm long) were made, coated with laminin and poly-l-lysine, filled with Schwann cells and then inserted into optic tract lesions (reviewed in Harvey *et al.*, 1995). The tubes were oriented rostrocaudally such that they formed a bridge between the dorsal lateral geniculate nucleus rostrally and the superior colliculus caudally. Concentrated Schwann cell conditioned medium and basic fibroblast growth factor were also applied to the lesion site. Regrown retinal ganglion cell and other axons (the latter identified by immunostaining with the antibody RT97, which recognizes 200kD neurofilaments) were consistently seen in association with grafted Schwann cells within the tubes. Retinal axons regrew for up to 1mm but did not reach the distal (tectal) end of the implants. Importantly however, the greatest number of axons was seen after short survival times (11–19 days post-implantation). At longer survival times (82–117 days), implants still contained viable Schwann cells but there was less cellular material and fewer axons inside the tubes. There appeared to be a collapse of the cellular structure within the tubes and a dense cord of tissue (200-250μm thick) was formed which extended the length of the implant. This was somewhat similar to the tissue cords described previously (Montgomery & Robson, 1993). We obtained ultrastructural evidence for oligodendroglial migration into the polycarbonate tubes, but unlike Montgomery & Robson (1993) we did not see evidence of either peripheral or central myelination of axons within the intracranial bridging implants. The reasons for these differences in cellular organization within the tubes are unclear, but they may relate to differences in the initial packing density of the Schwann cells and/or to the intracranial versus extracranial positioning of the implants. In the study of Montgomery & Robson (1993) the extent to which non-CNS cells migrated into the extracranial opening of the implants, and whether or not other peripherally derived axons also made up some of the population of myelinated fibres within the polycarbonate tubes, was not determined.

Because of the apparent instability of the cellular architecture within our intracranial polycarbonate tubes, we added an additional supporting extracellular matrix for the transplanted Schwann cells within the lumen (preliminary data reviewed in Harvey *et al.*, 1995; Plant & Harvey in preparation). Cultured Schwann cells were first seeded on sheets of lens capsule-derived extracellular matrix prior to insertion into the polycarbonate tubes. We also used the tubes as a physical support and guide for pieces of predegenerate peripheral nerve implanted into lesioned rat optic tract (Harvey *et al.*, 1995; Plant & Harvey in preparation). To date, the most consistent regrowth of retinal ganglion cell and axons of other neuronal types within polycarbonate implants has been seen in tubes containing the Schwann cell-coated lens capsule extracellular matrix (Figure 1). In these implants, some regenerating axons become myelinated, either by grafted Schwann cells or more rarely by infiltrating host oligodendroglia.

The importance of some kind of supporting matrix within tubular polymer guidance channels has also recently been demonstrated in the injured adult rat spinal cord (Xu *et al.*, 1995; Oudega et al. 1997). In these studies, cultured Schwann cells are first suspended in Matrigel, a commercially available mixture of basal lamina components, and then inserted into semipermeable PAN/PVC (60:40 acrylonitrile:vinylchloride) copolymer tubes (2.6mm in diameter). In most studies from the Miami group, the proximal end of the tube is attached to the transected rostral part of the thoracic cord and the distal end is capped rather than attached to the caudal cord stump. This approach allows the implantation of long (10mm) guidance tubes and permits substantial elongation of axons within the channels; however capping of the tubes isolates any regrowing axons from possible target-derived trophic influences emanating from caudal spinal cord (see also Bunge, chapter 8, this volume).

Xu *et al.* (1995) examined these cell-seeded guidance channels one month after implantation into lesioned rat spinal cords. The cords were transected at T8 and the cord segments from T9-T11 were also removed prior to insertion of the guidance

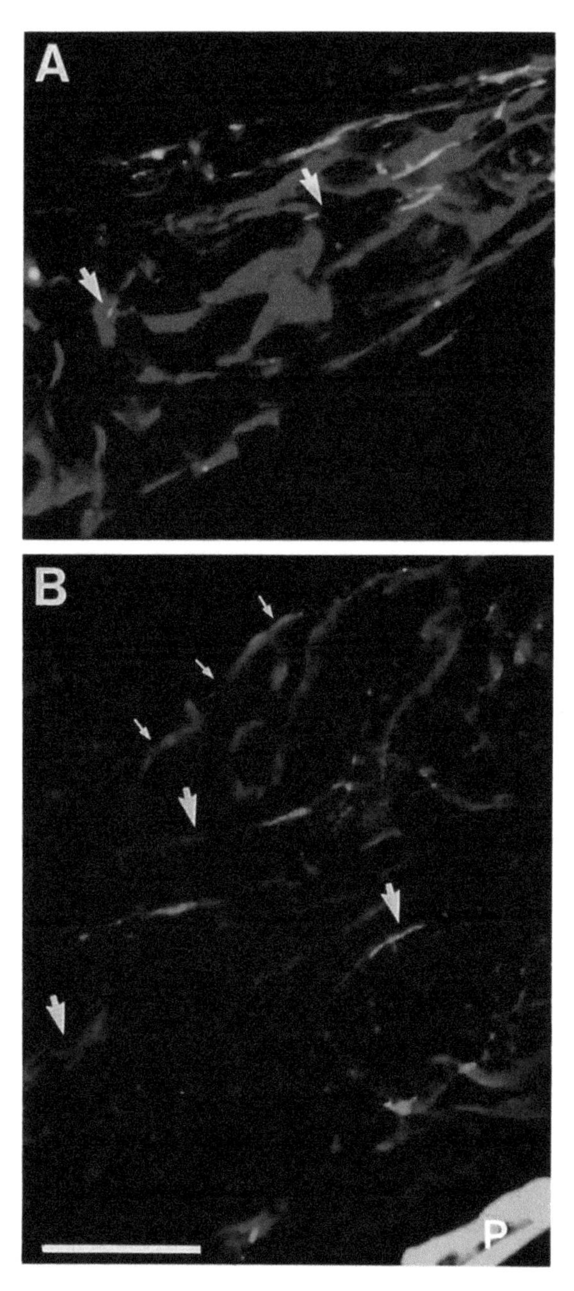

Figure 1 A, B: Confocal photomicrographs showing single or fasciculated CNS axons (green/yellow, large arrows) elongating within polycarbonate tubes filled with lens capsule extracellular matrix (red, small arrows in B) that had been coated with cultured Schwann cells. The cell/polymer bridges were implanted into the lesioned rat optic tract. Axons and lens capsule extracellular matrix were visualized using anti-neurofilament (RT97) and anti-laminin antibodies respectively. In B, a wall of the polycarbonate tube (P) can be seen. Scale bar for both photomicrographs = 100μm.

channels. Myelinated and unmyelinated axons were found within the vascularized tissue cable that was present within each of the tubes. Axons were myelinated by the engrafted Schwann cells. On average, about 500 myelinated axons were found 3 mm into the guidance channels, falling to a mean of 82 axons at 9mm. The reasons for this decline in numbers along the length of the implants are unclear. Consistent with many other bridging studies, far fewer regenerating axons were seen in implants lacking viable Schwann cells. Retrograde labelling studies revealed that neurons projecting axons into the tubes were mostly spinal cord interneurons with some fibres originating in dorsal root ganglia. The majority of spinal neurons were located within 3 segments of the implanted cell/polymer tubes, although occasional labelled cells were found as far rostral as C7. No supraspinal label was seen, highlighting the difficulty of inducing regenerative growth in the mature CNS when axons are damaged some distance away from the parent cell bodies.

In a later study, in an attempt to (i) promote more axonal regrowth, (ii) stimulate Schwann cell proliferation, and (iii) enhance myelination, insulin-like growth factor-1 (IGF-1) and platelet-derived growth factor (PDGF) were added to the Matrigel/ Schwann cell mixture prior to insertion into the PAN/PVC tubes (Oudega *et al.*, 1997). Addition of these factors did indeed increase the myelinated:unmyelinated fibre ratio and increased the thickness of the myelin sheaths. Overall however, in the presence of IGF-1 and PDGF there was less axonal ingrowth into the guidance tubes, perhaps as a result of increased secondary deterioration, cavity formation and axonal dieback in these animals. The study highlights problems that may arise with the exogenous application of growth factors to cell/polymer bridging structures in the CNS; because most of these factors have multiple actions and can initiate or be involved in a number of different biochemical cascades, gains or improvements in one facet of tissue regeneration may be offset by negative effects on another element of the repair process. Note also that Matrigel, which incidentally is of tumorogenic origin, tends to contract in volume, thus the matrix inside these PAN/PVC tubes may undergo some reorganization over time.

Very recently, these same Schwann cell-filled polymer guidance tubes have been used as bridges between the distal and caudal stumps after removal of a 4mm segment of thoracic spinal cord in adult rats (Ramón-Cueto *et al.*, 1998). In a novel attempt to promote axonal regeneration, ensheathing glia from the olfactory bulb (known to provide a permissive environment for axonal growth in the adult CNS) were also transplanted into the spinal cord stumps either side of the implanted bridge. The ensheathing glia promoted long-distance regrowth of both ascending and descending axons within the injured spinal cord, perhaps by facilitating the entry and exit of axons into the guidance channels. This interesting study points towards the development of a multiple implantation strategy, using certain cell/matrix/polymer components for support and guidance of axons across an injury site, and engraftment of other cell types into normal CNS tissue to ensure sustained growth of axons beyond the damaged region (see also Bunge, chapter 8, this volume).

Note that it may be important to ensure that different types of growth-promoting cell are confined to particular locations within and beyond the lesioned area. For example, although Schwann cells are particularly effective in promoting axonal elongation, we have obtained evidence that the presence of Schwann cells within CNS grey matter can, in some circumstances, induce axons to grow into inappropriate (non-target) regions, presumably by producing growth-promoting factors that mask or compete with signals released from the target neurons themselves (Harvey & Plant, 1995). Leakage or migration of Schwann cells out of bridging matrices into surrounding neuropil may therefore affect the selectivity with which axons regrow into deafferented areas, potentially compromising any functional recovery.

Hydrogel based three-dimensional matrices in neural tissue repair

While the use of relatively rigid tubular polymer/cell implants in peripheral nerve and spinal cord repair suits the particular geometry of these structures, lesions in the CNS are often irregular in shape, they affect intersecting fibre tracts, and there is variability in the extent and pattern of tissue loss. We have made progress in the development of effective polymer/matrix/Schwann cell bridges for implantation into the optic tract (Harvey *et al.*, 1995), however variation in the exact placement of such implants can have important consequences because axonal ingrowth and outgrowth is restricted to the proximal and distal openings of the tubes. What might eventually prove to be more appropriate for tissue repair and pathway reconstruction in many parts of the brain is a biocompatible, porous three-dimensional matrix that adheres to neural tissue, has physical/elastic properties similar to the brain, and thus can provide a flexible but stable scaffold. Hydrogels, which are hydrophilic polymer-based macromolecular networks swollen in water, are a class of material that appears to fulfil many of these requirements.

In a series of pioneering studies, Woerly and colleagues used *in vitro* and *in vivo* methods to assess the potential of different types of acrylic hydrogel materials for use in transplantation and brain tissue repair (Woerly *et al.*, 1990, 1993). Early studies showed that such polymers were biocompatible. The extent of *in vivo* tissue ingrowth depended on the hydrophilic and mechanical characteristics of the polymers, the geometry and size of the interconnecting pores and the presence of biological (biopolymer) macromolecules such as collagen to promote cell attachment to the implanted matrix (Woerly *et al.*, 1990). In a later *in vitro* study, Woerly *et al.* (1993) tested a variety of hydrophilic synthetic polymers in an attempt to find a hydrogel design that provided optimal adhesion and viability of dissociated embryonic neurons and supported long-distance neuritic outgrowth. Methacrylate (HEMA, GMA) or methacrylamide (HPMA) hydrogels (homo- and copolymer series) were used, either in their unmodified form or modified to contain collagen, basement membrane proteins (Matrigel) or the aminosugar glucosamine. Addition of carbohydrate or extracellular matrix proteins to HPMA hydrogels provided optimal conditions for neuronal viability and attachment, while addition of collagen stimulated neuritogenesis (Woerly *et al.*, 1993).

More recently the effects of adding peptides or aminosugars to HPMA hydrogels has been assessed *in vivo* by implanting the modified polymers into lesion cavities in the cerebral cortex or optic tract in rats (Plant *et al.*, 1997). In short, hydrogels containing the RGD (arginine-glycine-aspartic acid) peptide sequence, which mediates binding to integrin receptors, supported the greatest amount of host cellular infiltration (mostly astrocytes and macrophages) and the most extensive axonal ingrowth. *In vivo* implantation of macroporous hydrogels that contain relevant biological signalling sequences thus results in better incorporation of the polymer into surrounding brain tissue and increased prospects for tissue repair. We are currently testing the effects of adding adhesion molecules such as L1 and NCAM to implanted hydrogels.

Use of cell/hydrogel polymer hybrid structures in CNS repair

Optimal regenerative growth requires the activation and regulation of growth-associated genes in neurons with damaged processes, and this in turn requires not only supporting matrix structures but also the presence of cells that produce relevant diffusible neurotrophic factors. Thus in addition to the cross-linking of biological sequences or macromolecules to hydrogel polymers, it might be useful to incorporate appropriate primary or genetically engineered cells into the matrices prior to implantation into CNS injury sites.

Using HEMA based polymers, we have impregnated hydrogels with either collagen I (Harvey *et al.*, 1995) or collagen IV (Plant *et al.*, 1998) and then added cultured, Hoechst-labelled Schwann cells to the pre-formed macroporous network. We have tried various ways of incorporating cells into the pre-cut polymer pieces, the aim being to get large numbers of cells broadly distributed within the matrix. It is possible to almost dry the hydrogel and then rehydrate it in medium containing a high concentration of Schwann cells. Cells can also be injected directly into the centre of each piece or the polymers can be co-cultured with cells for a number of days prior to implantation (Plant *et al.*, 1998). We have found that large numbers of Hoechst-labelled Schwann cells can survive for many weeks within HEMA sponges implanted into the brain, particularly in hydrogels containing collagen IV (Plant *et al.*, 1998). Interestingly, the type of collagen incorporated into the gels also influenced Schwann cell differentiation and the host glial reaction to the implants. Thus, while both collagen I/HEMA and collagen IV/HEMA hydrogels contained regrown (RT97 immunopositive) axons, Schwann cell myelination was only seen in polymers containing collagen IV, and there was considerably less astrocytic infiltration in these same gels.

An alternative approach to cell incorporation into hydrogels is to build the polymers around the cells — to entrap and immobilize the cells during polymerization of the macroporous network. The advantage of this approach is that a known number of cells can be incorporated in a given volume of polymer and the cells will presumably be uniformly dispersed throughout the polymerized matrix. We have evidence that differentiated cells (astrocytes, Schwann cells or fetal neurons) can survive such immobilization within HPMA-based polymers for at least 6 days *in vitro* (Woerly *et al.*, 1996). Some toxicity during the polymerization procedure was evident, much of this presumably due to the generation of free radicals, heat production and changes in osmotic pressure. However, by using low catalyst concentrations, minimizing the cells' exposure to monomer reagents and by using hydrophilic gels (which permits the rapid release of toxic products) we were able to achieve a consistent level of cell viability within the polymer network.

In an effort to avoid cytotoxic effects associated with the cross-linking of acrylic polymers, cells have been immobilized within three-dimensional agarose hydrogel matrices (Bellamkonda *et al.*, 1995). Agarose is a clear, thermoreversible hydrogel made of polysaccharides and has physical properties similar to neural tissue. Bellamkonda et al. (1995) used a hydroxyethylated agarose that gels at 17°C, thus entrapment of cells in such polymers is a relatively benign process (Bellamkonda *et al.*, 1995). This method clearly has some advantages and deserves further investigation, but it may not necessarily be useful for long-term *in vivo* implantation strategies since the agarose may depolymerize to some extent at body temperature, adversely affecting the all-important cell/polymer matrix structure.

Where to from here?

Hydrogels clearly have some potential as templates for tissue repair and for use as bridges across lesion defects in the CNS. By incorporating biomaterials, signalling molecules, growth factors and/or cells into the polymer network there are considerable possibilities for manipulation of the implanted matrices which could facilitate their use in tissue engineering in the brain and spinal cord. Lamination of alternating nonpermissive, permissive, nonpermissive gel layers may permit the creation of regionally distinct, three-dimensional neural tracts (Bellamkonda *et al.*, 1995). To date, the orientation of the interconnected pores within hydrogels is random. Organization of the macroporous network into oriented guidance channels that are aligned with injured tracts in adjacent brain or spinal cord tissue is desirable and would be an important step forward. Potentially, hydrogels might be polymerized around a template which is itself removed or degraded prior to cell incorporation and implantation into wound sites.

Growth cones respond to electromagnetic fields and in culture they preferentially grow towards cathodes. Schmidt *et al.* (1997) recently described stimulation of neurite outgrowth from PC12 cells cultured on oxidized polypyrrole (PP), an electrically conducting polymer. The data suggest that increased neurite extension resulted primarily from passage of electronic current through the polymer material. PP appears to be biocompatible and elicits little adverse tissue responses *in vivo* (Schmidt *et al.*, 1997). Earlier studies in the peripheral nervous system also showed the potential of using polymers carrying surface electrical charges (reviewed in Fields *et al.*, 1989). Whether or not polymers such as PP can be used alone or in combination with other macroporous or tubular polymers as devices to promote CNS regeneration has yet to be evaluated. Growth of regenerating peripheral nerve axons through micromachined, perforated silicon chips has also recently been reported (Zhao *et al.*, 1997). In the peripheral nervous system, such nerve/polymer interactions could theoretically be used in association with specially designed electronic circuits to control limb prostheses. To my knowledge, the potential use of such axon/sieve electrode interactions in white matter tracts in the CNS has not yet been explored.

This chapter has reviewed different types of cell/polymer hybrid structures that may prove useful as conduits for axonal regeneration in the adult CNS. Each generation of implants gives new insights into how best to synthesize the polymer networks and incorporate cells and relevant biological molecules. Within the bridges themselves, ways must also be found of (i) facilitating bi-directional axonal growth and (ii) encouraging axons that have grown into a favourable bridge environment to then grow out of that environment into a less supportive CNS milieu (Harvey *et al.*, 1995). In this regard, the study of Ramón-Cueto *et al.* (1998) shows that the transplantation of permissive olfactory ensheathing glia to normal spinal cord tissue either side of a Schwann cell/polymer bridge can enhance regrowth of axons into and beyond the implants. Only by ensuring the continuation of regrowth beyond bridging implants will axons have the opportunity of re-entering specific grey matter areas and perhaps re-establishing appropriate functional connections with target neurons. Effective CNS repair clearly requires a multifactorial approach, but the construction of effective supporting scaffolds for regenerative growth across lesion sites will be a critical part of the future neurosurgeons armamentarium.

Key points

Requirements for future progress:

- Further development of hydrogels by incorporation of biomaterials, signalling molecules and growth factors or cells.
- Development of oriented gels to provide more accurate guidance channels for growing neurites.
- Use of electric fields to improve orientation of growing neurites.
- Development of interventions that promote growth of neurites beyond the confines of the bridge.

References

Bellamkonda R Ranieri J P Bouche N and Aebischer P (1995) Hydrogel-based three-dimensional matrix for neural cells. Journal of Biomedical and Materials Research 29, 663–671

Cadelli D and Schwab M E (1991) Regeneration of lesioned septohippocampal acetylcholinesterase-positive axons is improved by antibodies against the myelin-associated neurite growth inhibitors NI-35/250. European Journal of Neuroscience 3, 825–832

Chen M Harvey A R and Dyson S E (1991) Regrowth of lesioned retinal axons associated with the transplantation of Schwann cells to the brachial region of the optic tract. Restorative Neurolology and Neuroscience 2, 233–248

Deglon N Heyd B Tan S A Joseph J M Zurn A D and Aebischer P (1996) Central nervous system delivery of recombinant ciliary neurotrophic factor by polymer encapsulated differentiated C2C 12 myoblasts. Human Gene Therapy 7, 851–860

Fields R D Le Beau J M Longo F M and Ellisman M H (1989) Nerve regeneration through artificial tubular implants. Progress in Neurobiology 33, 87–134

Harvey A R Fan Y Connor A M Grounds M D and Beilharz M W (1993) The migration and intermixing of donor and host glia on nitrocellulose polymers implanted into cortical lesion cavities in adult mice and rats. International Journal of Developmental Neuroscience 11, 569–581

Harvey A R and Plant G W (1995) Schwann cells and fetal tectal tissue cografted to the midbrain of newborn rats: fate of Schwann cells and their influence on host retinal innervation of grafts. Experimental Neurology 134, 179–191

Harvey A R Plant G W & Tan M M. L (1995) Schwann cells and the regrowth of axons in the mammalian CNS: a review of transplantation studies in the rat visual system. Clinical and Experimental Pharmacology and Physiology 22, 569–579

Kliot M Smith G M Siegal J D and Silver J (1990) Astrocyte-polymer implants promote regeneration of dorsal root fibers into the adult mammalian spinal cord. Experimental Neurology 109, 57–69

Montgomery C T and Robson J A (1993) Implants of cultured Schwann cells support axonal growth in the central nervous system of adult rats. Experimental Neurology 122, 107–124

Oudega M Xu X M Guénard V Kleitman N and Bunge M B (1997) A combination of insulin-like growth factor-1 and platelet-derived growth factor enhances myelination but diminishes axonal regeneration into Schwann cell grafts in the adult rat spinal cord. Glia 19, 247–258

Plant G W Chirila T V and Harvey A R (1998) Implantation of collagen IV/poly(2- hydroxyethyl methacrylate) hydrogels containing Schwann cells into the lesioned rat optic tract. Cell Transplantation 7, 381–391

Plant G W Woerly S W and Harvey A R (1997) Hydrogels containing peptide or aminosugar sequences implanted into the rat brain: influence on cellular migration and axonal growth. Experimental Neurology 143, 287–299

Ramón-Cueto A Plant G W Avila J and Bunge M B (1998) Long-distance axonal regeneration in the transected adult rat spinal cord is promoted by olfactory ensheathing glia transplants. Journal of Neuroscience 18, 3803–3815

Sautter J Tseng J L Braguglia D Aebischer P Spenger C Seiler R W Widmer H R and Zurn A D (1998) Implants of polymer-encapsulated genetically modified cells releasing glial cell line-derived neurotrophic factor improve survival, growth and function of fetal dopaminergic grafts. Experimental Neurology 149, 230–236

Schmidt C E Shastri V R Vacanti J P and Langer R (1997) Stimulation of neurite outgrowth using an electrically conducting polymer. Proceedings of the National Academy of Sciences USA 94, 8948–8953

Schreyer D J and Jones E G (1987) Growth of corticospinal axons on prosthetic substrates introduced into the spinal cord of neonatal rats. Developmental Brain Research 35, 291–299

Silver J and Ogawa M Y (1983) Postnatally induced formation of the corpus callosum in acallosal mice on glia-coated cellulose bridges. Science 220, 1067–1069

Tan S A and Aebischer P (1996) The problems of delivering neuroactive molecules to the CNS. Ciba Foundation Symposium 196, 211–236

Winn S R Lindner M D Lee A Haggett G Francis J M and Emerich D F (1996) Polymer-encapsulated genetically modified cells continue to secrete human nerve growth factor for over one year in rat ventricles: behavioral and anatomical consequences. Experimental Neurology 140, 126–138

Woerly S Maghami G Duncan R Subr V and Ulbrich K (1993) Synthetic polymer derivatives as substrata for neuronal adhesion and growth. Brain Research Bulletin 30, 423–432

Woerly S Marchand R and Lavallée C (1990) Intracerebral implantation of synthetic polymer/biopolymer matrix: a new perspective for brain repair. Biomaterials 11, 97–107

Woerly S W Plant G W and Harvey A R (1996) Neural tissue engineering: from polymer to biohybrid organs. Biomaterials 17, 301–310

Xu X M Guénard V Kleitman N and Bunge M B (1995) Axonal regeneration into Schwann cell-seeded guidance channels grafted into transected adult rat spinal cord. Journal of Comparative Neurology 351, 145–160

Zhao Q Drott J Laurell T Wallman L Lindström K Bjursten L M Lundborg G Montelius L and Danielsen N (1997) Rat sciatic nerve regeneration through a micromachined silicon chip. Biomaterials 18, 75–80

10

Neural stem cells: Regulation and potential use in neuronal regeneration

P F Bartlett, G J F Brooker, C H Faux, A M Turnley and T J Kilpatrick

Introduction

The concept of a stem cell whose properties include the ability to give rise to a multitude of cell types and to self-renew, has been well established in many systems, particularly the haematopoietic system, and yet only recently has it been deemed applicable to the central nervous system. This was partly because of the inability to grow stem cells *in vitro* or to monitor their progeny *in vivo*, but a more important impediment to the acceptance of the stem cell concept has been the unwillingness to embrace the property of self-renewal, mainly because this implied an ongoing presence of stem cells in the mature nervous system.

Cogent support for the stem cell concept, however, has accumulated over the last few years: populations of stem cells were successfully grown *in vitro* (Murphy, Drago & Bartlett, 1990); individual stem cells were cloned and their multipotentiality and self renewal properties formally demonstrated (Bartlett *et al.*, 1988; Kilpatrick & Bartlett 1993; Davis & Temple, 1994); stem cells with multipotential capacity were demonstrated *in vivo* using retroviral markers (Walsh&Cepko, 1988; Reid *et al.*, 1995); and finally, stem cells in the brains of animals at times beyond the neurogenic period (Kilpatrick & Bartlett, 1995) and into adulthood (Weiss & Reynolds, 1992; Richards *et al.*, 1992) were identified — confirming the self renewal capacity of the stem cell population.

It is not the intention of this review to cover all areas of stem cell biology, instead it will focus on recent results which address the mechanisms regulating the differentiation of stem cells in the forebrain of embryonic and adult mice and discuss how this knowledge may be used to stimulate repair of the damaged nervous system.

Embryonic stem cell regulation

Proliferation

The first suggestion that the fibroblast growth factor (FGF) family may influence stem cell growth came from *in vitro* studies. FGF-2 and FGF-1 were shown, in the presence of serum and insulin-like growth factor (IGF-1), to be the most effective agents in stimulating cell division in populations of neuroepithelial cells obtained from the embryonic day 10 (E10) mouse forebrain (Murphy *et al.*, 1991; Drago *et al.*, 1992). Although both neurons and astrocytes were reported to be generated from dividing cells, no conclusion could be drawn about the nature of the precursor or its potential since the experiments were performed at high cell density.

The subsequent demonstration that FGF-2 stimulated single cells from E10 neuroepithelium to produce clones consisting of several thousand cells which, in the presence of a glial-derived conditioned medium, could produce neurons in addition to astrocytes confirmed the suspicion that FGF-2 could stimulate the proliferation of stem cell populations (Kilpatrick & Bartlett, 1995). The frequency of precursors with the ability to give rise to clones of a significant size (>100 cells) was, on average, about 1 in 20 of the population. A similar frequency of cortical clones was observed by Qian *et al.* (1997) under culture conditions which did not include serum. Close examination of the growth curves obtained from bulk cultures stimulated with FGF-2 over a 5 day period (Murphy *et al.*, 1992) also suggested that the majority of cells arose from a small sub-population of cells no larger than 10%. Thus, the number of cells in the E10 cortical neuroepithelium with the ability to generate a large number of progeny appears to be about 10% of the population, even though it is clear that >99% of cells are dividing at this stage of development. The remaining dividing cells are limited in their capacity to divide and therefore generate only a small number of progeny — as is evident *in vivo* using retroviral markers to identify cortical clones (Walsh & Cepko, 1988). Of course, there is almost certainly a heirachy of precursor cells within the developing forebrain according to their proliferative and lineage potential, nevertheless, since a hallmark of true stem cells is their ability to generate large numbers of progeny it is these which concern us here.

In addition to generating large clones, the FGF-2 responsive cells have the property of self renewal, with >80% of the clonal progeny cells giving rise to new clones (Kilpatrick & Bartlett, 1993). Recent evidence from FGF-2 deficient mice (Dono *et al*, 1998) shows that FGF-2 may also be the relevant growth factor regulating stem cell proliferation *in vivo* since these mice show a decrease in cortical thickness and up to a 50% loss of neurons (Vaccarino *et al.* unpublished observations). In addition, the infusion of FGF-2 into the ventricles of E14 rats has been shown to increase the number of neurons in the cortex by up to 80% without interfering with the overall cytoarchitecture of the cortex (Rhee *et al.*, 1998); the mechanism appears to be an increase in number of divisions the precursor goes through and an increase in the neurogenic period. Both these *in vivo* studies support the concept that FGF-2 is a major regulator of stem cell proliferation in the developing forebrain.

Key points

- **The proliferation of stem cells in the developing forebrain appears to be regulated by FGF-2 in association with protoeglycans.**

Differentiation

Although the initial studies in high-density cultures indicated that FGF-2 could stimulate the generation of neurons, especially at higher concentrations of FGF-2, subsequent clonal examination revealed that FGF-2 stimulation in the presence of serum rarely produced neurons, although astrocytes did arise. In fact, studies revealed that neuronal differentiation could be inhibited by FGF-2, reinforcing the idea that FGF-2 appears to predominantly drive precursor proliferation. Earlier, however, it was observed that immortalised precursors from the E10 forebrain differentiated in response to FGF-2 (Bartlett *et al.*, 1988) and subsequent studies in serum-free medium by Qian *et al.* (1997) have shown that cortical clones generated in high doses of FGF-2 contain both neurons and oligodendrocytes. The latter authors also reported that clones generated in low levels of FGF (0.1ng/ml) only contained neurons as did the small number of clones arising without FGF-2. This suggests several things: first, that serum inhibits neuronal generation and second, that the level of FGF-2 determines the cell-type of the constituents of the clone. Closer examination, however, reveals that these results are similar to those obtained with serum-generated clones because the number of neurons generated per clone, regardless of FGF-2 concentration and size of clone, is small (average of 15) and the size of clones generated with low doses of FGF is, in the main, below 16 cells. Thus, like the serum-derived clones, FGF-2 at the higher concentrations, which generate large clones, appears to inhibit further neuronal production by the stem cell. Whether low doses of FGF-2 actually induce neuronal differentiation seems, as the authors state, unlikely since FGF-free clones also contain neurons. In addition, similar to the FGF-2 clones generated in serum, the large serum-free clones contain a majority of glial cells. There is, however, one major difference: the serum-plus clones contain astrocytes with virtually no oligodendrocytes whereas the serum-free clones are almost exclusively comprised of oligodendrocytes. Astrocyte production requires additional factors which, as we will discuss below, may function by stimulation through the leukaemia inhibitory factor (LIF) receptor complex. It is well established that oligodendrocyte production is enhanced by serum free conditions by a mechanism that is still unclear (Raff *et al.*, 1983).

Thus, the evidence suggests that the production of neurons within a large clone requires additional factors to FGF-2. Ghosh & Greenberg (1995) showed neurotrophin-3 (NT-3) could stimulate neurogenesis in FGF-2-stimulated cultures of E14 forebrain and we have shown that conditioned medium from an astrocyte cell line can result in a significant number of FGF-2 expanded clones producing neurons after FGF-2 withdrawal (Kilpatrick & Bartlett, 1992). One strong candidate for providing a neurogenic stimulus is FGF-1, which we demonstrated was expressed at E11 just as neurogenesis begins in the mouse cortex. Whereas FGF-2 was present much earlier at E9.5, prior to the commencement of neuronal production, but coinciding temporally with the onset of proliferation. However, we were not able to demonstrate a differential action of FGF-1 compared to FGF-2 in our cultures regardless of presence or absence of heparin. Nevertheless, Guillemot & Cepko (1992) had shown that FGF-1 was far more potent than FGF-2 in promoting the differentiation of retinal ganglion cells.

Role of heparan sulphate proteoglycans in FGF responsiveness

In the process of isolating FGF-1 and FGF-2 from the neuroepithelial cells from mouse forebrain, it was discovered that the majority of the FGF was bound to a single dominant heparan sulphate proteoglycan (HSPG) (Nurcombe *et al.*, 1993), which we have since identified as a splice variant of perlecan (Joseph *et al.*, 1996). The most interesting finding, however was the binding specificity of this HSPG isolated from the developing forebrain at different times; HSPG from E10 brains (HS-2) predominantly bound FGF-2, whereas from E12 brains (HS-1) it preferentially bound FGF-1 (Nurcombe *et al.*, 1993). This shift was associated with an increase in the number of sulfated domains and increased heparan sulphate glycosaminoglycan (HS) side chain length (Brickman *et al.*, 1988). It is known that the domains of charge created by sulfation are critical to FGF-1 and FGF-2 binding and also are thought to influence interaction of FGF with its cognate receptor(s). One current hypothesis which we favour is that HS serve to couple FGF to specific HS-binding regions on specific FGF receptors (FGFR) to form an activated signalling complex of FGF/HS/FGFR.

It was found that precursor proliferation in high density cultures was significantly enhanced when FGF-1 was used with HS-1, or FGF-2 with HS-2 (Nurcombe *et al.*, 1993) confirming the importance of this type of presentation mechanism. This provides a mechanism by which cell activation can be regulated without the requirement for changes in FGF concentration or receptor down regulation, and is probably used by a number of the heparin-binding growth factors. Recently, we have used FGF-1 in combination with HS-1 and shown that >80% of the clones generated contain large numbers of neurons (>100), whereas less than 3% of clones have neurons when heparin is used (Bartlett *et al.*, 1998).

The precise mechanism by which neuronal signalling occurs is not known, but it probably involves the differential signalling through one or more of the FGFRs on the stem cell's surface. We and others have shown that isoforms of FGFR 1, 2 and 3 are expressed on the precursor population in developing cortex during this period (Brickman *et al.*, 1995; Qian *et al.*, 1997), so the question remains as to which receptor signals neurogenesis and which signals proliferation.

Factor regulation of astrocyte differentiation

As discussed above, there is good evidence for the bi-potential stem cell's fate being determined, at least in part, by environmental factors such as growth factors. Previously we had shown, *in vitro*, that leukemia inhibitory factor (LIF) could stimulate precursors from the E10 spinal cord to express glial fibrillary acidic protein (GFAP) (Richards *et al.*, 1996). In addition, this study showed that antibodies to the LIFR significantly reduced the number of astrocytes which developed in the absence of exogenous growth factors, suggesting that endogenous ligands acting through the LIFR influence astrocyte development. Other ligands which signal through the LIFR complex — a heterodimer composed of LIFR and gp130 — such as ciliary neurotrophic factor (CNTF), also have been shown to promote GFAP expression in CNS precursor populations (Johe *et al.*, 1996). Thus, the *in vitro* results strongly suggest that ligands which signal through the LIFR complex may have a role in regulating astrocyte differentiation.

In support of LIFR's role in regulating astrocyte production was the demonstration that E19 embryonic mice with a targeted disruption of the low affinity LIF receptor gene, which appear to have normal CNS development, have a deficiency of GFAP positive cells in the developing hindbrain (Ware *et al.*, 1995). Unfortunately, since these animals die at E19 — which is just 2 days after the first appearance of GFAP (Abney *et al.*, 1981) — it was difficult to determine whether this astrocyte deficiency was due to general retardation in development or to a failure in astrocyte generation due to lack of signalling through the LIF receptor. To explore these possibilities further, the properties of precursor cells from the forebrain of LIFR deficient mice were examined *in vitro* (Koblar *et al.*, 1998). It was shown that precursors from the forebrains of mice homozygous for the LIFR null mutation (LIFR–/–) failed to generate significant numbers of GFAP positive cells even after 3 weeks *in vitro*. To determine if the lack of GFAP expression in LIFR–/– precursors fully reflected a failure in astrocyte development, an assay to assess astrocyte function was performed. Previously, it has been shown that astrocytes promote neuronal differentiation and / or survival in a number of systems (Kilpatrick *et al.*, 1993; Cohen *et al.*, 1986) thus, the ability of established monolayers derived from LIFR +/+, +/–, and –/– forebrains to support the neuronal differentiation was tested. No difference was found in the number of neurons produced on the LIFR+/+ or +/– monolayers, however there was approximately 10 fold fewer neurons found on the LIFR–/– monolayers. Thus, signalling through the LIFR was required for the generation of functional astrocytes not just for the expression of GFAP. This is an important point, since it has recently been shown that one of the downstream signalling pathways activated by signalling through LIFR – the JAK-STAT pathway – can directly activate the GFAP gene. STAT 3 has been shown to directly bind to a consensus site in the promoter region of the GFAP gene (Bonni *et al.*, 1997). Thus, the regulation of GFAP expression can be directly regulated through the LIFR complex — both LIFR and gp130 appear to be required for this signal (Bonni *et al.*, 1997).

Key points

- **The differentiation of stem cells into astrocytes requires factors which signal through the LIF receptor complex.**

It was subsequently shown that the precursor population in the LIFR–/– forebrain was in fact present. Stimulation of LIFR–/– forebrain cells *in vitro* with BMP-2, a member of the transforming growth factor (TGF-β) growth factor family, previously shown to stimulate GFAP expression in astrocytes, resulted in the generation of a significant percentage of GFAP positive cells after 10 days in culture. In addition, long term passaging *in vitro* (>5 weeks) revealed significant numbers of GFAP cells in LIFR–/– cultures which supported neuron generation and/or survival (Koblar *et al.*, 1998).

We also found that there was no decrease in the total number of neural clones generated from the LIFR–/– mouse forebrain precursors with FGF-2; this strongly suggests that LIFR signalling was not essential for the maintenance of precursor cells. As mentioned above, we had shown that FGF-2 stimulated forebrain precursors have the ability to generate two types of clones: clones which contain both neurons and glia, or clones restricted to astrocytes. However, since the frequency of neuron-containing clones generated with FGF-1 and HSPG-1 is also unaltered in the LIFR–/– population, it suggests that there is no change in the relative frequency of either the bipotential or astrocyte-restricted clones in these animals.

The question arises as to whether signalling through the LIF receptor instructs a precursor to become committed to the astrocyte pathway. Several pieces of evidence support such an hypothesis: first, in the presence of LIF >80% of precursors become

GFAP positive *in vitro* (Richards *et al.*, 1996); second, STAT-3, which is directly activated by LIFR signalling, can bind to the promoter region of the GFAP gene and regulate its expression (Bonni *et al.*, 1997); and third, stimulation with LIF can significantly inhibit the neuronal differentiation of clones (Brooker & Bartlett, unpublished observations). This latter finding is also true of clones derived from adult subventricular zone. Although this confirms the idea that signalling through the LIFR may actively promote astrocyte differentiation, an additional interpretation is that LIFR signalling may inhibit neuronal differentiation leading to astrocyte production by default. Thus, LIFR signalling may actively keep the precursor in an undifferentiated, or stem cell state — as it does for pluripotential embryonic stem cells — and neurogenesis may result from individual stem cells overcoming this inhibition. A candidate for mediating this type of action is the recently discovered suppressors of cytokine signalling (SOCS) family which have been shown to inhibit signalling through the LIFR (Starr *et al.*, 1997).

It is not known which LIFR ligand mediates these effects. We have shown that mice with a targeted deletion in the LIF gene have a reduction in the number of astrocytes in the hippocampus but it is incomplete (Koblar *et al.*, 1998). Since ciliary neurotrophic factor (CNTF) also has been shown to promote astrocyte formation, it may play a part, but again, the CNTF–/– mouse has not been reported to have an astrocyte deficiency. Other ligand-receptor pathways may replace LIFR at later stages of development masking the early defect. The finding of GFAP positive cells in longterm cultures from LIFR–/– mice supports this idea as do recent experiments in which portions of LIFR–/– brains were transplanted to a syngeneic recipient and shown to contain GFAP cells several weeks after transplantation (Bartlett & Harvey, unpublished observations).

Neuronal differentiation results from disinhibition

The concept raised above, in which neurogenesis results from overcoming signals that favour the maintenance of a stem cell-state, is best exemplified by the action of the neurogenic genes *Delta* and *Notch,* which code for a cell surface ligand and receptor respectively, and through a process of lateral inhibition prevent adjacent precursors from differentiating. This process has been well demonstrated to regulate neurogenesis in drosophila and *xenopus,* and more recently it has been shown to play a role in mammalian retinal differentiation (Austin *et al.*, 1995).

The key step in this phenomenon is the ability of a single precursor to express more of the ligand Delta than its neighbours thereby activating the neighbour's Notch receptor signalling pathway which inhibits neurogenesis by inhibiting the production of the helix-loop-helix transcriptional regulators Neurogenin and Neuro-D; which in turn regulate Delta levels. To investigate whether the action of the growth factors FGF-2 and FGF-1 could influence this pathway, we have begun to examine neuronal production in high cell density conditions where, as we have previously shown, FGF-1 and FGF-2 promote proliferation rather than neuronal differentiation (Murphy *et al.*, 1991). When Notch-1 expression is reduced by the addition of antisense oligonucleotides to the cultures a significant increase in the number of neurons was found in the presence of FGF-1, but not FGF-2 (Faux, Turnley & Bartlett, unpublished observations). Again this demonstrates the superior ability of FGF-1 to promote neuronal differentiation compared to FGF-2 (at similar concentration). It also suggests that neurogenesis *in vivo* may require both inhibition of Notch signalling and activation of

FGF receptor signalling; although there maybe a common mechanism whereby Notch expression is further reduced by FGF-1 signalling to levels below the threshold for inhibition.

Key points

- **Neuronal differentiation requires stem cells to overcome the affects of inhibitory factors which signal through receptors like Notch and the LIF receptor.**

Stem cells from the adult forebrain

So far we have only mentioned the regulation of embryonic stem cells, however, it was shown several years ago that a population of precursors exist within the adult brain of mice which could be stimulated *in vitro* to produce neurons in response to FGF-2 (Richards *et al.*, 1992) and also epidermal growth factor (EGF) (Reynolds & Weiss, 1992). Subsequently, it was shown that the subventricular layer of the lateral ventricle was rich in stem cells (Lois & Alvarez-Buylla, 1993) capable of generating neurons when they migrated to the olfactory bulb (Lois & Alvarez-Buylla, 1994). More recently, it has been shown that EGF- and FGF-responsive stem cells exist throughout the ventricular neuroaxis of the brain and spinal cord (Weiss *et al.*, 1996), and FGF responsive stem cell have been isolated from various other regions of the brain not associated with ventricular regions such as hippocampus and striatum (Gage *et al.*, 1995). Thus stem cells may be widely spread throughout the adult nervous system of rodents. Recently, neuronal replacement has been shown to occur in granule cells of the dentate gyrus in the adult marmoset (Gould *et al.*, 1998), indicating that similar precursors almost certainly exist in all primates including humans.

Key points

- **Neural stem cells appear to be widespread in the adult CNS.**
- **Generation of neurons is actively inhibited in the adult CNS except in areas such as olfactory bulb and hippocampus.**

Is there a separate EGF- and FGF-responsive stem cell in the adult brain?

In the last few years there has been continued intensive study of the relative effects of two growth factors, EGF and FGF-2 upon precursor cells. Weiss and coworkers have reported extensively that EGF stimulates the proliferation of quiescent stem cells, that reside within the adult lateral ventricle and striatum, to form non-adherent neurospheres *in vitro*. The spheres have been dissociated and the progeny cloned and subjected to multiple passages. The percentage of cells that has the capacity to generate spheres is actually increased in secondary as opposed to primary spheres and cells which stain for neuron specific enolase, O4 and GFAP have been identifiable at each passage. The findings have been interpreted to indicate that an EGF-responsive multipotential precursor has the capacity to undergo both self renewal and differentiation into multiple lineages. It has also been reported that EGF treated precursors upregulate the expression of the FGF-2 receptor. This implies that FGF might act upon a more differentiated precursor and it has been suggested that, in the presence of serum constituents, FGF stimulates the proliferation of either a committed neuroblast or a bipotential progenitor with the capacity to differentiate into either a neuron or an astrocyte.

In an attempt to assess whether EGF-responsive progenitors have the capacity to generate neurons *in vivo*, Craig *et al.* (1996) infused EGF into the lateral ventricle of adult mice over six days. After 7 weeks, it was stated that 25% of the labelled cells had differentiated into astrocytes without apparent accompanying astrogliosis and 3% of the cells had both migrated into the parenchyma (cortex, striatum and septum) and differentiated into neurons.

Similar *in vivo* studies have been undertaken by Kuhn *et al.* (1996) in the rat. In these experiments, EGF induced polyp-like hyperplasias in the lateral ventricles that regressed within four weeks of cessation of therapy. Labelled cells were observed to migrate into the cortex, septum and around the cannular tract but although bromodeoxy uridine (BrDU) labelled, S100 positive glia were identified, NeuN positive neurons were not detected. The reason for the discrepant results reported by Craig *et al.* (1995) and by Kuhn *et al.* (1996) is unclear. Certainly there may be species differences but it is also possible that stereological techniques are required, as employed by Kuhn *et al.* (1996) to absolutely exclude the possibility of including overlapping cells in the assessment. Kuhn *et al.* (1996) have also indicated that gliogenesis accompanies the deficit in the olfactory migratory stream exhibited in the EGF treated rodents, raising the possibility that the primary effect of EGF relates to a change in lineage specification rather than due to a primary alteration in migratory capacity of the precursor cell.

Like Richards *et al.* (1992), Gritti *et al.* (1996) have recently used FGF-2 to generate neurospheres from adult mouse striatum using FGF-2 and, like EGF responsive cells, the cells generated with FGF appeared to exhibit multipotentiality. It was of note that neurons imunoreactive for GABA, Substance P, ChAT and glutamate were all delineated and cells with neuronal morphologies were also shown to exhibit electrophysiological properties characteristic of neurons. The proportion of cells isolated from the striatum that exhibited FGF responsiveness was similar to that previously documented to exhibit EGF responsiveness. The parsimonious view would be that the same precursor responds to both EGF and FGF.

Recently, work from our laboratory has shown that the FGF-2-responsive stem cell and the EGF-responsive cell are most probably the same cell. We have used flow cytometry to sort cells on the basis of size and cell-surface markers and then stimulated subpopulations with either FGF-2 or EGF in the presence of serum.

A stem cell with the cell-surface phenotype of peanut agglutinin negative (PNA-ve) and A2B5 -ve, which represents approximately 3% of the population, responds equally well to both EGF and FGF-2. In a clonal assay, the same number of clones arises with either factor and the frequency is unchanged when both factors are added simultaneously (Figure 1). The one difference is that the FGF-2 stimulated clones can give rise to neurons in addition to astrocytes when switched into FGF-1, as described for embryonic-derived clones above, whereas we have been totally unable to induce neuronal differentiation in EGF stimulated clones regardless of the growth factor used. Thus, in the presence of serum EGF appears to restrict a common precursor to the production of glial cells. A similar finding occurred with E16 neuroepithelial cells: FGF-

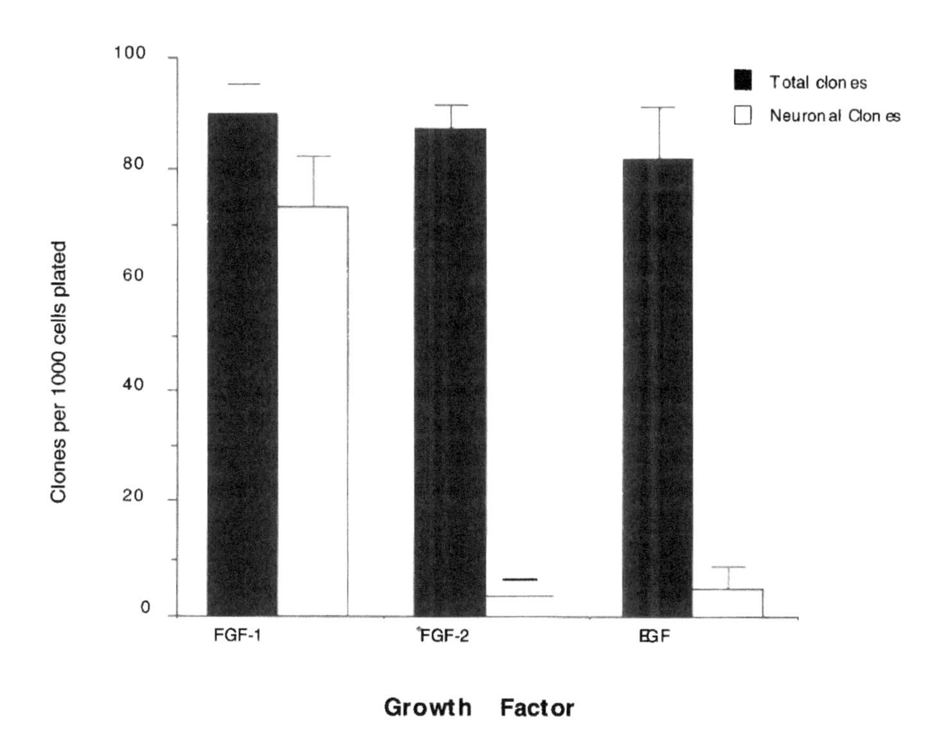

Figure 1 Response of stem cells selected on the basis of surface markers A2B5 and PNA showing that they respond equally to FGF-2, FGF-1 and EGF, but only FGF-1 results in clones containing neurons.

2 stimulated neuronal containing clones whereas EGF only gave rise to astrocyte-containing clones (Kilpatrick & Bartlett, 1995).

How does one reconcile these findings with the neurosphere data? It would appear that under serum free conditions EGF may be able to stimulate a common stem cell to proliferate and maintain its multipotentiality whereas in serum, EGF stimulation drives the precursor toward the astrocyte lineage. The role of serum, however, should not be overestimated since the number of neurons generated from individual neurospheres is still very low in EGF generated neurospheres in the absence of co-stimulation with FGF-2. Thus, EGF's primary action may be to restrict the differentiation capability of stem cells to the glial lineage. The data of Kuhn *et al.* (1996; see above) also suggests EGF is an astrocyte stimulus, and so do the experiments in which EGF-generated neurospheres were injected into the brains of embryonic rats (Winkler *et al.*, 1998).

Use of stem cells in neuronal regeneration

Since stem cells have the capacity to generate large numbers of neurons as well as other cell types and appear to be able to differentiate into all major neuronal cell types

Key points

- EGF stimulates the proliferation of quiescent stem cells that reside within the adult lateral ventricle and striatum to form non-adherent neurospheres *in vitro.*
- FGF-2 has also been used to generate neurospheres from adult mouse striatum; like EGF responsive cells, the cells generated with FGF appeared to exhibit multipotentiality.
- The FGF-2-responsive stem cell and the EGF-responsive cell are probably the same cell.
- In the presence of serum, EGF appears to restrict a common precursor to the production of glial cells.
- Under serum-free conditions EGF may be able to stimulate a common stem cell to proliferate and maintain its multipotentiality whereas in serum, EGF stimulation drives the precursor toward the astrocyte lineage.

depending on environmental signals, they have the potential to replace neurons lost or damaged as the result of neurodegeneration or injury. In principle, there are two approaches to stem cell therapy: either to stimulate endogenous stem cells or to transplant *in vitro*-generated stem cells.

Stimulation of endogenous stem cells

There appear to be two major problems with this approach: first, the migration of precursors to the appropriate site, and second, the differentiation of stem cells into neurons rather than astrocytes.

Stem cells and their progeny appear to migrate only if they are placed in the rostral migratory stream of the adult brain, suggesting that the factors in this structure which are required for migration are absent in other brain regions. The other area where neurogenesis occurs in the adult is the gentate gyrus and here the radial glia are retained into adult life (Kosaka *et al.*, 1986). However, lack of radial glia or rostral migratory stream factors may not be an insurmountable barrier to repair since stem cells appear to be present in most areas of the brain and spinal cord and thus extensive migration may not be required.

The second problem appears to be of much greater difficulty. Outside of what appears to be the normal physiological generation of neurons in the adult olfactory bulb and hippocampus, there have been no reports of neuronal replacement after injury in the adult brain. Since stem cells appear likely to be present in the damaged area it suggests that neuronal differentiation is inhibited.

We have recently gained evidence for an active inhibitory mechanism affecting precursors in the adult brain. As reported by Lois & Alvarez-Bullya (1993), dividing cells within the explants of subventricular zone grown *in vitro* are unable to generate neurons. However, we have shown that the dividing cells within the explant do have the propensity to give rise to neurons when dissociated and replated at clonal or low cell density (Dutton *et al.* unpublished observations). Replating dissociated cells at high cell density leads to inhibition of neurogenesis. The results suggest an inhibitory mechanism similar to lateral inhibition previously described. Both Delta and Notch are expressed in the adult subventricular zone (Faux & Bartlett, unpublished observations). Such inhibitory mechanisms may restrict the ability of precursors within the adult subventricular zone to generate neurons apart from those destined for the olfactory bulb. It could be postulated that the olfactory stream provides signals that may reduce these inhibitory affects.

In addition to the Delta-Notch system, it now appears that growth factors can also negatively regulate the production of neurons in both the embryonic and adult brain. As we have discussed above, FGF-2 can inhibit neuronal production by way of its potent stimulation of proliferation and, since it is produced by neuroepithelial stem cells, it is possible that it acts as an active inhibitor of neurogenesis. Factors acting through the LIF receptor may also inhibit neurogenesis and promote astrocyte formation as they do in their embryonic counterpart. The addition of LIF to clones derived from adult adult subventricular zone results in a marked reduction in neuron containing clones (Brooker & Bartlett, unpublished observations).

Another important inhibitor of neuronal production in the adult may be the release of inflammatory factors associated with disease or injury. Work from the Snyder laboratory (Snyder *et al.*, 1997) indicates that transplanted stem cells can generate neurons in the cortex of adult mice only under conditions in which the adult neurons are killed by targeted photolysis which produces little inflammatory response. When kainic acid was used to produce the cortical lesion, only astrocytes were produced. Thus, the production of neurons from endogenous stem cells may require a combination of stimulatory factors and the suppression of inhibitory factors. The transplantation of donor stem cells may help to define these conditions.

Key points

- Two approaches to stem cell therapy: either stimulate endogenous stem cells or to transplant *in vitro*-generated stem cells.
- Stimulation of endogenous stem cells requires migration to appropriate site and differentiation into neirons rather than astrocytes.
- Widespread distribution of stem cells may mean migraion is not a major problem.
- Inhibition of precursor differentiation to neurons may be more of a problem.

Transplantation of stem cells

The first hint that exogenous stem cells could be used to replace neurons came from transplantation studies using precursors immortalised with retrovirus or large T-antigen (Snyder *et al.*, 1992; Renfranz *et al.*, 1991). Surprisingly, when these cells were transplanted into embryonic recipients they were able to replace an array of neuronal phenotypes regardless from which area of the brain the stem cells originated. This plasticity has also been shown to be true of transplanted embryonic or adult stem cells generated with FGF-2 (Kilpatrick *et al.*, 1994, Gage *et al.*, 1995). Thus, it appears that signals in the host's local environmental are sufficient to induce the appropriate differentiation in transplanted stem cells. The phenomenon even extends across species barriers: stem cells isolated from human fetal telencephalon were able to populate a wide range of areas of the mouse and rat brain with phenotypically normal neurons when transplanted during development (Flax *et al.*, 1998; Brüstle *et al.*, 1998).

Although stem cells from both the adult and embryo have been shown to be capable of giving rise to tissue-specific neurons when transplanted into the adult CNS of mice, this capacity is largely restricted, although not entirely (Harvey *et al.*, 1997), to areas in which there is ongoing neurogenesis: the adult subventricular zone – rostral migratory stream – olfactory bulb system and the dentate gyrus (Suhonen *et al.*, 1996; Gage *et al.*, 1995). This indicates that the same restrictions/inhibitions are placed on transplanted stem cells as are placed on the endogenous stem cells in the adult. It is further cogent evidence that the focus of future research must be on defining these regulatory signals if stem cell-mediated regeneration is to be successful in the adult nervous system.

Key points

- **Stem cells from adult and embryo are capable of giving rise to tissue-specific neurons when transplanted into the adult CNS of mice; this is largely restricted, to areas in which there is ongoing neurogenesis: eg adult subventricular zone.**
- **Same restrictions/inhibitions are placed on transplanted stem cells as on endogenous stem cells in adult.**
- **Focus of future research must be on defining these regulatory signals, if stem cell-mediated regeneration is to be successful in the adult nervous system**

Immunological considerations

Although the CNS was originally thought to be an immunological privileged site it is now clear that allogeneic or xenogeneic neural tissue transplanted into the brain, even during embryogenesis, will eventually be rejected (Kerr & Bartlett, 1989). We demonstrated that neural cells express major histocompatibility antigens (MHC) in response to inflammatory cytokines such as interferon-γ and thus can be targeted by cytotoxic T cells (Wong *et al.*, 1984). Initiation of the immune response is probably due to dendritic cells which reside within the brain parenchyma.

It was found, however, that neurons do not express MHC in response to interferon-γ and therefore, if not surrounded by allogeneic astrocytes, could be successfully transplanted. Furthermore, we found a lack of MHC-inducibility on progenitors committed to become neurons, but not on the stem population (Bailey *et al.*, 1994). Thus, MHC -ve progenitors selected from forebrain neuroepithelium could be transplanted to allogeneic hosts without rejection. Thus, under certain circumstances it may be preferable to transplant progenitors restricted to the neuronal lineage rather than stem cell populations.

Key points

- CNS is less immunologically privileged than previously thought.
- Neural cells, but not neurons themselves, express major histocompatibility antigens (MHC) in response to inflammatory cytokines such as interferon-γ.
- Neuronal progenitors also lack MHC and may therefore be preferred candidates for transplantation, rather than stem cells, which will also give rise to MHC +ve astrocytes.

Acknowledgments
This work was supported by National Health and Medical Research Council of Australia, the Collaborative Research Centre for Cellular Growth Factors, and the Australasian Spinal Research Trust.

References

Abney E R Bartlett P F and Raff M C (1981) Astrocytes, ependymal cells and oligodendrocytes develop on schedule in dissociated cell cultures of embryonic rat brain. Developmental Biology 83, 301–310

Austin P A Feldman D E Ida J A and Cepko C L (1995) Vertebrate retinal ganglion cells are selected from competent progenitors by the action of Notch. Development 121, 3637–3650

Bailey K A Drago J D and Bartlett P F (1994) Neuronal progenitors identified by their inability to express class I histocompatibility antigens in response to interferon γ. Journal of Neruoscience Research 39, 166–177

Bartlett P F Brooker G J Faux C H Dutton R Murphy M Turnley A and Kilpatrick T J (1998) Regulation of neural stem cell differentiation in the forebrain. Immunology and Cell Biology 76, 414–418

Bartlett P F Reid H H Bailey K A and Bernard O (1988) Mouse neuroepithelial cells immortalized by the *c-myc* oncogene. Proceedings of the National Academy of Sciences USA 85, 3255–3259

Bonni A Sun Y Nadal-Vicens M Frank D A Rozovsky I Stahl N Yancopoulos G D and Greenberg M E (1997) Regulation of gliogenesis in the central nervous system by the JAK-STAT signaling pathway. Science 278, 477–482

Brickman Y G Ford M D Gallagher J T Nurcombe V Bartlett P F and Turnbull J E (1998) Structural modification of fibroblast growth factor-binding heparan sulfate at a determative stage of neural development. Journal of Biological Chemistry 273, 4350–4359

Brickman Y G Ford M D Small D H Bartlett P F and Nurcombe V (1995) Heparan sulphates mediate the binding of basic fibroblast growth factor to a specific receptor on neural precursor cells. Journal of Biological Chemistry 270, 24941–24948

Brüstle O Chaundhary K Karram K Hüttner A Murray K Dubois-Dalcq M and McKay R D G (1998) Chimeric brains generated by intraventricular transplantation of fetal human brain cells into embryonic rats. Nature Biotechnology 16, 1040–1044

Cohen J Burne G F Winter J and Bartlett P F (1986) Retinal ganglion cells lose response to laminin with maturation. Nature 322, 465–467

Craig C G Tropepe V Morshead C M Reynolds B A Weiss S and Van der Koop, D (1996) *in vivo* growth factor expansion of endogenous subependymal neural precursor cell populations in the adult mouse brain. Journal of Neuroscience 16, 2649–2658

Davis A A and Temple S (1994) A self-renewing multipotential stem cell in the embryonic rat cerebral cortex. Nature 372, 263–266

Dono R Texido G Dussel R Ehmke H and Zeller R (1998) Impaired cerebral cortex development and blood pressure regulation in FGF-2-deficient mice. EMBO 17, 4213–4225

Drago J Murphy M Carroll S Harvey R P and Bartlett P F (1991) FGF-mediated proliferation of CNS precursors depeneds on endogenous production of IGF-I. Proceedings of the National Academy of Sciences USA 88, 219–2203

Flax J D Aurora S Yang C Simonin C Wills A M Billinghurst L L Jendoubi M Sidman R L Wolfe J H Kim S U and Snyder E Y (1998) Engraftable human neural stem cells respond to developmental cues, replace neurons, and express foreign genes. Nature Biotechnology 16, 1033–1039

Gage F H Coates P W Palmer T D Kuhn H G Fisher L J Suhonen J O Teterson D A Suhr S T and Ray J (1995) Survival and differentiation of adult neuronal progenitor cells transplanted to the adult brain. Proceedings of the National Academy of Sciences USA 92, 11879–11883

Gould E Tanapat P McEwen B S Flugge G and Fuchs E (1998) Proliferation of granule cell precursors in the dentate gyrus of adult monkeys is diminished by stress. Proceedings of the National Academy of Sciences USA 95, 3168–71

Ghosh A and Greenberg M E (1995) Distinct roles for bFGF and NT-3 in the regulation of cortical neurogenesis. Neuron 15, 1–10

Gritti A Parati E A Cova L Frolichsthal P Calli R Wanke E Faravelli L Morassutti D J Roisen F Nickel D D and Vescovi A (1996) Multipotential cells from the adult mouse brain proliferate and self-renew in response to basic fibroblast growth factor. Journal of Neuroscience 16, 1091–1100

Guillemot F and Cepko C L (1992) Retinal fate and ganglion cell differentiation are potentiated by acidic FGF in an *in vitro* assay of early retinal development. Development 114, 743–754

Harvey A R Symons N A Pollett M A Brooker G J F and Bartlett P F (1997) Fate of adult neural precursors grafted to adult cortex monitored with a Y chromosome marker. Neuroreport 8, 3939–3943

Johe K K Hazel T G Muller T Dugich-Djordjevic M M and McKay R D (1996) Single factors direct the differentiation of stem cells from the fetal and adult central nervous system. Genes and Development 10, 3129–3140

Joseph S J Ford M D Barth C Portbury S Bartlett P F Nurcombe V and Greferath U (1996) A Proteoglycan that activates fibroblast growth factors during early neuronal development is a perlecan variant. Development 122, 3443–3452

Kerr R S C and Bartlett P F (1989) The immune response to intraparenchymal foetal CNS transplants. Transplant Proceedings 21, 3166–3169

Kilpatrick T J and Bartlett P F (1993) Cloning and growth of multipotential precursors: requirements for proliferation and differentiation. Neuron 10, 255–265

Kilpatrick T J and Bartlett P F (1995) Cloned multipotential precursors from the mouse cerebrum require FGF-2, whereas glial restricted precursors are stimulated with either FGF-2 or EGF. Journal of Neuroscience 15, 3653–3661

Kilpatrick T J Cheema S S Koblar S A Tan S S Bartlett P F (1994) The engraftment of transplanted primary neuroepithelial cells within the postnatal mouse brain. Neuroscience Letters 181, 129–133

Kilpatrick T J Richards L R and Bartlett P F (1995) The regulation of neuronal precursor cells within the mammalian brain. Molecular and Cellular Neuroscience 6, 2–15

Kilpatrick T J Talman P T and Bartlett P F (1993) The differentiation and survival of murine neurons *in vitro* is promoted by soluble factors produced by an astrocytic cell line. Journal of Neuroscience Research 35, 147–161

Koblar S A Dutton R Ware C B and Bartlett P F (1998) Neural precursor differentiation into astrocytes requires signaling through the leukemia inhibitory factor receptor. Proceedings of the National Academy of Sciences USA 95, 3178–3181
Kosaka T and Hama K (1986) Three-dimensional structure of astrocytes in the rat dentate gyrus. Journal of Comparative Neurology 249, 242–260
Kuhn H G Winkler J Kempermann G Tha L J and Gage F H (1997) Epidermal growth factor and fibroblast growth factor-2 have different effects on neural progenitors in the adult rat brain. Journal of Neuroscience 17, 5820–5829
Lois C and Alvarez-Buylla A (1993) Proliferating subventricular zone cells in the adult mammalian forebrain can differentiate into neurons and glia. Proceedings of the National Academy of Sciences USA 90, 2074–2077
Lois C and Alvarez-Buylla A (1994) Long distance neuronal migration in the adult mammalian brain. Science 264, 1145–1148
Martinez-Serrano A and Björklund A (1997) Immortalized neural progenitor cells for CNS gene transfer and repair. Trends in Neuroscience 20, 530–538
Morshead C M Reynolds B A Craig C G McBurney M W Staines W A Morassutti D Weiss S and van der Kooy D (1994) Neural stem cells in the adult mammalian forebrain: a relatively quiescent subpopulation of subependymal cells. Neuron 13, 1070–1082
Murphy M Drago J and Bartlett P F (1990) Fibroblast growth factor stimulates the proliferation and differentiation of neural precursor cells *in vitro*. Journal of Neuroscience Research 25, 463–475
Nurcombe V Ford M D Wildschut J A and Bartlett P F (1993) Developmental regulation of neural response to FGF-1 and FGF-2 by heparan sulfate proteoglycan. Science 260, 103–106
Palmer T D Takahashi J and Gage F H (1997) The adult rat hippocampus contains primordial neural stem cells. Molecular and Cellular Neuroscience 8, 389–404
Qian X Davis A A Goderie S K and Temple S (1997) FGF2 concentration regulates the generation of neurons and glia from multipotent cortical stem cells. Neuron 18, 81–93
Raff M C Miller R H and Noble M (1983) A glial progenitor cell that develops *in vitro* into an astrocyte or an oligodendrocyte depending on culture medium. Nature 303, 390–396
Ray J Peterson D A Schinstine M and Gage F H (1993) Proliferation, differentiation, and long-term culture of primary hippocampal neurons. Proceedings of the National Academy of Sciences USA 90, 3602–3606
Refranz P J Cunningham M G and McKay R D G (1991) Region-specific differentiation of the hippocampal stem cell line HiB5 upon implantation into the developing mammalian brain. Cell 66, 713–729
Reid C B Liang I and Walsh C (1995) Systemic widespread clonal organization in cerebral cortex. Neuron 15, 299–310
Reynolds B A and Weiss S (1992) Generation of neurons and astrocytes from isolated cells of the adult mammalian central nervous system Science 255, 1707– 1710
Reynolds B A Tetzlaff W and Weiss S (1992) A multipotent EGF-responsive striatal embryonic progenitor cell produces neurons and astrocytes. Journal of Neuroscience 12, 4565–4574
Rhee J S Raballo R Schwartz M S and Vaccarino F M (1998) Lineage and non-lineage effects of FGF-2 on progenitor cells in the developing cerebral cortex. Society for Neuroscience Abstracts 24, 281
Richards L J Kilpatrick T J and Bartlett P F (1992) De novo generation of neuronal cells from the adult mouse brain. Proceedings of the National Academy of Sciences USA 89, 8591–8595
Richards L J Kilpatrick T J Dutton R Tan S S Gearing D P Bartlett P F and Murphy M (1996) Leukemia inhibitory factor or related factors promote the differentiation of neuronal and astrocytic precursors within the developing murine spinal cord. European Journal of Neuroscience 8, 291–299

Snyder E Y Deitcher D L Walsh C Arnold-Aldea S Hartweig E A and Cepko C L (1992) Multipotent cell lines can engraft and participate in the development of mouse cerebellum. Cell 68, 1–20

Snyder E Y Yoon C Flax J D and Macklis J D (1997) Multipotent neural precursors can differentiate toward replacement of neurons undergoing targeted apoptotic degeneration in adult mouse neocortex. Proceedings of the National Academy of Sciences USA 94, 11663–11668

Starr R Willson T A Viney E M Murray L J Rayner J R Jenkins B J Gonda T J Alexander W S Metcalf D Nicola N A and Hilton D J (1997) A family of cytokine-inducible inhibitors of signalling. Nature 387, 917–921

Suhonen J O Peterson D A Ray J and Gage F H (1996) Differentiation of adult hippocampus-derived progenitors into olfactory neurons *in vivo*. Nature 383, 624–627

Turner D L and Cepko C L (1987) A Common progenitor for neurons and glia persists in rat retina late in development. Nature 328, 131–136

Walsh C and Cepko C L (1988) Clonally related cortical cells show several migration patterns. Science 241, 1342–1345

Ware C B Horowitz M C Renshaw B R Hunt J S Liggitt D Koblar S A Gliniak B C Mckenna H J Stamatoyannopoulos G Thoma B Donovan P J Peschon J J Bartlett P F Willis C R Wright B D Carpenter M K Davison B L and Gearing D P (1995) Targeted disruption of the leukemia inhibitory factor receptor b gene causes placental, skeletal, neural and metabolic defects and results in perinatal cell death. Development 121, 1283–1299

Weiss S Dunne C Hewson J Wohl C Wheatley M Peterson A C and Reynolds B A (1996) Multipotent CNS stem cells are present in the adult mammalian spinal cord and ventricular Neuroaxis. Journal of Neuroscience 16, 7599–7609

Winkler C Fricker R A Gates M A Olsonn M Hammang J P Carpenter M K and Björklund A (1998) Incorporation and glial differentiation of mouse EGF- responsive neural progenitor cells after transplantation into the embryonic rat brain. Molecular and Cellular Neuroscience 11, 99–116

Wong G H Bartlett P F Clark-Lewis I Battye F and Schrader J W (1984) Inducible expression of H-2 and Ia antigens on brain cells. Nature 310, 688–691

11

The low affinity neurotrophin receptor, p75: A multifunction molecule with a role in nerve regeneration?

I A Ferguson, J J Lu, X F Zhou and R A Rush

Introduction and overview

Neurotrophic factors play important roles in the normal development, maintenance and repair of the nervous system (review: Lewin & Barde, 1996). One family of neurotrophic factors, the neurotrophins, has been described from their relationship with the well characterised nerve growth factor (NGF). The neurotrophins can bind specifically to two types of cell surface protein: the trk family of transmembrane receptor tyrosine kinases (Barbacid, 1995) and p75, the low affinity neurotrophin receptor (Chao & Hempstead, 1995). The high affinity neurotrophin receptors (trkA, trkB and trkC) mediate the biological effects of neurotrophins in responsive neurons; however, the function of the p75 receptor is less clear. Studies with some null p75 mice suggest p75 is not essential for mediating the action of neurotrophins on neurons, but these studies are not conclusive. Accumulating evidence has shown that p75 may have functional roles including mediating the binding of NGF to the trkA receptor and directly or indirectly promoting apoptosis. The observation that Schwann cells dramatically upregulate their expression of p75 after axotomy led to the hypothesis that p75 played an important and perhaps critical role in peripheral nerve regeneration (Johnson, 1988). Although initial studies using null p75 mice indicated that this was not the case, more recent studies support the notion that p75 is involved in facilitating the directional growth of neurotrophin responsive neurons in association with glia expressing high levels of p75. This selective overview of the p75 literature focused on the roles of p75 particularly after nerve injury, but also includes a brief outline of its properties and diverse functions to aid the non-specialist reader. This approach also seems appropriate to help appreciate the many changes in focus which have resulted from each new wave of research into this enigmatic molecule. Readers are referred to a number of excellent reviews for specific details of neurotrophins, trks and p75 (details below).

Background: the p75 molecule

Neurotrophins and their receptors

The mammalian neurotrophin family comprises four structurally related molecules: nerve growth factor (NGF), brain derived neurotrophic factor (BDNF), neurotrophin-3 (NT-3), and neurotrophin-4/5 (NT-4/5) (review: Lewin & Barde, 1996). The neurotrophins are highly basic proteins of around 120 amino acids in length and each is processed from a larger precursor which is subsequently cleaved to yield mature neurotrophins. Each neurotrophin is a homodimer, with the monomer characterised by three disulphide bridges flanked by regions of high homology among all the neurotrophins. The molecular domains on the neurotrophin molecule involved in binding to either the trk or p75 molecules have been identified. The regions involved in binding to the trk receptors differ significantly between the different members of the neurotrophin family. The regions implicated in binding to p75 have been less rigorously studied but it is known that the V1 loop of NGF is involved.

In addition to stimulating neuron growth, the interaction of neurotrophins with the trk receptors promotes neuron survival by preventing apoptosis: elimination of either the trk receptor via deletion of the trk gene, or elimination of the neurotrophin ligand for the trk receptor leads to neuronal cell death (reviews: Barbacid, 1995; Bothwell, 1995). The trk family of receptors were discovered and identified as high-affinity (K_d about 10^{-11} M) neurotrophin receptors in the early 1990s (review: Chao & Hempstead, 1995). They have a single transmembrane spanning region and exhibit tyrosine kinase activity following neurotrophin binding. Three members of the trk family have been identified and named trkA, trkB and trkC. While some cross-talk is apparent, in general, trkA functions as a high affinity receptor for NGF, trkB functions as the high affinity receptor for BDNF and NT-4/5, and trkC functions as the high affinity receptor for NT-3 (review: Barbacid, 1995). The neurotrophin responsiveness of a particular neuronal type correlates with its expression of trkA, B or C. In contrast to trkA, which is expressed by relatively few neuronal cells in the central nervous system (CNS), transcripts for trkB and trkC are widely distributed throughout the brain and many CNS neurons express both trkB and trkC as well as alternatively spliced trkB or trkC receptors of different lengths (Barbacid, 1995; Bothwell, 1995).

p75 differs from trk receptors in terms of both the dissociation constant (Kd of approx. 10^{-9} M) and the kinetics of receptor association. Indeed, p75 was previously termed the fast NGF receptor because NGF dissociates from the p75 receptor at a considerably faster rate than from the high affinity (also termed slow) receptor. The p75 gene from a number of different species was cloned in the mid 1980's and in all these species, p75 is a single peptide chain of approximately 400 amino acid residues, with a single membrane spanning domain separating a slightly longer extracellular domain from a shorter cytoplasmic domain. The extracellular domain of p75 can be proteolytically clipped from cells, but retains its capacity to bind neurotrophins (DiStefano & Johnson, 1988).

Cellular origin of p75

As might be expected of a molecule able to mediate the physiological effects of neurotrophins, p75 is expressed by a wide range of neurons in the CNS and peripheral nervous system (PNS) at different times during development and into maturity (Koh *et al.*, 1989). For example, p75 immunoreactivity can be detected in spinal motor neurons between embryonic day 15 and post-natal day 10, but not in the adult rat (Koh *et al.*, 1989). Within the postnatal rat brain, p75 is widely expressed within individual neurons in different areas including neurons of the basal forebrain (cholinergic neurons), caudate/putamen, ventral premammillary nucleus, mesencephalic trigeminal nucleus, prepositus hypoglossal nucleus, raphe nucleus, nucleus ambiguous and Purkinje cells of the cerebellum (Koh *et al.*, 1989). In the mature human spinal cord at cervical level, p75 immunoreactivity is not observed in the spinal motor neurons but can be observed within the dorsal root and Lissauer's tract, within substantia gelatinosa (lamina II), to a lesser extent in laminae III-IV, and in the spinal cord homologue of the descending spinal nucleus of nerve V (Mufson *et al.*, 1992). After spinal cord injury, p75 mRNA levels in the injury region increase about 7-fold over laminectomy control, with a maximal expression at about 7 days post injury (Brunello *et al.*, 1990).

In addition to its expression by many neuronal populations, p75 has been identified in a variety of non-neuronal cells of neuroectodermal origin, usually in the absence of trk receptors. These include Schwann cells, melanocytes, meningeal and glial cells, as well as other non-neuronal cells of various developmental lineages such as basal keratinocytes, epithelial cells in the ducts and acini of mammary and prostatic glands and spleen cells and in skin mesenchyme, somites and muscle anlage, testes and kidney (see e.g., Wyatt *et al.*, 1990). Since many of these cells and tissues do not respond to neurotrophins, it is unclear what function p75 serves in these tissues.

Within the non-neuronal cells of the peripheral nervous system, the pattern of expression of p75 by Schwann cells is distinctive: Schwann cells strongly express p75 during development or in culture but down-regulate their expression as the nervous system matures (DiStefano & Johnson, 1988). Separation of the Schwann cell from the functional nerve axon (eg, following nerve injury or removal into culture) stimulates the upregulation of p75 expression by Schwann cells; p75 expression down-regulates again once contact with the functional nerve axon is re-established. In contrast, the CNS glial cells, astrocytes and oligodendrocytes, do not express significant levels of p75, either during development or in response to axonal contact (see Gai *et al.*, 1996). The expression of high levels of p75 by peripheral glia but its absence from most central glia, has prompted speculation about the possible functions of p75 in contributing to the more permissive, regenerative environment of peripheral nerves compared with white matter tracts of the CNS.

p75: What does it do?

Overview of approaches to study of p75 function

The history of significant advances in attempts to determine the physiological functions of p75 correlates with the development of new tools to probe the function of the molecule. During the late 1970s and early 1980s, the availability of radiolabelled NGF of high specificity provided a method of tagging and identifying the p75 molecule for biochemical, kinetic and *in vivo* retrograde tracer studies. Information was generated about the size and distribution of p75 and the kinetics of NGF binding. These studies strengthened the conclusion that there was a high affinity receptor which was different to the low affinity NGF receptor, p75. The generation of a monoclonal antibody to rat p75, 192-IgG, by Chandler *et al.* (1984) provided the next leap in understanding about p75. The antibody enabled immunohistochemical studies of the distribution and changes in expression of p75 within the nervous system as well as reinforcing conclusions made using radiolabeled NGF. The hypothesis that p75 plays an important role in nerve regeneration emerged as a result of these studies.

The cloning of the gene encoding for p75 from a number of different species in 1986 and 1987 and the subsequent availability of the p75 gene sequence for use with molecular biology techniques provided the next advance in understanding. *In situ* hybridization studies followed which reinforced the data from immunohistochemical studies. Recombinant expression of the p75 gene reinforced the conclusion that p75 was not a signal transducing receptor for NGF, although this view has subsequently been modified (see below). The unexpected discovery in the early 1990s that the trk receptor was the high affinity and signal transducing NGF receptor, coupled with the cloning of NT-3 and NT-4 neurotrophins and the trkB and trk C receptors ushered in a shift in focus towards the trk family of neurotrophin receptors. This shift in scientific interest was reinforced by the generation of transgenic null p75 mice (Lee *et al.*, 1992) which initially suggested that p75 had relatively minor effects compared with dramatically defective nervous system development observed in the transgenic null trk mice. Thus, throughout much of the early 1990s, p75 appeared to be a molecule searching for a function. More recently, the availability of tools for studying apoptosis has led to the demonstration that p75 can stimulate apoptosis, renewing scientific interest in p75 and its function. Below we provide a selective overview of some of these studies and their primary conclusions.

Can p75 modulate the binding of neurotrophin to trk receptors?

Compared with trk receptors, p75 is typically expressed by neurons at higher copy numbers. Co-expression of low and high affinity receptor classes can be expected to concentrate and facilitate transfer of ligand to high affinity receptors. Indeed, p75 has been shown to enhance neurotrophin binding to the trk proteins and increase neurotrophin responsiveness and function (Davies *et al.*, 1993; Verdi *et al.*, 1994). In addition to these simple ligand concentration effects, some studies indicate that p75 may specifically modulate the function of trk receptors. For example, coexpression of both p75 and trk results in the appearance of high affinity NGF binding sites not seen in the absence of coexpression (Chao & Hempstead, 1995). Other studies support this conclusion: the number of NGF high affinity binding sites in PC-12 cells and in sensory neurons is decreased by adding antibodies to p75 or BDNF (because, at these concentrations, BDNF binds to p75 but not trkA; Barker & Shooter, 1994). Also, transfection of the p75 gene into mutant PC-12 cells which normally lack p75, restores high affinity binding and NGF-induced (trk-mediated) protein phosphorylation (Pleasure *et al.*, 1990). An 8 fold higher level of tyrosine phosphorylation by trkA in response to NGF binding in sympathoadrenal cells is seen when these cells co-express trkA and p75 (Verdi *et al.*, 1994). Mutant NGF, which binds to and activates trkA but not p75, is as potent as wild type NGF in stimulating fibroblasts transfected with trkA alone, but 3 to 4 fold less potent in stimulating sensory or sympathetic neurons indicating that p75 enhances the binding of NGF to trkA at low concentrations in neurons (Ryden *et al.*, 1997).

Is p75 involved in the retrograde transport of neurotrophins?
Receptor mediated internalisation and retrograde axonal transport of trophic factors from innervated target tissues is a fundamental feature of the theory of neurotrophic factor action. It is unclear whether p75 has either a critical, or simply facilitatory role in the internalisation and retrograde transport of the NGF-trkA complex. Involvement of p75 in the phenomenon of retrograde transport of neurotrophic factors is suggested by the reduction in retrograde transport of neurotrophins when neutralising antibodies to p75 are administered with the neurotrophins (Curtis *et al.*, 1995), although reduced transport may be the result of decreased internalisation at the nerve terminals. Furthermore, p75 can be retrogradely transported. Retrograde transport of radiolabelled antibody to p75 to the spinal motor neurons can be demonstrated following injection into the lower limb and this ability of spinal motor neurons to retrogradely transport ^{125}I-192-IgG is lost as the animal matures and the spinal motor neurons to cease express p75 (Yan *et al.*, 1989).

While suggestive, these studies do not allow the conclusion that p75 is involved in retrograde axonal transport of neurotrophins because of characteristics of the antibody to p75 used in the study. In contrast to antibodies which block neurotrophin binding to its receptor, the monoclonal antibody, IgG-192, can actually augment retrograde transport of NGF within sympathetic nerves (see Johnson, 1988). This antibody differs from many others in that 192-IgG was selected by its ability to increase, not block, the binding of ^{125}I-NGF to p75 (Chandler *et al.*, 1984). The ability of 192-IgG to increase the strength of binding of ^{125}I-NGF to p75 may indicate that the bi-functional monoclonal antibody leads to the formation of p75 dimers with a consequent decrease in the rate of dissociation of ^{125}I-NGF from a pair of p75 molecules. Note that a decrease in the rate of dissociation of NGF from p75 would manifest as an increase in the rate of binding because NGF has both a fast rate of association with p75 (on-rate) and a fast rate of dissociation (off-rate). The binding of ligands to receptors leading to receptor dimerisation and then receptor-ligand internalisation is a phenomenon common to many growth factors and their receptors. Ligand induced p75 dimerisation by ^{125}I-192-IgG may explain the internalisation and retrograde transport to the spinal motor neuron cell body observed by Yan *et al.* (1989).

p75 is conserved during evolution
Discovery of the p75 gene sequence enabled study of the amino acid sequence for clues to its possible functions. The p75 sequence is well conserved during evolution, suggesting that p75 subserves some physiologically important functions. 71% of the amino acid residues of the p75 sequence are identical in chicken, rat and the human molecule. This is comparable with the greater than 80% conservation of sequence in the p75 ligands, NGF, BDNF and NT-3 between human, chicken and rat. The most strongly conserved region of the p75 molecules (95% between chicken, rat and human) is the transmembrane sequence and the sequences immediately flanking it: the last 19 residues of the extracellular region and the first 46 residues of the cytoplasmic region;

the 22 hydrophobic residues spanning the membrane are conserved precisely. This suggests that this region may be critical for the biological function(s) of p75.

Transgenic animal studies

Genetic deletion studies can provide clues about physiological function of the deleted gene. The argument that p75 subserves functions as important as the trk receptors appears not to be supported by p75 knockout studies which demonstrate that, unlike deletions of any of the trk receptor genes which are lethal, deletion of the p75 gene is not lethal (Lee *et al.*, 1992). The first null p75 mice appeared superficially normal. Closer (later) examination revealed that up to 50% of the trkA positive, NGF responsive sensory neurons are lost in the null p75 mice (Stucky & Koltzenburg, 1997). These neurons mediate temperature and pain sensation and the null p75 mice lack the normal sensory innervation of footpad skin by calcitonin gene related peptide (CGRP) and substance P immunoreactive (trkA expressing) fibres. They also have low heat sensitivity. The fact that this abnormal phenotype can be corrected by crossing a human transgene encoding p75 into the p75–/– animals (Lee *et al.*, 1992) implicates p75 in mediating the normal development of these trkA positive sensory neurons.

The conclusion that p75 can modulate binding of NGF to trkA is supported by transgenic mice studies. Primary sensory neurons from null p75 mice require approximately 3–4 times higher levels of NGF to elicit a survival response than neurons which express p75 (Davies *et al.*, 1993). Lee *et al.* (1992, 1994) showed that a two fold greater level of NGF is required to stimulate sensory neurons removed from dorsal root ganglia of null p75 mice at the peak of naturally occurring cell death period (embryonic day 15) compared with ganglia from normal mice. Interaction of p75 with trkA to enhance NGF binding and signal transduction would explain the substantial loss of trkA positive neurons in the null p75 animal (Davies *et al.*, 1993); in the absence of p75, the greater levels of naturally occurring neuronal cell death occurs where NGF synthesis rate is unaltered.

The null p75 mouse studies have also implicated p75 in normal function of other classes of neurons including the trkC expressing sensory neurons and the trkA expressing sympathetic neurons. The large diameter sensory neurons in dorsal root ganglia are typically trkC expressing, NT-3 responsive and immunoreactive, and mediate proprioception and mechanoception. Mechano-transduction by sensory C fibres and D-hair receptors is abnormal in the null p75 mice. Like many small diameter sensory neurons, sympathetic neurons also express trkA and are NGF responsive. As with trkA expressing sensory neurons (see Ryden *et al.*, 1997), three fold greater levels of NGF are required to stimulate sympathetic neurons removed from the superior cervical ganglia of null p75 mice at the peak of naturally occurring cell death period for sympathetic neurons (postnatal day 3).

The possible involvement of p75 in the phenomenon of retrograde transport of neurotrophic factors also has been studied using null p75 mice. For example, Curtis *et al.* (1995) showed that similar levels of radiolabelled BDNF are retrogradely transported within the spinal motor neurons of null p75 and normal mice but that the amount of radiolabelled NGF, BDNF, or NT-4 retrogradely transported to the dorsal root ganglia was markedly reduced in null p75 mice. The low numbers of sensory neurons in p75 knockout mice make it difficult to determine whether the low transport rates are the result of absence of p75 mediated retrograde transport or a function of the reduced number of neurons transporting the labelled neurotrophic factors.

Limitations of the null p75 transgenic mice studies

The null p75 transgenic mice studies of Lee *et al.* (1992, 1994) have been used as evidence for a non-critical role of p75 in the normal development and maintenance of neurotrophin dependent neurons. However, this conclusion assumes that the gene deletion strategy ablates all p75 specific functions. This may not be a valid assumption given that only the extracellular neurotrophin binding domain was deleted in these mice leaving intact the promoter, transmembrane domain and intracellular domain of p75 gene. The gene sequence encoding the highly conserved region of the p75 molecule (the transmembrane and flanking sequences) remains intact in these mice. Given that the promoter region remains intact, the possibility remains that p75 molecules devoid of an extracellular binding domain but with functional transmembrane and intracellular domains may be expressed in these null-p75 mice. It is interesting to note that proteolytic cleavage of the extracellular domain of p75 from the transmembrane and intracellular domain can occur as a normal consequence of p75 expression (DiStefano & Johnson, 1988) and that deletion of extracellular ligand binding domain of growth factor receptors such as the epidermal growth factor (EGF) receptor can result in a constituitively active receptor. It remains to be determined whether expression of p75 lacking the extracellular domain has any biological effect on normal neuronal or non-neuronal cell function.

p75 expression and apoptosis?

Apoptosis, occuring primarily during the period of naturally occuring neuron cell death, plays an important role in the normal development of the nervous system and there is evidence that p75 can mediate apoptosis. The first evidence of p75 involvement in apoptosis was reported in 1993. Transfection of a neural cell line with p75 induced neural cell death constituitively in the absence of a p75 ligand (NGF or antibody), whereas addition of the p75 ligand prevented neural cell death (Rabizadeh *et al.*, 1993). This is analogous to CD40 induced cell death, in which binding of monoclonal antibody or ligand binding to the CD40 inhibits activation of apoptotic pathways (see Itho & Nagata, 1993). Application of p75 antisense oligonucleotide to the proximal stump of transected peripheral nerve effectively reduces p75 protein levels in sensory neurons located in the dorsal root ganglia and reduces the loss of these axotomized sensory neurons (see chapter 10 by Bartlett *et al.*, this volume). A pro-apoptotic effect of p75 has also been observed in primary cultures of sensory neurons. Early in development, endogenous NGF causes the death of retinal neurons that express p75, but not neurons which also express trkA (Frade *et al.*, 1996). Also within the CNS, approximately 25% of the trkA expressing basal forebrain cholinergic neurons normally die in mice between postnatal days 6 and 15, but fail to die in p75 knockout mice or in mice injected with a peptide which inhibits NGF binding to p75 (Van der Zee *et al.*, 1996).

The possibility that p75 may mediate apoptosis is supported by observations of sequence similarity with other molecules known to mediate apoptosis such as the receptors to tumour necrosis factors I and II (TNFRI and TNFRII), the human cell surface antigen Fas (Apo-1) and the B cell antigen CD-40 (see Itoh & Nagata, 1993). While

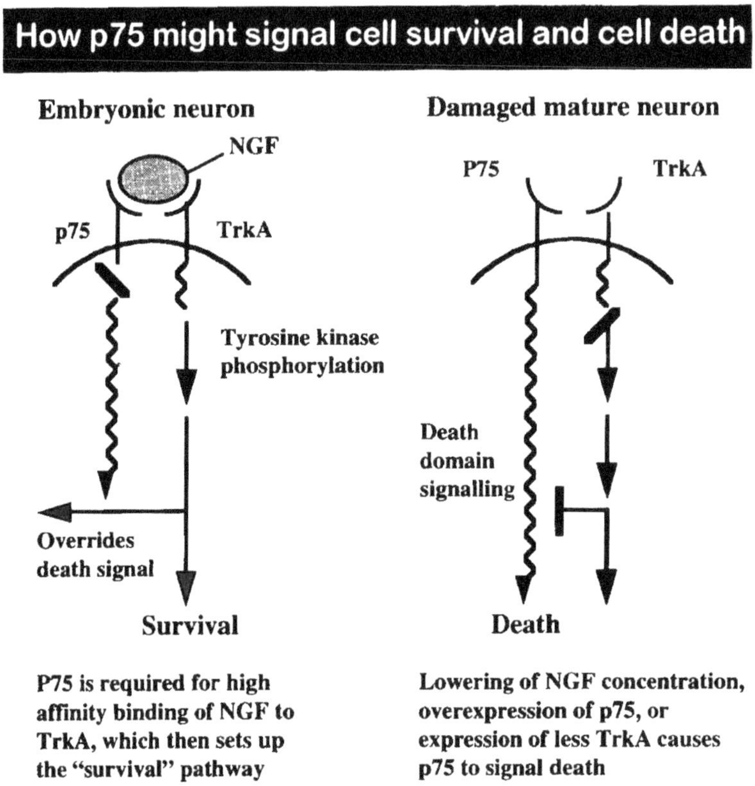

Figure 1 Model to explain a possible relationship between neurotrophin signalling via p75 and trkA in some neurons. According to this model, NGF binding to p75 activates an apoptotic pathway leading to death of the cell, but, when trkA is present, NGF interaction with trkA activates a counter-balancing pathway which prevents apoptosis and stimulates nerve growth. Redrawn from: Finkel (1996).

p75 does not activate tyrosine kinases (Barbacid, 1995; Chao & Hempstead, 1995), NGF can signal in non-transformed cells in the absence of trkA through p75 as evidenced by the nuclear translocation of the transcription factor NF-κB (Carter *et al.*, 1996). The second messenger involved in p75 mediated signal transduction may be the lipid messenger, ceramide. Transient (not sustained) increase in intracellular levels of ceramide has been observed in 3T3 cells transfected with p75 following addition of NGF, BDNF, NT-3 or NT4 (Dobrowsky *et al.*, 1994). This transient increase in ceramide levels is not associated with stimulation of apoptosis in these 3T3 cells transfected with p75 but sustained increase in ceramide and c-Jun amino-terminal kinase activity has been reported to lead to apoptosis following administration of NGF to oligodendrocytes cultured to express p75, but not trkA (Casaccia-Bonnefil *et al.*, 1996).

It is unclear whether all neurotrophins can stimulate apoptosis after binding to p75. While Casaccia-Bonnefil *et al.* (1996) reported that only NGF, and not BDNF, NT-3 or NT-4, can induce apoptosis in these p75 expressing oligodendrocytes, Bamji *et al.* (1998) showed that BDNF, which can activate p75 but not trkA, can induce apoptosis in sympathetic neurons (which express p75 and trkA). Bamji *et al.* (1998) suggested that p75 may be the mediator of naturally occurring neuronal cell death in the sympathetic nervous system. Finkel (1996) presented a model (Figure 1) to explain how p75 may lead to activation of apoptotic pathways in some neurons but not in others. According to this model, NGF binding to p75 activates an apoptotic pathway leading to death of the cell, but, when trkA is present, NGF interaction with trkA activates a counter-balancing pathway which prevents apoptosis and stimulates nerve growth.

p75 expression changes in response to injury

In many ways the response to injury in the nervous system recapitulates key events in the development of the nervous system. For example, within the PNS following axotomising injury, neurons commence regrowing towards their effector tissues, and Schwann cells deprived of contact with functioning nerve axons, de-differentiate and express high levels of p75.

Such changes in p75 expression during development and following injury are well illustrated in the spinal motor neuron. During development, p75 is expressed in spinal motoneurons at relatively high levels at the time of naturally occurring cell death and this expression then decreases to a relatively low level in adult. A unilateral crush lesion of the sciatic nerve resulted in an 8-fold increase in p75 mRNA in adult rat spinal cord motoneurons by 3 days after lesion which decreases to normal levels following motor nerve reinnervation of muscle and restoration of motor function. Similarly, p75 mRNA in facial motoneurons is also transiently increased following either facial nerve crush or transection (Ferri *et al.*, 1998). Armstrong *et al.* (1991) speculated that the upregulation of p75 in the motor neurons of the hypoglossal nucleus following injury (Figure 2) may be mediated by factor(s) released at the site of injury rather than by loss of a repressing signal from the innervated peripheral field, but this proposal has not been tested further. p75 is also elevated in spinal motor neurons in motor neuron disease (amyotrophic lateral sclerosis) (Seeburger *et al.*, 1993). Modulation of the p75 expression in injured motor neurons, perhaps with the use of anti-sense oligonucleotides may provide a mechanism to prevent p75 induced death of motor neurons following injury or in disease (see also Finkel, 1996).

The situation with motor neurons contrasts with that observed in sensory or sympathetic neurons. For example, both sensory (Zhou *et al.*, 1996) and sympathetic neurons (Miller *et al.*, 1994) down-regulate their expression of p75 following axotomy injury. This reduced expression of p75 has been suggested to protect these neurons from p75 mediated apoptosis associated with the interruption of the retrograde flow of NGF (Miller *et al.*, 1994).

Peripheral nerve injury results in a prominent upregulation of p75 expression by Schwann cells following loss of axonal contact which is maintained until the injured nerve re-establishes contact with the Schwann cell (Taniuchi *et al.*, 1986). This phenomenon, coupled with proteolytic cleavage of the p75 extracellular domain from the Schwann cell, results in elevated levels of truncated p75 in the blood and urine of patients with disorders such as peripheral neuropathy or motor neuron disease (amyotrophic lateral sclerosis) (see Lindner *et al.*, 1993).

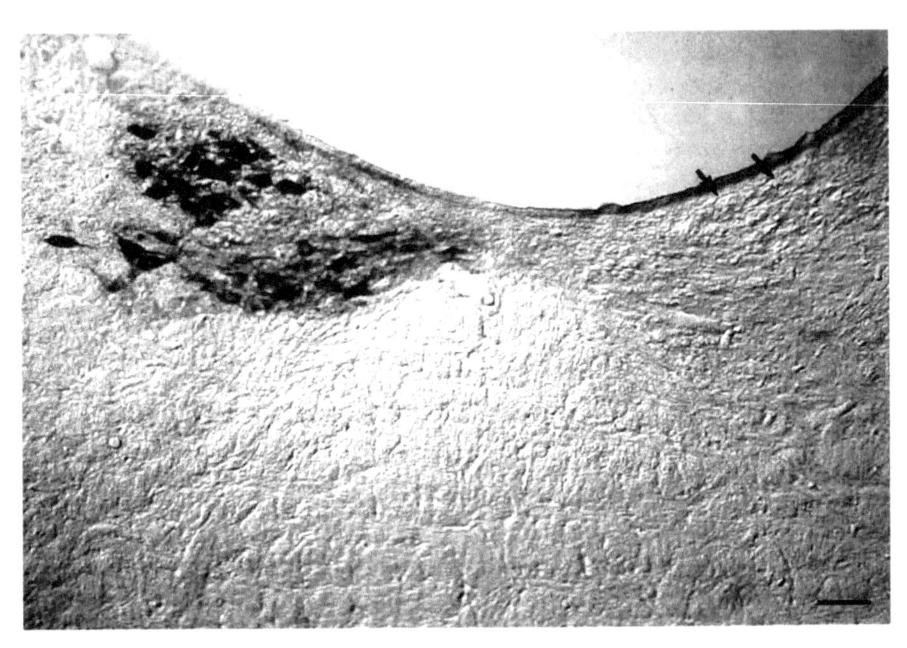

Figure 2 p75 is upregulated in motor nerves following peripheral nerve injury. p75 immunoreactivity in motor neurons of the hypoglosal nucleus following crush of the facial nerve. Note internal control (arrows): absence of p75 in facial motoneurons projecting through uninjured facial nerve. Scale bar = 10 μm.

The up-regulation of p75 expression by Schwann cells following peripheral nerve injury and axonal degeneration has been hypothesised to facilitate regeneration of the regrowing nerve along the p75 rich Schwann cells and back to their target tissue (Johnson, 1988). NGF is concentrated on the p75 expressed by Schwann cells following axonal degeneration (Figure 3). It therefore has been proposed (Figure 4) that the increased synthesis of p75 that occurs in peripheral nerve following injury endows the Schwann cell with the ability to concentrate and present neurotrophins to the high affinity receptors expressed on the outgrowing nerve growth cone (Taniuchi *et al.*, 1986). That is, p75 mediates a net transfer of NGF from the Schwann cell, where it is abundant, to the higher affinity receptors expressed on the regrowing neuron and that this transfer is facilitated by the kinetic differences between the two receptors: NGF dissociates much more rapidly from the p75 receptor than from the high affinity receptor.

Is p75 critical for nerve regeneration?

The Taniuchi hypothesis proposes that p75 plays a critical role in the concentration of neurotrophin molecules on the surface of Schwann cells and their presentation to the trk receptors expressed on the surface of the regenerating neuron's growth cone. This would predict that while nerve regeneration still might be expected to occur in the absence of p75, all other things being equal, the regenerative responsive might be expected to be less robust (due to less concentrated neurotrophin source) and with

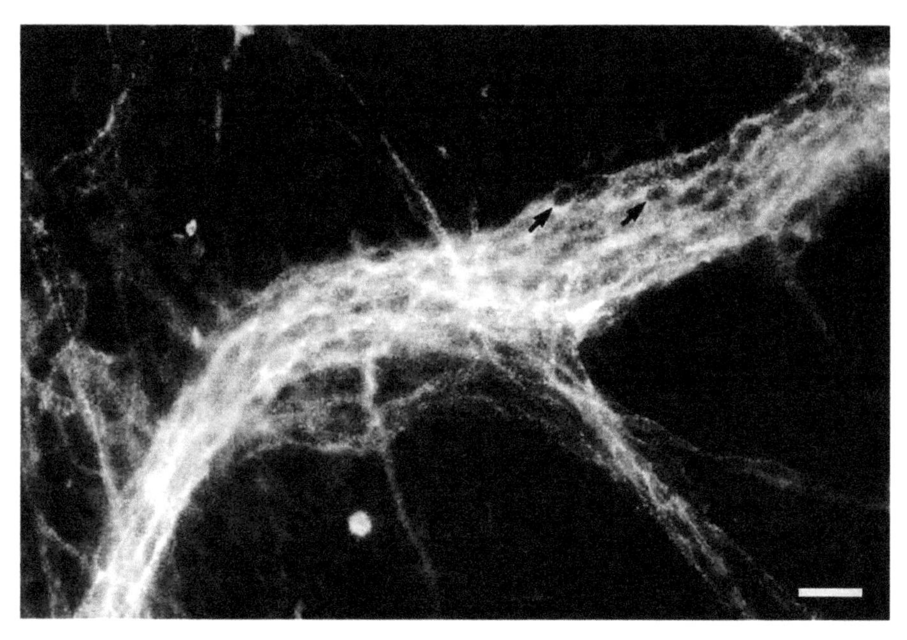

Figure 3 NGF is concentrated on the p75 expressed by Schwann cells following axonal degeneration. NGF immunoreactivity on Schwann cells within peripheral nerves of the iris denervated by removal into tissue culture. Note Schwann cell nuclei (arrows). From: Rush (1984). Scale bar = 10 μm.

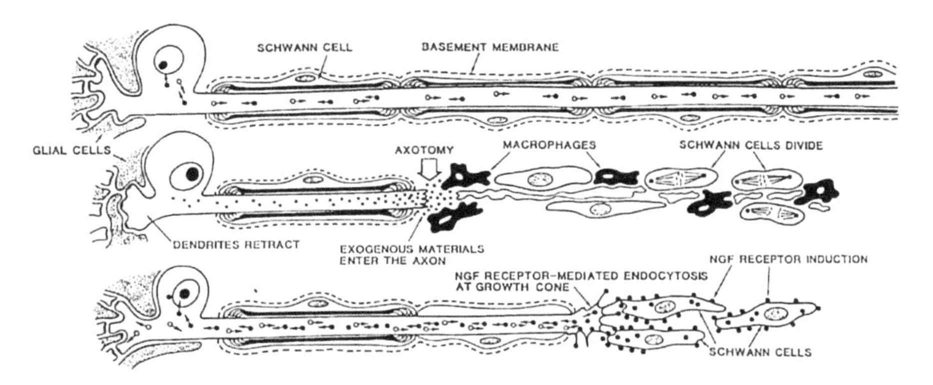

Figure 4 Hypothesised relationship between regenerating peripheral nerves and p75 expressed on Schwann cells. Increased synthesis of p75 by Schwann cells distal to a peripheral nerve injury endows the Schwann cell with the ability to concentrate and present neurotrophins to the high affinity receptors expressed on the outgrowing nerve growth cone. Adapted from: Taniuchi *et al.* (1986).

outgrowth perhaps less directed along the Schwann cells. In fact, it appears that p75 expression is not critical for the normal regeneration of the injured motor nerves. Indeed, functional recovery of whisker movement following a facial nerve crush occurred slightly earlier in adult null p75 mice than in wild type mice and by 25 days following axotomy, the survival profile in the adult p75(–/–) mice was significantly improved compared to p75(+/+) mice (Ferri *et al.*, 1998). Observations that treatment of lesioned sciatic nerve with NGF has no effect on either the length of regenerating axons, their number or their ability to reinnervate effector tissues provide further arguments against a role for glial p75 in facilitating nerve regeneration (Wyatt *et al.*, 1990).

However, a different model system provides evidence to support the notion that p75 mediates the directional growth of neurotrophin responsive axons on p75 expressing glial cells. Following peripheral nerve injury, the sympathetic nerves which innervate the blood vessels penetrating the dorsal root ganglia, sprout and grow towards large sensory neurons and form baskets around some of these cells (McLachlan *et al.*, 1993; Zhou *et al.*, 1996). Analogous to the up-regulation of p75 by Schwann cells deprived of axonal contact, this phenomenon correlates with an up-regulation in the levels of p75 expression by satellite glial cells within the dorsal root ganglia (McLachlan *et al.*, 1993; Zhou *et al.*, 1996). Sympathetic sprouting occurs in null p75 mice with normal (Ramer & Bisby, 1997) or enhanced levels of NGF expression (Walsh *et al.*, 1999) indicating that upregulation of p75 by satellite cells is not essential for sympathetic sprouting. However, the resulting pattern of sympathetic axon ingrowth in the null p75 mice is different to that in p75 expressing mice: the basket formation is less robust (Ramer & Bisby, 1997) and the ingrowth of sympathetic axons is random and imprecise (Walsh *et al.*, 1999). These data support the notion that while p75 is not essential for regeneration, its presence facilitates the outgrowth of neurotrophin responsive axons so that they grow towards and maintain contact with p75 expressing glial cells.

While null p75 mice differ from normal mice in aspects of the response to injury, interpretation of the role of p75 in nerve regeneration in null p75 mice is complicated by the possibility that p75 is also involved in induction of neuronal cell death (see chapter 10 by Bartlett *et al.*, this volume). Delayed but eventually robust sensory and sympathetic sprouting occurs into the cerebellum of both p75(–/–) and p75(+/+) mice with CNS glial cells which over-express NGF, indicating p75 is not essential for mediating sprouting and innervation in these neurons (Coome *et al.*, 1998). More sensory innervation in tooth pulp can be found in aged p75–/– mice than the age matched wild type mice, suggesting p75 might be involved in the aging induced neuronal degeneration (Coome *et al.*, 1998). The level of expression of p75 in motor neurons increases following injury. Recently Ferri *et al.* (1998) showed using the newborn facial nerve crush model, that the level of injury induced motor neuron death is greater in animals which express p75 compared to null p75 animals suggesting that p75 expression may increase apoptosis in motor neurons following injury.

p75 in Schwann cell migration

Nerve regeneration is dependent on the presence of Schwann cells and following traumatic injury, nerves will not grow across a gap until the gap is first bridged by Schwann cells migrating from the cut nerve stumps (Fields *et al.*, 1989). Schwann cells move more rapidly in the presence than in the absence of NGF and blocking antibodies

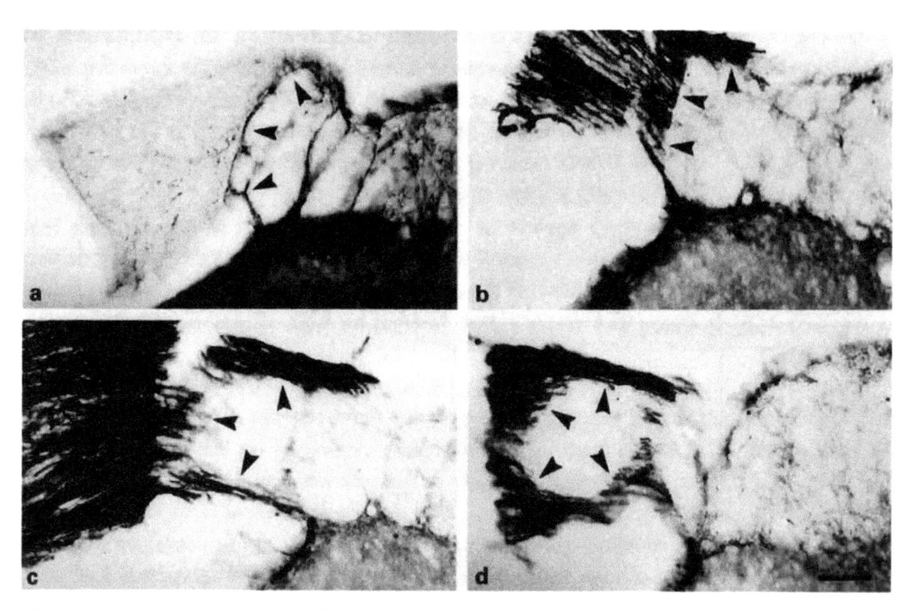

Figure 5 p75 is expressed by Schwann cells but not CNS following nerve injury. p75 expression in the dorsal root providing sensory input into the spinal cord. Arrowheads point to the transition between the glia of the peripheral nervous system (Schwann cells) and the glia of the CNS. From: Gai *et al.* (1996).

to either p75 or NGF inhibit this enhanced migration rate (Anton *et al.*, 1994). On the basis of these studies, these authors suggests that Schwann cells can use p75 to migrate in an NGF rich environment and speculate that poor Schwann cell migration following injury may in part account for the poor sensory re-innervation associated with the ulcers which develop in null p75 mice.

Unlike Schwann cells which upregulate p75 expression following nerve injury, CNS glia do not upregulate p75 expression following loss of contact with the nerve axon. This is clearly illustrated following injury to the dorsal root in which Schwann cells within the dorsal root synthesise high concentrations of p75 receptor, whereas oligodendrocytes immediately adjacent to these reactive Schwann cells and myelinating the same degenerating sensory axons, fail to produce detectable levels of p75 (Figure 5; Gai *et al.*, 1996). Neurotrophins are up-regulated in both central glia and peripheral Schwann cells (see eg, Funakoshi *et al.*, 1993). The failure of CNS glia to upregulate p75 expression raises the possibility that neurotrophins secreted from CNS glia diffuse away from these cells and that the absence of significant p75 expression by CNS glia may contribute to the relatively poor regeneration of CNS, compared to PNS, neurons.

Discussion and conclusions

There is evidence that p75 can modulate neurotrophin activity in both a trk dependent and independent manner. p75 can interact with trkA to form high affinity sites and to regulate trkA signalling. p75 can also act independently of trk by regulating apoptosis (Finkel, 1996) and the migration of Schwann cells (Anton *et al.*, 1994).

The neurotrophins can signal through p75 leading to stimulation of sphingomyelinase and subsequently increased ceramide production (Dobrowsky *et al.*, 1994), induction of apoptosis, as well as activation of gene transcription via NF-κB. Interestingly, while all neurotrophins can stimulate ceramide production via p75 binding, unlike NGF, NT-3 or BDNF binding to p75 does not stimulate NF-κB nuclear translocation in Schwann cells (Carter *et al.*, 1996).

While p75 can bind all neurotrophins, p75 gene deletion studies indicate that p75 influences NGF responsive neurons more than BDNF, NT-3 or NT-4 responsive neurons, presumably via modulation of the interaction of NGF with trkA. One possibility is that p75 may facilitate trkA dimerisation needed for high affinity binding (Carter & Lewin, 1997).

The high level of p75 expressed by Schwann cells when deprived of axonal contact was suggested as a mechanism to facilitate nerve regeneration by concentrating and presenting NGF to the high affinity receptors expressed on the regrowing nerve (Taniuchi *et al.*, 1986). Evidence from study of the p75 knockout mouse indicates that p75 is not critical for nerve regeneration. This does not necessarily indicate that the lack of a role for p75 in nerve guidance or function. Rather, it may indicate that p75 function is redundant because more than one molecule can subserve this function. Indeed, many neurotrophic factors including the glial cell line derived neurotrophic family act through both high and low affinity receptors and may act as parallel factors to the neurotrophins in many neuronal populations. Thus, further studies are required before the hypothesised mechanism can be dismissed.

The conservation during evolution of both low and high affinity receptors for several families of neurotrophic factors, suggests that low affinity neurotrophic factor receptors do subserve important physiological functions. Goodhill (1998) provided a theoretical model which indicates that the presence of a substrate bound gradient can increase the spatial limit for nerve axonal guidance due to a target diffusible gradient from approximately 1 mm, in the absence of substrate binding, to approximately 1 cm in the presence of substrate binding. It can be predicted that a low affinity receptor such as p75 can act as a substrate binding agent for neurotrophic factors and reduce the number of nerves growing to inappropriate targets during development or regeneration. The use of the p75 null mouse might allow these predictions to be tested (for example, by cross-transplantation of PNS tissue from p75(+/+) and p75(–/–) animals). Given the relative absence of p75 in the mature CNS compared to PNS, transfection of CNS glial cells might provide another approach for testing a possible effect of p75 on nerve regeneration.

Key points

- p75, the low affinity receptor for the neurotrophins, has alternatively been considered a central, or relatively unimportant, component of the neuronal mechanism required for neurotrophin signal transduction leading to survival, axonal outgrowth and a host of other responses.
- The view that p75 plays only a minimal role in regulating neuronal function was enhanced particularly by the initial findings from mutant animals lacking the extracellular domain of the receptor.
- p75 knockout animals survive with only minor functional consequences, but trk receptor deletion leads to a severe phenotype. Hence, the high affinity trk receptors again became the primary focus of intensive research.
- Recent work has re-instated the importance of p75, particularly in regulating cell death and in a manner which is independent of the activation of trk receptors.
- This chapter highlights the hypothesis, posed more than a decade ago, that p75 may act to encourage nerve regeneration, by creating a reservoir of neurotrophins ahead of regrowing nerve fibres.
- It is argued that little work has been done to directly test the hypothesis.
- The finding of other neurotrophic families which also have both high and low affinity receptors is highlighted as one possible mechanism involving overlapping activities which may mask the effect of gene deletion.

References

Anton E S Weskamp G Reichardt L F and Matthew W D (1994) Nerve growth factor and its low-affinity receptor promote Schwann cell migration. Proceedings of the National Academy of Sciences USA 91, 2795–2799

Armstrong D M Brady R Hersh L B Hayes R C and Wiley R G (1991) Expression of choline acetyltransferase and nerve growth factor receptor within hypoglossal motoneurons following nerve injury. Journal of Comparative Neurology 304, 596–607

Bamji S X Majdan M Pozniak C D Belliveau D J Aloyza R Kohn J Causing C G and Mille F D (1998) The p75 neurotrophin receptor mediates neuronal apoptosis and is essential for naturally occurring sympathetic neuron death. Journal of Cell Biology 140, 911–923

Barbacid M (1995) Neurotrophic factors and their receptors. Current Opinion in Cell Biology 7, 148–155

Barker P A and Shooter E M (1994) Disruption of NGF binding to the low affinity neurotrophin receptor p75LNTR reduces NGF binding to TrkA on PC12 cells. Neuron 13, 203–215

Bothwell M (1995) Functional interactions of neurotrophins and neurotrophin receptors. Annual Review of Neuroscience 18, 223–253

Brunello N Reynolds M Wrathall J R and Mocchetti I (1990) Increased nerve growth factor receptor mRNA in contused rat spinal cord. Neuroscience Letters 118, 238–240

Carter B D and Lewin G R (1997) Neurotrophins live or let die: Does p75NTR decide? Neuron 18, 187–190

Casaccia-Bonnefil P Carter B D Dobrowsky R T and Chao M V (1996) Death of oligodendrocytes mediated by the interaction of nerve growth factor with its receptor p75. Nature 383, 716–719

Chao M V and Hempstead B L (1995) P75 and trk – a two-receptor system. Trends in Neurosciences 18, 321–326

Chandler C E Parsons L M Hosang M and Shooter E M (1994) A monoclonal antibody modulates the interaction of nerve growth factor with PC12 cells. Journal of Biological Chemistry 259, 6882–6889

Coome G E Elliott J and Kawaja M D (1998) Sympathetic and sensory axons invade the brains of nerve growth factor transgenic mice in the absence of p75NTR expression. Experimental Neurology 149, 284–294

Curtis R Adryan K M Stark J L Park J S Compton D L Weskamp G Huber L J Chao M V Jaenisch R Lee K F Lindsay R M and DiStefano P S (1995) Differential role of the low affinity neurotrophin receptor (p75) in retrograde axonal transport of the neurotrophins. Neuron 14, 1201–1211

Davies A M Lee K-F and Jaenisch R (1993) p75-deficient trigeminal sensory neurons have an altered response to NGF but not to other neurotrophins. Neuron 11, 565–574

DiStefano P S and Johnson E M Jr (1988) Nerve growth factor receptors on cultured rat Schwann cells. Journal of Neuroscience 8, 231–241

Dobrowsky R T Werner M H Castellino A M Chao M V and Hannun Y A (1994) Activation of the sphingomyelin cycle through the low-affinity neurotrophin receptor. Science 265, 1596–1599

Fields R D Le Beau J M Longo F M and Ellisman M H (1989) Nerve regeneration through artifical tubular implants. Progress in Neurobiology 33, 87–134

Ferri C C Moore F A and Bisby M A (1998) Effects of facial nerve injury on mouse motoneurons lacking the p75 low-affinity neurotrophin receptor. Journal of Neurobiology 34, 1–9

Finkel E (1996) p75 blockade – can it prevent neuronal death. Lancet 347, 1684

Frade J M Rodrigueztebar A and Barde Y A (1996) Induction of cell death by endogenous nerve growth factor through its p75 receptor. Nature 383, 166–168

Funakoshi H Frisén J Barbany G Timmusk T Zachrisson O Verge V M K and Persson H (1993) Differential expression of mRNAs for neurotrophins and their receptors after axotomy of the sciatic nerve. Journal of Cell Biology 123, 455–465

Goodhill G J (1998) Mathematical guidance for axons. Trends in Neuroscience 21, 226–231

Gai W P Zhou X F and Rush R A (1996) Analysis of low affinity neurotrophin receptor (p75) expression in glia of the CNS-PNS transition zone following dorsal root transection. Neuropathology and Applied Neurobiology 22, 434–439

Itoh N and Nagata S (1993) A novel protein domain required for apoptosis. Mutational analysis of human Fas antigen. Journal of Biological Chemistry 268, 10932–10937

Johnson E M Jr (1988) Regulation of nerve growth factor receptor expression on Schwann cells. Progress in Brain Research 78, 327–331

Kerkhoff H Jennekens F G Troost D and Veldman H (1991) Nerve growth factor receptor immunostaining in the spinal cord and peripheral nerves in amyotrophic lateral sclerosis. Acta Neuropathologica 81, 649–656

Koh S Oyler G A and Higgins G A (1989) Localization of nerve growth factor receptor messenger RNA and protein in the adult rat brain. Experimental Neurology 106, 209–221

Lee K-F Li E Huber L J Landis S C Sharpe A H Chao M V and Jaenisch R (1992) Targeted mutation of the gene encoding the low affinity NGF receptor p75 leads to deficits in the peripheral sensory nervous system. Cell 69, 737–749

Lee K-F Davies A M and Jaenisch R (1994) p75-deficient embryonic dorsal root sensory and neonatal sympathetic neurons display a decreased sensitivity to NGF. Development 120, 1027–1033

Lewin G R and Barde Y A (1996) Physiology of the neurotrophins. Annual Reviews of Neuroscience 19, 289–317
Lindner M D Gordon D D Miller J M Tariot P N McDaniel K D Hamill R W DiStefano P S and Loy R (1993) Increased levels of truncated nerve growth factor receptor in urine of mildly demented patients with Alzheimer's disease. Archives of Neurology 50, 1054–1058
McLachlan E M Janig W Devor M and Michaelis M (1993) Peripheral nerve injury triggers noradrenergic sprouting within dorsal root ganglia. Nature 363, 543–545
Miller F D Speelman A Mathew T C Fabian J Chang E Pozniak C and Toma J G (1994) Nerve growth factor derived from terminals selectively increases the ratio of p75, to trkA NGF receptors on mature sympathetic neurons. Developmental Biology 161, 206–217
Mufson E J Brashers-Krug T and Kordower J H (1992) p75, nerve growth factor receptor immunoreactivity in the human brainstem and spinal cord. Brain Research 589, 115–123
Pleasure S J Reddy U R Venkatakrishnan G Roy A K Chen J Ross A H Trojanowski J Q Pleasure D E and Lee V M-Y (1990) Introduction of nerve growth factor (NGF) receptors into a medulloblastoma cell line results in expression of high- and low- affinity NGF receptors but not NGF-mediated differentiation. Proceedings of National Academy of Science USA 87, 8496–8500
Rabizadeh S Oh J Zhong L Yang J Bitler C M Butcher L L and Bredesen D E (1993) Induction of apoptosis by the low-affinity NGF receptor. Science 261, 345–348
Ramer M and Bisby M (1997) Reduced sympathetic sprouting occurs in dorsal root ganglia after axotomy in mice lacking low-affinity neurotrophin receptor. Neuroscience Letters 228, 9–12
Rush R A (1984) Immunohistochemical localisation of endogenous nerve growth factor. Nature 312, 364–367
Ryden M Hempstead B & Ibanez C F (1997) Differential modulation of neuron survival during development by nerve growth factor to the p75, neurotrophin receptor. Journal of Biological Chemistry 272, 16322–16328
Seeburger J L Tarras S Natter H and Springer J E (1993) Spinal cord motoneurons express $p75^{NGFR}$ and p145trkB mRNA in amyotrophic lateral sclerosis. Brain Research 621, 111–115
Stucky C L and Koltzenburg M (1997) The low affinity neurotrophin receptor p75, regulates the function but not the selective survival of specific subpopulations of sensory neurons. Journal of Neuroscience 17, 4398–4405
Taniuchi M Clark H B and Johnson E M (1986) Induction of nerve growth factor receptor in Schwann cells after axotomy. Proceedings of the National Academy of Sciences USA 83, 4094–4098
Van der Zee C E E M Ross G M Riopellè R J and Hagg T (1996) Survival of cholinergic forebrain neurons in developing p75NGFR- deficient mice. Science 274, 1729–1732
Verdi J M Birren S J Ibáñez C F Persson H Kaplan D R Benedetti M Chao M V and Anderson D J (1994) p75LNGFR Regulates Trk signal transduction and NGF-induced neuronal differentiation in MAH cells. Neuron 12, 733–745
Walsh G S Krol K M and Kawaja M D (1999) Absence of the p75, neurotrophin receptor alters the pattern of sympathosensory sprouting in the trigeminal ganglia of mice overexpressing nerve growth factor. Journal of Neuroscience 19, 258–273
Wyatt S Shooter E M and Davies A M (1990) Expression of the NGF receptor gene in sensory neurons and their cutaneous targets prior to and during innervation. Neuron 4, 421–427
Yan Q Snider W D Pinzone J J and Johnson E M (1989) Retrograde transport of nerve growth factor (NGF) in motoneurons of developing rats: assessment of potential neurotrophinc effects. Neuron 1, 335–343
Zhou X F Rush R A and McLachlan E M (1996) Differential expression of the p75, nerve growth factor receptor in glia and neurons of the rat dorsal root ganglia after peripheral nerve transection. Journal of Neuroscience 16, 2901–2911

12

Regeneration in the peripheral nervous system

Mark A Bisby

Introduction

My friends who study the problem of axon regeneration in the central nervous system regard my fascination with peripheral nervous system regeneration as similar to the attitude of the late-night reveller who was asked why he was looking for his lost keys on the side of the street opposite to where he dropped them: his explanation was that "Its lighter over here." There is some truth to this. It is undeniably more satisfying to study regeneration when it does occur, and it has provided insights into fundamental changes in the nervous system necessary to support regeneration, as well as the components of a sustaining environment for axon growth. Yet, studied on its own, it is unlikely to yield the keys to central nervous system regeneration. There is no need to make excuses for studying peripheral nerve regeneration: it remains an important health issue. In humans, peripheral nerve regeneration is slow and apparently much less successful than in small laboratory animals, and so there is much to be learned about fostering peripheral regeneration, not only after traumatic nerve injury, but also following resolution of toxic and metabolic neuropathies. In addition, insights gained from regeneration studies are also illuminating the understanding of neuropathic pain.

Overview

Successful peripheral nerve regeneration requires that a number of essentials are fulfilled (see box p. 240). In this review, I will focus on a number of significant and provocative recent papers which provide new information, or raise novel questions about these key requirements. For a more comprehensive discussion, I refer the reader to two recent reviews (Bisby, 1995; Fu & Gordon, 1997).

Signaling between axon and cell body

Injury to the axon results in major changes both in the cell body and proximal axon of the injured neuron, and in its axon distal to the injury. The distal changes are referred to as "Wallerian degeneration," and will be discussed later. The proximal changes are supposed, teleologically, to allow the neuron to switch its pattern of macromolecular synthesis from maintenance of the axon to that necessary to support regrowth. The signal(s) which convey the information that the axon has been injured to the cell body and effect the subtle changes in gene expression which follow are still undefined. There are two classes of possible signals: positive, where a factor/event produced by the injury

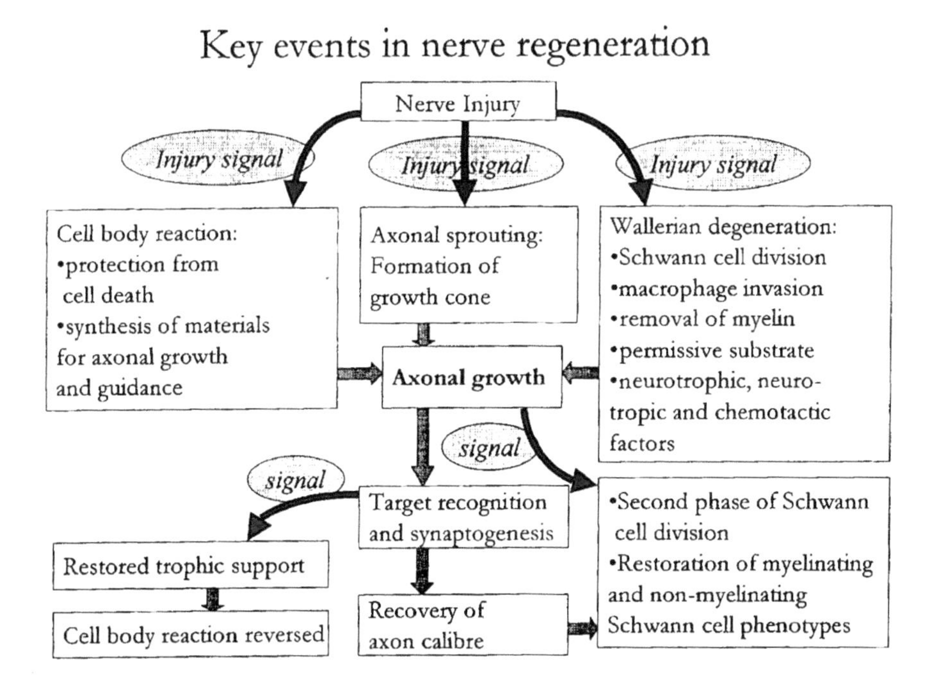

is involved, and negative, where a signal normally conveyed from the distal axons is interrupted by the axon injury. There is a good deal of evidence for the latter type of signal, specifically the interruption in retrograde transport of target-derived neurotrophic factors. Many injury-induced changes can be mitigated by direct application of these factors to the injured neuron cell bodies.

Before proceeding, its worth reviewing briefly two of the fundamental properties of neurons: first, their highly-developed internal transport systems; and second, their dependence on neurotrophic factors for survival and determination of phenotype (especially during development). Neurons are enormous cells: some of your spinal motoneurons have axons over a metre in length, and have to move macromolecules synthesized in their cell bodies to the ends of the axons in order to maintain this extended structure. To do this, they have developed systems of anterograde axonal transport which convey proteins and lipids at several different velocities, ranging from 400 to about 1 mm/day. Neurons recycle degraded terminal components back to the cell body, through retrograde axonal transport, and this retrograde system is also used by the neuron to acquire information about the environment of its axon and terminals. For an update on these transport mechanisms, which also exist in dendrites, see Sheetz *et al.* (1998).

There is a large number of neurotrophic factors produced by cells in the neuronal environment (as well as by some neurons themselves) which are required for the survival of neurons and determination of their phenotype. For example, normal survival and development of sympathetic neurons is absolutely dependent on nerve growth factor (NGF). A number of different neurotrophic factors will be encountered in this chapter, but it is important at the outset to distinguish between neurotrophic factors in general, and the neurotrophins, one specific family of related proteins which have neurotrophic activity. The members of this family include the nerve growth factor (NGF, discovered in the 1950s by Rita Levi-Montalcini and Stanley Cohen, for which they received the Nobel prize in 1986), as well as brain-derived neurotrophic factor (BDNF), neurotrophin-3 (NT-3) and neurotrophin 4/5 (NT-4/5). The neurotrophins exert their actions through two types of receptors: trk-family receptors, which are relatively specific, for example, NGF binds to trkA but not to trkB or C; and the p75 receptor which is promiscuous, binding all members of the neurotrophin family with roughly equal affinity. Most of the survival- and growth-promoting activities of the neurotrophins are mediated through trk receptors, and a key initial event in the signal transduction pathway is the autophosphorylation of the trk receptor when it binds its neurotrophin partner. The role of the p75 receptor is mysterious, but in some instances it seems to be involved not in survival, but in cell death, with an antagonistic relation between activation of trkA and p75 receptors determining cell fate (Yoon *et al.*, 1998; see also chapter 11, Ferguson *et al.*, this volume).

Conventional neurotrophic signaling requires the retrograde axonal transport of a signal molecule from the distal axons to the cell body where the signal stimulates a change in neuronal properties through alteration in gene transcription. This signal molecule might be the neurotrophin itself, a complex between the neurotrophin and its internalized surface receptor, or a second messenger produced as a result of interaction of neurotrophin with the receptor in the distal axon.

To distinguish between these possibilities, Riccio *et al.* (1997) used the clever compartmented culture chamber developed by Campenot (see later) which allows cell bodies and distal axons to be exposed to different environments. Within one compartment they placed neonatal sympathetic neurons, which extended axon processes into the adjacent compartments. After boosting the concentration of NGF in the axon compartment, they were able to detect that it had affected the cell body with a time-course consistent with signaling through conventional retrograde axonal transport. The indicator of the cell body effect which they used was the phosphorylation of a protein called cyclic AMP response element binding protein (CREB) which had previously been identified as a target of the NGF signaling pathway. (It is a transcription factor which binds to a cyclic AMP response element on DNA: phosphorylation by cyclic AMP-activated protein kinase A turns on its DNA-binding ability). Next, they applied NGF bound to microspheres to the axon compartment. They showed that this immobilized NGF indeed bound to axonal trkA receptors, as shown by the normal ligand-induced autophosphorylation of the trkA receptors, but this failed to induce CREB phosphorylation in the cell body compartment. They concluded that the signaling entity was not some second messenger induced by NGF binding to trkA in the distal axon, but a complex between NGF and internalized trkA. This was supported by the appearance of autophosphorylated trkA in the cell bodies after NGF application to the axon compartment. However, if an inhibitor of trkA autophosphorylation (K-252a) was applied to the cell bodies, NGF application to the axons did not cause phosphorylation of CREB. Thus, the signal which turns on phosphorylation of CREB is a complex of retrogradely-transported NGF and autophosphorylated trkA receptor. After axotomy, the absence of this complex arriving at the cell body would presumably fail to maintain the phosphorylation of CREB leading to inactivation of those genes with cyclic AMP response elements in NGF-sensitive neurons. It is reasonable to suppose that this model would apply to retrograde signaling by all members of the neurotrophin family, acting on the various classes of neurotrophin-sensitive peripheral neurons.

In the case we have just considered, a transcription factor, CREB, is regulated indirectly through interruption in neurotrophin supply as a result of axotomy. However, there is an intriguing possibility for more direct involvement of a transcription factor in signaling axon injury. The transcription factor NF-κB exists in cytoplasm complexed with an inhibitory factor IκB. Activation of proteases releases NF-κB which then moves from cytoplasm into the nucleus and regulates transcription. In theory, calcium ion entry at the site of axon injury could activate proteases such as calpain, resulting in the mobilization of NF-κB as a "positive" signal, followed by its retrograde transport to the cell body and its translocation to the nucleus. Povelones *et al.* (1997) studied a NF-κB homologue in *Aplysia* neurons and found that rather than being activated by axon injury, there was a loss of DNA binding activity in the axon within five minutes, and later in the cell bodies, as measured using an electrophoretic mobility shift assay on extracts of cytoplasm extruded from the neurons. This assay uses an oligonucleotide probe that recognizes the DNA sequence to which the transcription factor of interest binds. If the transcription factor is bound to the DNA, the electrophoretic mobility of the DNA-transcription factor complex is lower than of the DNA alone, and the labeled band on the electrophoretic gel shifts upwards. This loss of activity was not due to an injury-induced re-association of NF-κB with IκB since treatment of nerve extracts with deoxycholate, which dissociated IκB from NF-κB did not restore DNA-binding activity. Pavilions et al. speculated that axotomy caused the destruction of NF-κB through Ca^{++}-activated proteolytic pathways. The loss of existing NF-κB activity following axotomy is consistent with a study by Doyle & Hunt (1997) who found a reduction in the proportion of primary sensory neurons of the cat dorsal root ganglion expressing nuclear NF-κB immunoreactivity at 7h, but not at 24h, following sciatic nerve injury. However, by seven days following a partial, but not a complete, sciatic nerve injury, we found an increase in nuclear immunoreactive neurons (Ma & Bisby, 1998), perhaps due to the effect of cytokines or other factors produced by Wallerian degeneration on the neurons with spared axons following partial injury. Overall, then, while there is evidence for a change in NF-κB binding after injury, the response is not, as theorized above, a simple axonal activation of this transcription factor. Indeed, evidence is accumulating that NF-κB activation is one of the effects of trophic factor application which protects neurons from cell death. Cell death following withdrawal of NGF from cultured sympathetic neurons (mimicking axotomy) could be prevented by microinjecting a plasmid expressing NF-κB (Maggirwar *et al.*, 1998).

Another candidate for a positive axon injury signal is the cytokine, ciliary neurotrophic factor (CNTF). Unlike other neurotrophins and cytokines, this is present in Schwann cells of normal peripheral nerve at high levels, but does not appear to be secreted. Because physical disruption of the Schwann cells could liberate cytoplasmic CNTF into the extracellular space where it can be taken up by axons and retrograde transported to cell bodies, it seems well-suited to be an injury signal. Some experimental support for this hypothesis was provided by Sendtner *et al.* (1997). They used the *pmn* mouse mutant which suffers from a rapid progressive motoneuron death, and found that a facial nerve lesion at day 28 of life reduced the extent of motoneuron cell death when measured on day 42: normally about 50% of facial motoneurons have died by this time, but on the lesioned side, only about 16% had died. That this protective effect of axon injury might be due to the trophic effects of injury-liberated CNTF was demonstrated by crossing *pmn* mutants with CNTF null mutants. In these animals, facial nerve injury had no significant protective effect on motoneuron loss. While this study supports the idea that CNTF released from injured Schwann cells can have a trophic effect on injured motoneurons, it does not provide evidence that CNTF acts as an injury signal initiating the cell body reaction. Nor does it exclude the possibility that CNTF is acting indirectly via other mediators which actually perform the signaling function. Moreover, it is not yet apparent that the cell body reaction is defective in CNTF null mutants, as the hypothesis would predict.

While most changes in cell body gene expression following injury occur with a latency consistent with retrograde transport of an injury signal (positive or negative), occasional reports appear of changes much too rapid for the relatively stately pace of ~400 mm/day for fast axonal transport. One example is the induction, within 45 minutes of axon injury, of gap junction protein in astrocytes surrounding axotomised motoneurons (Rohlmann *et al.*, 1994). A provocative paper with evidence for a much faster axon-to-cell body communication system was published by Senger & Campenot (1996), again using the compartmented chamber perfected by Campenot, and sympathetic neurons from neonatal rats. When NGF was applied to the axon compartment, autophosphorylated trkA appeared in the cell body within one minute. Retrograde transport of NGF occurred much more slowly: using ^{125}I-NGF as a probe, they showed that very little had reached the cell body an hour after application to the distal axon compartment. Thus, the trkA phosphorylation signal is conveyed by a much faster mechanism than retrograde axonal transport, and Senger & Campenot (1998) suggested that they were observing the phosphorylation of existing trkA in the cell body, rather than translocation of phosphorylated trkA from the distal axon. Further work is urgently required to establish that this phenomenon exists *in vivo.*

The cell body reaction — fishing with new tackle

A powerful rationale for studying the cell body reaction to injury is that proteins/genes more strongly expressed after injury are presumably important for axon regeneration. Comparison of proteins synthesized by injured cell bodies and transported into the proximal axons, using two-dimensional gel electrophoresis, permitted the identification of protein GAP-43 (Growth-Associated Protein with apparent molecular weight of 43 kilodaltons), which based on its localization within growth cones, and effects of transfection experiments, was supposed to be involved in axon outgrowth. However, GAP-43 null mutant mice show defects in pathfinding at the optic chiasm during development of the retinotectal pathway, rather than reduced axonal outgrowth (Strittmatter *et al.*, 1995).

The discovery of GAP-43 was essentially the result of "fishing expeditions" to identify unknown proteins selectively expressed by regenerating neurons. This approach continues to be fruitful, though nowadays the fishing tackle consists of molecular biology techniques, such as differential display PCR (Livesey & Hunt, 1996). Such techniques sometimes catch new fish: for example, Nakayama *et al.* (1995) extracted RNA from the pooled facial nuclei of 40 rats, 24h after unilateral facial nerve section. They identified nine cDNA clones which were induced by axotomy, and confirmed by *in situ* hybridization those expressed in the injured facial nucleus. These clones included those encoding the familiar up-regulated GAP-43 and α-tubulin, providing confidence in the technique. A single, novel cDNA was identified as suppressed by nerve injury. The protein product was christened neurodap-1, and found to contain a characteristic RING-H2-finger motif. In a complete molecular workout, a GST-neurodap-1 fusion protein was expressed and used to generate a rabbit polyclonal antibody. The antibody was used to localize neurodap-1 at light and EM levels. It was associated with the postsynaptic region of axosomatic synapses. Its downregulation following nerve injury is thus consistent with the process of "synaptic shedding" long-recognized in axotomised motoneurons.

Using a similar technique, Su *et al.* (1997) discovered a strongly up-regulated PCR product in samples from axotomised mouse hypoglossal motoneurons, and confirmed up-regulation using *in situ* hybridization with the product as probe. The PCR product was sequenced and found to be highly homologous to the kinesin light chain. This was an exciting result because kinesin is a motor for the fastest type of anterograde axonal transport, consisting of a tetramer of two light and two heavy chains, with the light chains possibly involved in regulating heavy chain activity, (and thus transport velocity), or binding of transported proteins, (thus altering capacity). All light chain isoforms were up-regulated after injury, but no heavy chain isoforms increased. However, an earlier study on a family of kinesin-related proteins, using northern hybridization exclusively, reported that all detected members were modestly down-regulated after injury (Takemura *et al.*, 1996). Since different kinesin-related proteins and combinations of the various kinesin light and heavy chains may transport different organelles within the axon, the differences in the two studies may not be contradictory, but rather indicate a shift in the types of organelles conveyed by axonal transport during regeneration. In both studies, the retrograde motor, dynein, was found to be up-regulated, suggesting that there is an increase in retrograde transport capacity following injury, perhaps to accommodate increased membrane recycling during the remodeling of the axon, perhaps to increase the sensitivity of neurotrophin signaling. Three other recently-discovered proteins which are expressed by peripheral neurons after axon injury, TOAD-64 (Minturn *et al.*, 1995), ninjurin (Araki & Milbrandt, 1996) and Reg-2 (Livesey *et al.*, 1997) likely play a role in axon growth and guidance, and we will return to them later.

Of course, not all changes in gene expression in neurons after axotomy are likely to be supportive: upregulation of inappropriate genes might interfere with adequate regeneration, or failure to turn on a regenerative program may be one reason for lack of regeneration in the central nervous system (Tetzlaff *et al.*, 1991). Comparison of the response to axotomy of adult rat Purkinje cells, with that of neurons of the inferior olive, lateral reticular nucleus, and deep cerebellar nuclei, shows that the expression of markers of a robust cell body reaction, such as c-jun (see below) and GAP-43 is correlated with the ability of the axons of these neurons to regenerate when offered a growth-permissive substrate. However, the Purkinje cells can be persuaded to express c-Jun, both *in vivo* and in organotypic cerebellar cultures, if neutralizing antibodies against one component of central myelin (NI-250) are applied (Zagrebelsky *et al.*, 1998). This component of myelin has a strong inhibitory effect on axonal growth which is due both to local effects at the growth cone, and, as this report shows, to suppression of the cell body response to axotomy. Stimulation of the cell body response by application of trophic factors may be one way of promoting axon regeneration. Rubrospinal neurons undergo atrophy after axotomy, but this can be prevented by application of BDNF, which also stimulates GAP-43 expression and increases the regenerative ability of the rubrospinal neurons when they are offered the supportive substrate of a segment of peripheral nerve (Kobayashi *et al.*, 1998).

C-jun as a key step in the cell body reaction to axotomy

We have already seen that axon injury alters the activity of transcription factors in the cell body, an obvious necessary step in changes in the pattern of gene expression. The best studied change in transcription factors involves c-jun: increased nuclear expression of this transcription factor is an early and universal marker of the axotomised peripheral neuron (Herdegen *et al.*, 1997). But what stimulates this change?

Kenney & Kocsis (1998) hypothesized that in axotomised neurons c-jun protein kinases (also known as stress-activated protein kinases) would stimulate jun activation by serine phosphorylation, as has been demonstrated for cultured non-neuronal cells in response to various stressors such as x-ray radiation or tumour necrosis factor (Herdegen *et al.*, 1997). Using an immune complex kinase assay, they found an increase in c-jun protein kinase activity in dorsal root ganglion within 30 minutes of proximal (within 1 cm) nerve transection, but not until 3h after distal (~4 cm) nerve transection. This was sustained for up to 30 days. The increase in c-jun protein kinase activity preceded the previously reported increase in c-jun (three hours after proximal lesions), and was not associated with increases in levels of c-jun protein kinase protein, as assessed by Western blotting. Immunocytochemistry showed increased phosphorylated c-jun in dorsal root ganglion neurons, and electrophoretic mobility shift assays showed increased AP-1 binding to an Hmt IIA promoter from 12 hours to 30 days after axotomy. Treatment of dorsal root ganglion extracts used for electrophoretic mobility shift assay with antibodies against the DNA-binding domain of c-jun reduced complex formation showing that the AP-1 binding complexes actually contained c-jun. In dorsal root ganglion extracts obtained after nerve crush, which permitted regeneration and functional recovery, both c-jun levels and AP-1 binding activity fell by 30 days.

This key paper showed that axotomy first induces c-jun protein kinase activity, with a time course consistent with a retrograde transported signal, and this is followed by n-terminal phosphorylation of c-jun which allows c-jun both to act as a transcription factor (Herdegen *et al.*, 1997), as shown by formation of AP-1 binding complexes in the Kenney & Kocsis (1998) paper, and to induce its own synthesis through binding as a heterodimer with ATF-2 (whose phosphorylation by c-jun protein kinase also increases its transcriptional ability) to the c-jun promoter.

Of course, important questions remain: are there reasonable links between putative axotomy signals, such as trophic factor deprivation, and activation of c-jun protein kinase? What genes, involved in the cell body response to axotomy, are regulated by c-jun? Recent work by Eilers *et al.* (1998) provides an affirmative answer to the first question: NGF withdrawal from cultured neonatal sympathetic neurons rapidly induced c-jun protein kinase activity, jun phosphorylation and activation of the c-jun promoter.

The pathway by which NGF withdrawal induces c-jun protein kinase activity (or conversely, by which NGF suppresses it) is becoming clearer. Two kinases involved in NGF-signaling cascades mediated by the trkA receptor, Akt protein kinase, a putative effector of phosphotidylinositol-3 kinase, are likely involved. Inhibitors of phosphotidylinositol-3 kinase block NGF-mediated survival of cultured sympathetic neurons and induce c-jun, and conversely expression of constitutively active Akt or phosphotidylinositol-3 kinase in neurons protects against death after NGF withdrawal. (Crowder & Freeman, 1998). In cultured dorsal root ganglion neurons, the same inhibitors of phosphotidylinositol-3 cause c-jun phosphorylation, as does withdrawal of NGF (Vogelbaum *et al.*, 1998). In adult dorsal root ganglion neurons, NGF withdrawal does not cause apoptotic cell death, as it does in immature dorsal root ganglion neurons. Treatment with the specific inhibitor of phosphotidylinositol-3 kinase also resulted in apoptosis in immature but not mature dorsal root ganglion neurons but induced c-jun phosphorylation irrespective of whether the cells survived or died. However, the ratio of expression of the death-promoting gene bax to the survival-promoting gene bcl-xL was many times higher in immature than mature dorsal root ganglion neurons. These results suggest that the pathway from NGF binding to trkA extends through phosphotidylinositol-3 kinase to suppress c-jun phosphorylation. In the absence of NGF, this pathway is ineffective, and c-jun phosphorylation occurs. Whether this causes the death of the neuron depends on steps downstream from c-jun, and involves differential expression of members of the bcl-2 gene family, well known as key players in the regulation of cell death.

Genes potentially regulated by c-jun include GAP-43, which contains an AP-1 site, but there is no consensus on the *in vivo* importance of this site in transcriptional regulation, although c-jun expression following axotomy in neurons appropriately precedes GAP-43 expression. Other candidate genes include galanin and vasoactive intestinal polypeptide, which are strongly expressed in dorsal root ganglion neurons after axotomy. However, microinjection of c-jun antisense oligonucleotides into dorsal root ganglion neurons axotomised by placing them in culture reduced expression of vasoactive intestinal polypeptide and neuropeptide Y, but not galanin (Mulderry & Dobson, 1996). It would have been useful if these authors had also examined GAP-43 expression, but they did show, through electrophoretic mobility shift assay and transfection of cultured dorsal root ganglion neurons with plasmid containing multiple copies of the cyclic AMP response element to out-compete transcription factors binding to cyclic AMP response element, that vasoactive intestinal polypeptide indication by c-jun was mediated through the cyclic AMP response element, rather than through AP-1 promoter sites. Thus, a very large pool of genes could be affected by c-jun (Herdegen *et al.*, 1997).

The application of molecular biological techniques to attack the old question "what is the signal for the cell body reaction?" is providing partial answers at an exciting pace. The final answer(s) are of more than intellectual interest. Many central nervous system neurons do not respond following a distal axotomy with the same changes in regeneration-associated gene expression as do peripheral neurons: this failure may reduce their regenerative potential (see chapter 3 by Anderson & Lieberman, this volume). In addition, in many neurons, axotomy leads to death, again with c-jun as an intermediary (Herdegen *et al.*, 1997). Understanding of the c-jun mediated pathways which lead to either survival and regeneration or to cell death will likely be of therapeutic importance in strategies to improve functional recovery from damage to nervous tissue.

Formation of the growth cone

A key, and poorly-understood step, in axonal regeneration is the conversion of the proximal axon stump into a growth cone, the flexible "battering ram", described by Santiago Ramón y Cajal, which is not only the site of axonal elongation, but also the sensor involved in directional guidance. Typically, growth cones do not emerge from the axon at the precise site of injury, but some distance proximal to it: in myelinated axons at the node proximal to the site of injury. Using *Aplysia* axons, Ziv & Spira (1997) studied the formation of growth cones following axotomy, and correlated intra-axonal calcium ion levels, measured with Fura-2 imaging, with the ultrastructural appearance of the axons. They found that nascent growth cones, identified by the formation of lamellipodia, formed not at the severed axon tip, but proximal to it, corresponding to a region where calcium ion concentration, which was in excess of 1500 μM at the tip, fell through the 300–500 μM range after axotomy (Figure 1). However, the growth cone did not emerge from this region until after calcium ion levels had fallen back to near-normal values (approx. 100 μM) following the rapid sealing of the cell membrane over the cut axon. The growth cone formation region also corresponded to a transition zone between more proximal normal axoplasm and axoplasm closer to the injured tip, which was highly disrupted. As a further demonstration of the important role of calcium ions, they made a local chemical axotomy, using the calcium ionophore ionomycin, and showed growth cone formation only from regions of the axon where induced calcium ion concentrations of 400–600 μM were achieved, again with the growth cone only forming after calcium ion concentrations had returned to normal.

It seems likely that calcium-activated proteases such as calpain result in an irreversible disruption of the axonal cytoskeleton at the tip of the injured axon, so that even as calcium ion levels are restored, material transported from the proximal axon cannot enter it. However, since inhibition of calpain activity abolishes growth cone formation, a certain level of Ca^{++}-dependent proteolysis seems essential for generation of a growth cone (Gitler & Spira, 1998).

The initial formation of the growth cone occurs too rapidly for it to be dependent on metabolic changes in the cell body, and in any event, isolated axons can develop growth cones. However, this initial outgrowth is slower than long distance elongation, which does seem to be dependent on cell body changes. This was nicely demonstrated by Smith & Skene (1997) who showed distinct initial modes of axon growth by adult dorsal root ganglion neurons obtained from intact animals, versus animals whose axons had been injured *in vivo* 2–7 days before culturing the dorsal root ganglion neurons. In freshly-injured dorsal root ganglion neurons, compact, highly branched axons formed, but in dorsal root ganglion neurons preconditioned by injury, there was rapid extension of long axons with few branches (Figue 2). However, by the second day in culture, the mode of axon growth from freshly-injured dorsal root ganglia began to resemble that of the preconditioned neurons. The transition in growth mode seemed

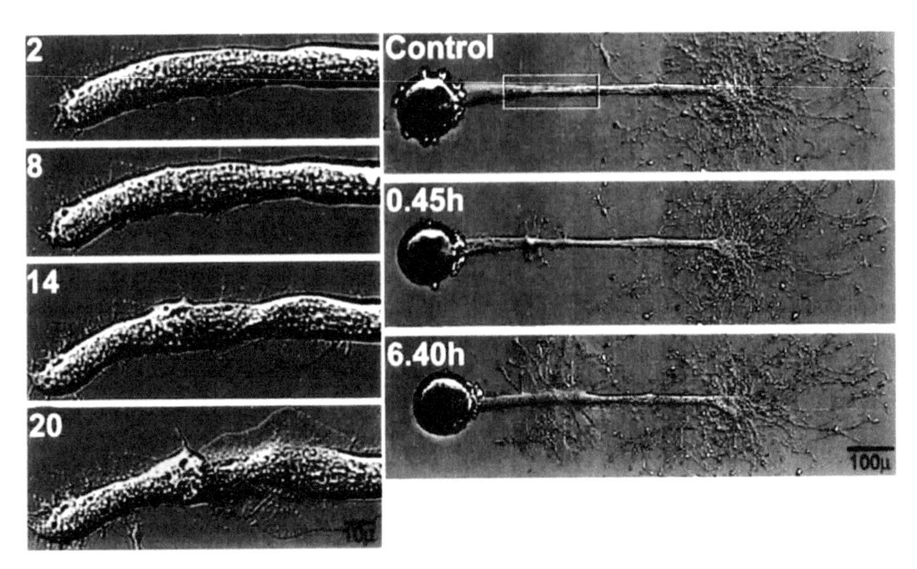

Figure 1 Growth cone formation from cut *Aplysia* axons. Left-hand column: the formation of lamellipodia at increasing times (in minutes) after axotomy. Note that the lamellipodia originate proximal to the cut axon tip, corresponding to the region where immediately after injury Ca^{++} levels increased to approximately 300 μM, compared to >1000 μM at the injured tip. Right-hand column: effects of focal application of the calcium ionophore, ionomycin, to the axon, within the region indicated by the white rectangle. Lamellipodia formed ectopically in the region where ionomycin induced a transient increase in Ca^{++} concentration to 300–600 μM. From Ziv & Spira (1997) Journal of Neuroscience 17, 3568–3579.

due to a gradual change in gene expression, and this was supported by placing the freshly-removed dorsal root ganglia in medium containing a transcriptional inhibitor for the first 16 hours of culture. This displaced the onset of the rapidly elongating mode of growth by a roughly equivalent period, but it had no effect on elongating axon outgrowth in preconditioned neurons, showing that the inhibitor was not interfering with elongation itself. Smith & Skene (1997) concluded that intact adult dorsal root ganglion neurons were able to support the highly branched mode of axon growth, but that the transition to the rapid phase of elongation required a change in gene transcription. Although this study was performed *in vitro*, its conclusions are consistent with *in vivo* studies showing that initial axon sprouting occurs before newly synthesized proteins would have time to arrive at the site of axon injury, and that initial axon outgrowth is slow, accelerating after the first 24–48 hours. Furthermore, it is clear that axonal sprouting from uninjured motor axons in response to partial denervation can occur without detectable changes in gene expression in motoneuron cell bodies.

Growth Cone Guidance

It is not enough to form a growth cone. Growth has to be appropriately directed, and in peripheral regeneration this direction is obviously into the distal stump. Growth cones from the ends of cut axons are reluctant to make a U-turn and grow back into the

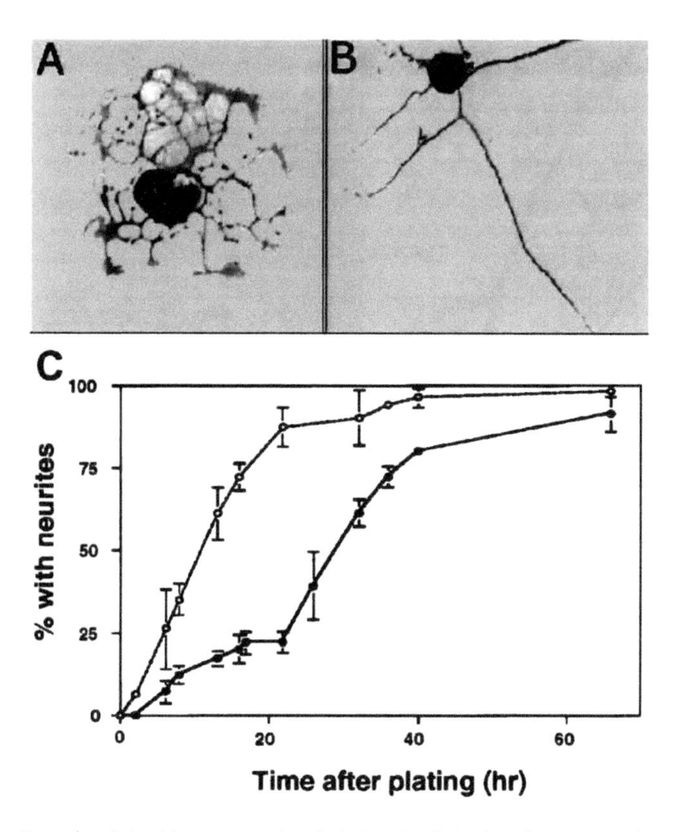

Figure 2 Growth of freshly- or previously-injured adult dorsal root ganglion neurons in dissociated cell culture, shown with immunocytochemistry for GAP-43. A: Freshly-injured dorsal root ganglion neuron after 16 hours in culture. Note short, highly-branched neurites with flattened growth-cones. B: Neuron previously axotomized for 7 days prior to culturing for 16 hours. Note long, straight, unbranched neurites. C: Time-course of neurite outgrowth. The percentage of neurons with neurites is shown at various times following culture. Broken line, neurons axotomized 7days prior to culture, showing high rate of neurite production. Solid line: neurons axotomized by the act of placing in culture. Note initial low rate of neurite production, followed by a transition to a higher rate after 20–30 hours. From Smith & Skene (1997) Journal of Neuroscience 17, 646–658.

proximal nerve stump, and also are directed towards the distal stump by neurotropic (nerve-attracting) substances secreted by the degenerating distal stump.

Agius & Cochard (1998) compared the growth of neurites from chicken dorsal root ganglion neurons on frozen sections of normal and 14 day-predegenerated rat sciatic nerve as culture substrates. While the percentage of neurons growing neurites was the same on both substrates, the total neurite length and number of neurites per neuron was higher on predegenerated sections. Axons clearly avoided regions containing myelin debris, and in normal axon sections, axons were constrained to travel along the endoneurial based lamina and never entered the Schwann cell tubes (Figure 3). However, after degeneration and removal of myelin, axons penetrated deep inside the

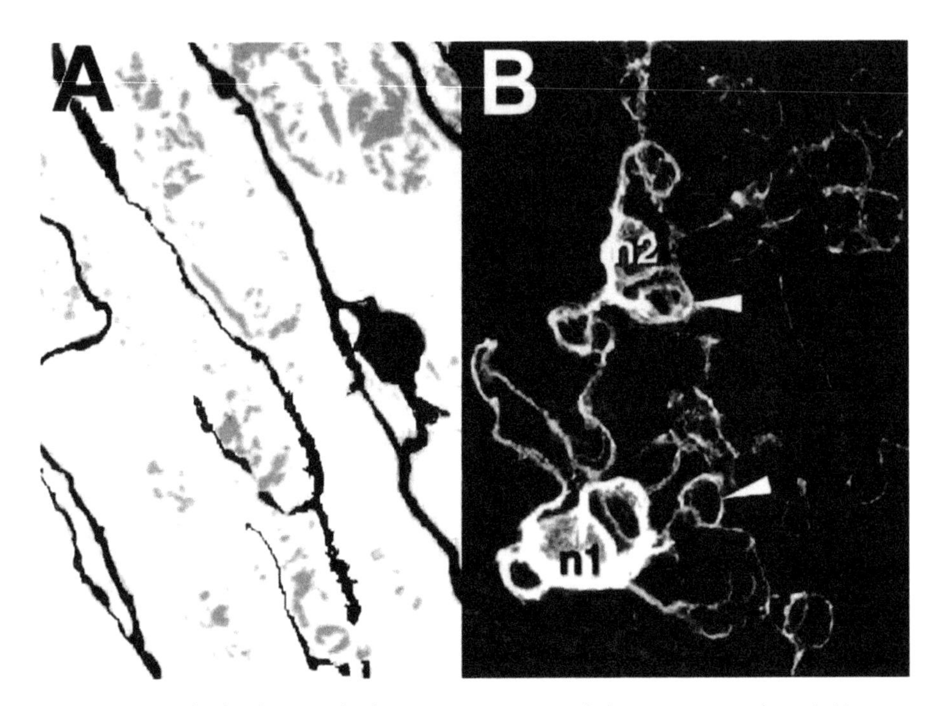

Figure 3 Growth of embryonic chick sensory neurons on sciatic nerve cryosections. A: Neurons detected by immunocytochemistry with anti-NCAM antibodies (black) growing over a longitudinal section of 15-day injured distal nerve. The neurites avoid remnant myelin fragments (grey areas). B: Two neurons (n1 and n2) growing on a transverse section of normal nerve. The neurites take serpentine routes, looping around myelinated axons (arrowheads), and growing along the endoneurial basal lamina. From Agius & Cochard (1997) Jounal of Neuroscience 18, 328–338.

Schwann cell tubes. Given the choice between degenerating nerve of the distal nerve stump and intact proximal nerve, it seems obvious that growth cones of regenerating axons would opt to grow distally.

The chemotactic property of the distal nerve stump, first inferred by Ramón y Cajal at the turn of the 19th century, was demonstrated directly by Perez *et al.* (1997). Conditioned medium obtained by placing frog sciatic nerve in culture for 7 days was concentrated and loaded into micropipettes, then injected by pressure pulses from the micropipettes placed adjacent to and at a defined angle to the longitudinal axis of growth cones extending from dissociated frog dorsal root ganglion neurons. Within minutes, the growth cones turned towards the pipette (Figure 4). When the conditioned medium was filtered through a 30 kilodalton cutoff filter, its chemotactic properties were abolished.

What are these distal neurotropic/chemotactic factors? Neurotrophins are one possibility. NGF and BDNF, both of which are synthesized in increasing quantities by Schwann and other endoneurial cells following nerve injury, have chemotactic actions on growth cones. A remarkable demonstration of the sensitivity of growth cones to

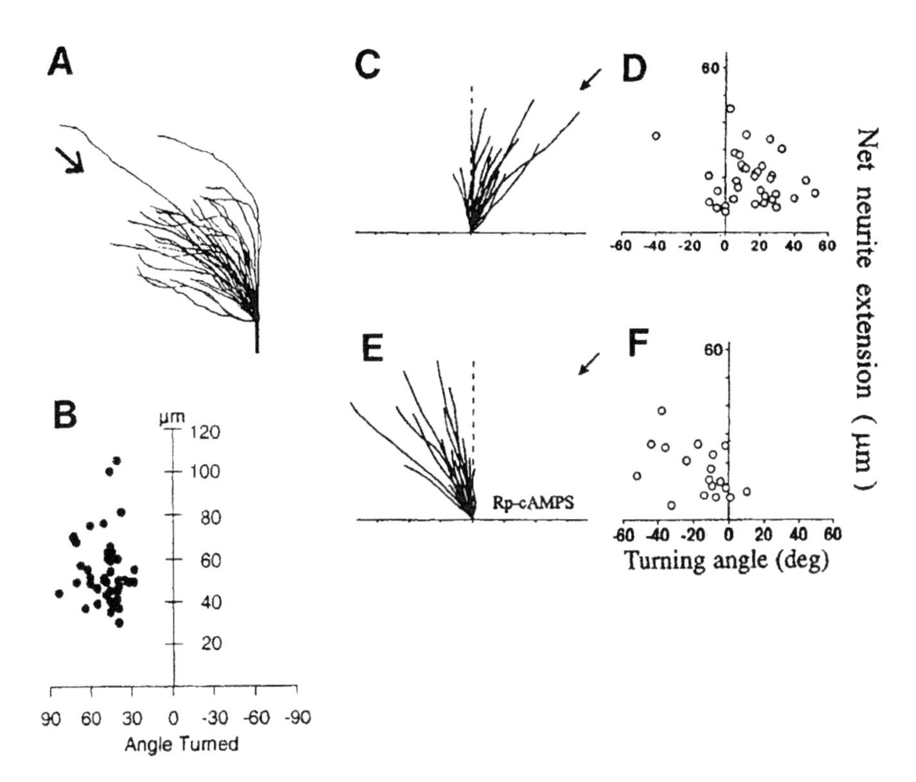

Figure 4 Growth cones respond to factors synthesized by degenerating peripheral nerve. A: The trajectory of 40 growth cones following ejection of conditioned medium obtained by incubating a segment of peripheral nerve in culture medium for 7 days. The position of the micropipette from which the conditioned medium was ejected is shown by the arrow. It was positioned at an average of 57° from the original axis of orientation of the growth cone. B: The data from (A), replotted, showing the relation between the outgrowth distance and the angle turned, averaging 49°C: Orienting of growth cones to similar application of BDNF from a micropipette, with D showing the data replotted as in (B). E, F: conversion of the chemotactic effects of applied BDNF to a chemorepulsive effect by reducing the effective cyclic AMP concentration through adding Rp-CAMPS, a non-hydrolysable analog competitor of cyclic AMP. A, B from Perez *et al.* (1997) Experimental Neurology 153, 196–202. C-F from Song *et al.* (1997) Nature 388, 275–279.

neurotrophic factors was provided by Song *et al.* (1997), using cultures of embryonic *Xenopus* spinal neurons. BDNF was injected from a micropipette placed 100 µm from a growth cone, and at a 45 degree angle to the current direction of neurite extension. After an hour, the direction and length of neurite outgrowth was measured. There was a dose-dependent turning of the growth cones towards the source of BDNF, apparently mediated through trk receptors, since an inhibitor of tyrosine kinase activity abolished the turning response. Calcium ions also seemed to be involved in the turning response, since it was abolished by reducing calcium concentration of the culture solution, without reducing neurite outgrowth. Most remarkably, the chemoattractant effects of BDNF could be converted to a chemorepulsive one if effective cyclic AMP levels were reduced by adding a non-hydrolysable competitive analogue (Figure 4), or if protein kinase A was inhibited. The growth cones also grew towards a source of neurotrophin-3, but this chemoattraction was not sensitive either to reductions in extracellular calcium concentration, or to reduced available cyclic AMP levels. These elegant experiments not only reveal clearly the chemoattractant properties of neurotrophins produced by the distal nerve stump, but they also begin to unravel the mechanisms involved, and suggest how growth cone attraction and repulsion can be modulated by local environmental factors operating through second messenger systems. Understanding how this modulation occurs could be an important step in overcoming the inhibition of axon growth within the central nervous system, as well as accelerating the relatively slowly regenerating human peripheral nervous system.

Rapid growth of regenerating axons and their guidance along the bands of Bungner formed by the chains of proliferating Schwann cells also depends on selective adhesion and repulsion between axonal growth cones and their substrates. Numerous studies show changes in production of cell adhesion molecules and extracellular matrix components during Wallerian degeneration (Fu & Gordon, 1997). One novel example of a protein likely to be involved in repulsive interactions is the protein known as TOAD-64 (Turned On After Division), examined by Minturn *et al.* (1995). This protein was first identified by comparing the two-dimensional gel electrophoresis patterns of rat membrane-associated proteins expressed by neuronal progenitors with those of post-mitotic neurons. Partial amino-acid sequencing of the protein extracted from a gel spot which was TOAD permitted production of antisera to localize the temporal and spatial expression of the protein, showing that it was one of the earliest expressed by post-mitotic neurons. In adult CNS, its expression is barely detectable, but after sciatic nerve injury it is re-expressed in motoneurons within one day of injury. Using various combinations of degenerate oligonucleotide primers, a PCR product was obtained and used to screen a rat brain cDNA library and obtain the complete TOAD-64 coding sequence. The sequence had homologies to the *unc-33* gene product of *C. elegans*: mutations in this gene result in defects in axonal trajectories, suggesting that it plays a key role in growth cone guidance, consistent with the concentration of TOAD-64 within lamellipodia and filopodia. Subsequently, rat TOAD-64 was identified as a member of a 4-gene family closely related to chick CRMP-62 (Collapsin Response Mediator Protein) which mediates inhibition of growth cone motility by collapsin (Wang & Strittmater, 1996).

The existence of a mediator for the collapsin/semaphorin family of proteins in regenerating axons suggests that this family is involved in peripheral nerve regeneration. Evidence for an inhibitory role for semaphorin III in peripheral nerve growth and maintenance was provided by Tanelian *et al.* (1997), using an *in vivo* rabbit cornea preparation. The cornea is innervated only by A and C sensory fibres, and since it is thin and transparent, plasmids encoding semaphorin III could be injected into the cornea using a gene gun, which uses a helium pressure pulse to blow DNA-coated gold particles from a small plastic cartridge directly into the target tissue. Plasmids encoding green fluorescent protein and NGF were used as controls. Expression of semaphorin III following transfection was determined indirectly through the expression of a Myc epitope tag contained in the semaphorin III expression vector. In normal corneas transfected with the semaphorin III vector, there was a significant reduction in innervation density in the corneal regions containing the gold particles injected by the gene gun. Following a wound in the corneal epithelium, uninjured collaterals sprout into the denervated area, and are later displaced as the injured axons regenerate. Expression of semaphorin III in areas where gene gun application overlapped the margins of the corneal wound inhibited the entry of collateral sprouts into the wound. These exciting results show that adult sensory nerves retain sensitivity to semaphorin III, and that axon growth can be inhibited by it, but they do not yet provide direct evidence for a physiological role. Tanelian *et al.* (1997) point out that treatment with semaphorin III might prevent inappropriate growth of small diameter afferent fibres following nerve injury and so may have a therapeutic role in treatment of some types of neuropathic pain.

Cell bodies of peripheral sensory and motor neurons express the mRNAs for both semaphorin III and its receptor neuropilin-1. After axotomy, neuropilin mRNA levels do not change, but there is a reduction in semaphorin III mRNA expression in axotomised motoneurons, which returns to normal after regeneration is completed (Pasterkamp *et al.*, 1998). These results suggest that mature peripheral neurons secrete their own semaphorin III which inhibits their growth through autocrine interaction with the neuropilin receptor. After axotomy, the reduction in semaphorin III may allow axonal growth to occur.

Another recently discovered protein, ninjurin, may be an example of a protein mediating adhesive interactions between substrate and growth cone during regeneration. This protein was identified by Araki & Milbrandt (1996) by differential screening of mRNA produced by the distal stump of injured sciatic nerve. After elucidating the sequence, synthetic peptides were used to generate antibodies, and ninjurin-immunoreactivity co-localized with S-100, showing it was associated with Schwann cells. Using cell-surface labelling techniques, they demonstrated that ninjurin was present in the plasma membrane. When Jurkat T cell leukemia cells or chinese hamster ovary cells were transfected with CMV-ninjurin plasmids, cell aggregates formed, suggesting that the protein is involved in homophilic adhesion. Ninjurin synthesis in dorsal root ganglion neurons was upregulated within 1 day after axotomy, and it was shown, using classical ligature techniques, to be anterograde transported into injured axons. When embryonic dorsal root ganglion neurons were co-cultured with Chinese hamster ovary cells expressing ninjurin, the average length of neurites extended over a 6 hour period was approximately double that when co-cultured with non-transfected Chinese hamster ovary cells. Thus, there is persuasive evidence that ninjurin synthesized by neurons and Schwann cells play a role in homophilic adhesion and promotion of axon growth during regeneration. It remains to be shown if neutralizing antibodies to ninjurin actually reduce axon regeneration, as has been shown for other molecules involved in adhesive interactions in peripheral nerve, such as laminin, L1 and N-cadherin (Fu & Gordon, 1997).

Recently, the human ninjurin gene was mapped to chromosome 13, within the candidate region for the autosomal dominant disorder hereditary sensory neuropathy Type I (Chadwick *et al.*, 1998), which is characterized by degeneration of small diameter sensory neurons subserving pain and temperature. Thus, ninjurin may play a role not only in outgrowth of dorsal root ganglion axons, but also in their maintenance during adult life.

Dialogue between axons and Schwann cells during degeneration and regeneration
So far, our focus has been on the regenerating neuron itself. However, the principle difference between the central nervous system and the peripheral nervous system is in the environment of the injured axon. While in the central nervous system this is inhibitory to regeneration, the environment created by Wallerian degeneration of the distal nerve stump is favourable. The importance of Wallerian degeneration to subsequent regeneration is demonstrated by examination of the *Wlds* mutant mouse (Coleman *et al.*, 1998), where Wallerian degeneration is slowed and muted. Peripheral nerve regeneration is also severely impaired (Brown *et al.*, 1994), even though the cell body reaction to axon injury appears normal (Bisby *et al.*, 1995). Wallerian degeneration provides an optimal environment in a number of ways: removal of inhibitory myelin, production of neurotrophic, neurotropic and chemotactic factors, provision of adhesive substrates for axonal elongation, and recycling of old axonal materials which can be used by the growing axons (Bisby, 1995).

It is clear that during Wallerian degeneration and subsequent regeneration there is an intimate molecular dialogue occurring between axon and Schwann cell, as well as with other endoneurial cells such as macrophages and fibroblasts. The initial trigger for Wallerian degeneration is unknown, but it is not simply due to exhaustion of the axon, deprived of transported components from the cell body. This assertion is backed by the *Wlds* mouse observations, where it has been shown that isolated axons remain structurally and functionally intact for days after the normal onset of Wallerian degeneration (see also chapter 13 by Hughes & Perry, this volume).

During subsequent regeneration in normal nerve, the arrival of the axons induces further changes in Schwann cells, and is followed by a gradual return to normal in their relationship with axons. The presence of Schwann cells is essential for rapid regeneration, and a graphic demonstration of this was provided by Torigoe *et al.* (1996). They sandwiched the proximal stump of a cut mouse sciatic nerve between two transparent pieces of film so that they could observe subsequent regenerative events. Regenerating axons emerged from nodes of Ranvier proximal to the end of the stump within 3 hours, but initially these were not accompanied by Schwann cells and outgrowth rates were low. From 2 days after axotomy (corresponding with the time of onset of Schwann cell proliferation), Schwann cells migrated from the proximal stump along the advancing axons, and eventually moved in front of the axons. Coinciding with the association of Schwann cells with the regenerating axons, the rate of outgrowth accelerated almost fourfold. This was confirmed by placing films which contained migrated Schwann cells on the distal ends of freshly-cut nerves. The distance grown by the regenerated axons over the film when examined two days later was greater if the film contained Schwann cells, and even greater if these were cells which originally had migrated onto the film from the distal, degenerating nerve stump. Clearly, Schwann cells from the distal stump stimulate axon outgrowth, and some explanations for this are provided below.

Some of the many molecules likely involved in the dialogue between axon and Schwann cell have been identified (Fu & Gordon, 1997), including neurotrophins, cytokines, transforming growth factors (TGFs) and glial-derived neurotophic factor, but

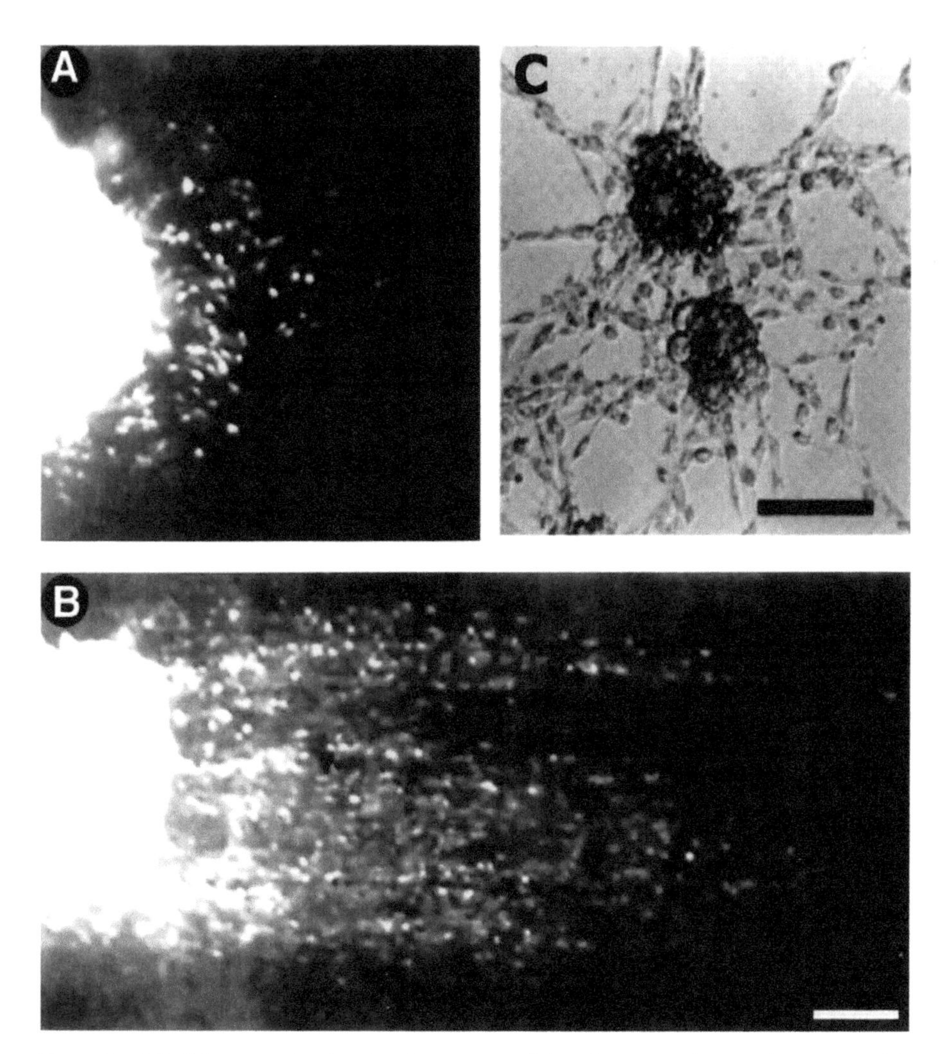

Figure 5 A,B. Schwann cell migration from neonatal dorsal root ganglion explants onto adult sciatic nerve cryosections, revealed by immunohistochemistry using S-100 antibodies. A: No neuregulins added. B: Glial Growth Factor 2, 2-ng/ml added. Scale bar = 100 μm. C: Schwann cell chemotaxis onto Sepharose beads coated with glial growth factor 2, 12 pg/bead. The beads were added to pure Schwann cell cultures. From Mahanthappa *et al.* (1996) Journal of Neuroscience 16, 4673–4683.

here I will focus on but one class as an example. Neuregulins have been identified as survival factors and mitogens for Schwann cells *in vitro* and during development, and they play a similarly important role during regeneration. Mahanthappa *et al.* (1996) showed that migration of Schwann cells from neonatal dorsal root ganglion explants onto frozen sections of adult sciatic nerve was enhanced by addition of the neuregulin glial growth factor-2 (GGF-2) (Figure 5), but with a maximal effect at a concentration about one tenth of that required for maximal Schwann cell proliferation: this stimulation of migration was not merely a result of increased proliferation. If Sepharose beads were coated with GGF-2, they induced migration of cultured Schwann cells towards the beads (Figure 5), after which the Schwann cells began to proliferate. In a third ingenious experimental preparation, they cultured superior cervical ganglia in polythene tubes containing a collagen matrix, and observed again that GGF-2 enhanced Schwann cell migration from the explant, independent of cell division and independent of neurite outgrowth, which occurred later. Most importantly, for relevance to peripheral nerve regeneration, conditioned medium harvested from Schwann cells grown in different concentrations of GGF-2 (and diluted to correct for increasing Schwann cell numbers), showed a concentration-dependent stimulatory effect on neurite outgrowth from ganglion explants, suggesting that GGF-stimulated Schwann cells produce increased amounts of neurotrophic/tropic factors. However, the GGF-2 concentrations required to produce this effect were greater than those required for the migratory or proliferative effects. From these results, they proposed an attractive model in which Schwann cells would react to a progressive spatial gradient of neuregulins secreted from the growth cone. Furthest away from the growth cone, Schwann cells would migrate towards the site of injury, and then would be stimulated to divide as they entered the region where GGF-2 concentrations were sufficiently high. Nearest to the growth cone, the highest GGF-2 concentrations would stimulate Schwann cells to emit their unidentified trophic factors for the benefit of regenerating axons. This model deals with neuregulins released from the regenerating axon, but do they also play a signalling role in the initial burst of Schwann cell mitosis following nerve injury, when the axon is degenerating? Carroll *et al.* (1997) found neuregulin mRNAs within the degenerating sciatic nerve distal to injury, representing splice variants different from those present in dorsal root ganglion and spinal cord. Clearly, these could not be neuronal in origin. The mRNAs were detectable within 3 days of axotomy, and increased up until 30 days post-axotomy.

Neuregulin-immunoreactive proteins increased in the nerve with the same time-course, corresponding to the onset of Schwann cell proliferation. The erbB2 neuregulin receptor was also upregulated, and localized to Schwann cells by immunocytochemistry (Figure 6). Thus, neuregulins produced by Schwann cells following nerve injury could act in an autocrine fashion during early Wallerian degeneration to stimulate their division. Confirmation of erbB2 activation during the period of post-axotomy Schwann cell mitosis was provided by Kwon *et al.* (1997), who used antibodies recognizing the autophosphorylated sites on the erbB2 receptor, and showed parallel changes in erbB2 phosphorylation and Schwann cell mitosis, as indicated by bromodeoxyuridine labelling, with both declining about two weeks following nerve injury. During this later period, when Schwann cell mitosis declines, neuregulin mRNA levels remain elevated, but the Schwann cells lose their responsiveness to neuregulins due to decreases in erbB2 receptor expression (Li *et al.*, 1997). Yet, when axons regenerate through the distal nerve during this period, they induce a second burst of mitosis. Perhaps the Schwann cells become refractory only to autocrine forms of neuregulins, and still respond to axonally-secreted forms, or perhaps other axonal mitogens are involved.

Reg-2 is a novel Schwann cell mitogen which, unlike neuregulins, is up-regulated in neurons following nerve injury. It was identified by Livesey *et al.* (1997) using differential display PCR on RNA extracted from normal and axotomised dorsal root ganglia, and found to correspond to a previously described pancreatic secreted protein.

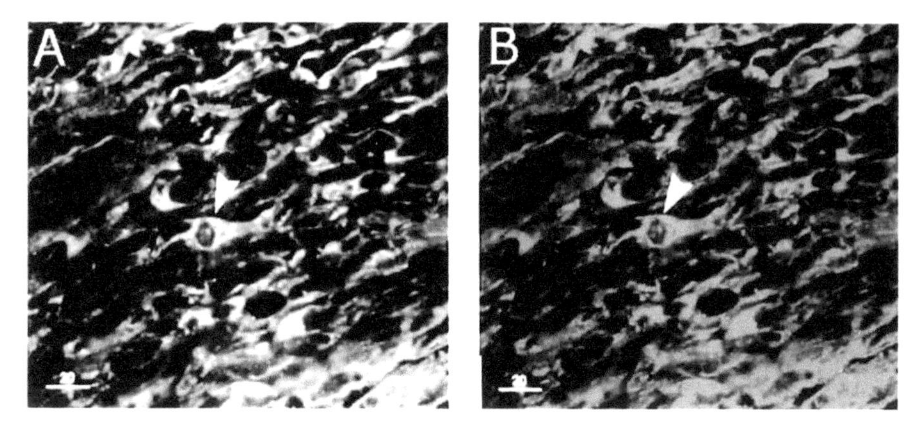

Figure 6 Localization of neuregulin immunoreactivity to Schwann cells in a five-day injured distal sciatic nerve. A: confocal image stained with a pan-neuregulin antibody. B: the same section showing localization of the Schwann cell marker S100. Arrowhead shows the same cell in both panels. There is an exact correspondence between neuregulin- and S100-immunoreactive structures. From Carroll *et al.* (1997) Journal of Neuroscience 17, 1642–1659.

In situ hybridization revealed the expression of the mRNA in motoneurons and a subpopulation of medium to large dorsal root ganglion neurons within 24 hours of sciatic nerve crush. Immunocytochemistry showed that Reg-2 was axonally transported along regenerating axons. A polyclonal antiserum, shown to block the mitogenic effects of Reg-2 on Schwann cells *in vitro,* was injected into the nerve immediately before injury, and 4 days later the number of Reg-2 containing axons distal to the injury was reduced, suggesting an inhibitory effect on axon regeneration. Livesey *et al.* (1997) further studied the regulation of Reg-2 gene expression. Since the Reg-2 gene contains interleukin-6 response elements, and since Interleukin-6 and several other cytokines including CNTF signal through the Leukemia Inhibitory Factor receptor, they examined Reg-2 expression in leukemia inhibitory factor receptor null mutants, and found that it was not expressed after nerve injury. A possible mechanism for the rapid increase in Reg-2 expression after injury is that physical damage to Schwann cells liberates CNTF (above), and through the leukemia inhibitory factor receptor, this induces Reg-2 in the injured neurons, which then acts as a mitogen at the site of injury, and subsequently adjacent to the growth cone. During regeneration, leukemia inhibitory factor and Interleukin-6, known to be produced by the distal injured nerve stump, maintain Reg-2 synthesis by regenerating neurons. This putative scheme emphasizes the existence of an intimate dialogue between axons and Schwann cells, achieved through reciprocal interactions between agents produced by the two cell types as a result of axotomy-induced changes in gene expression.

Target reinnervation

The final stage in regeneration is reinnervation of the peripheral target, accompanied by recovery of functions. The good recovery observed in experimental studies in small rodents contrasts with the generally less satisfactory outcome in humans, partly because we can appreciate residual defects much better in humans, partly because human peripheral nerve injuries are more variable in extent and location, and partly because regeneration seems to be poorer over the longer distances involved in humans and correspondingly longer periods of peripheral denervation.

In companion papers, Fu & Gordon (1995a,b) explored the reasons for poor motor functional recovery when there are long delays between nerve injury and reinnervation of the denervated muscles. The first paper focussed on the motoneurons, and the long-term effects of axotomy without reinnervation. The second paper examined the ability of long-term denervated muscle fibres and the distal nerve stumps to support axon regeneration. Their ingenious experimental paradigm was either to cross-suture the proximal tibial nerve stump, cut up to 12 months earlier, to the distal stump of the freshly-cut common peroneal nerve, or to cut the common peroneal nerve months before suturing to its distal stump the freshly-cut tibial nerve. In the former situation, chronically axotomised motoneurons were invited to reinnervate freshly-denervated distal nerve and muscle, and in the second case, acutely-axotomised motoneurons reinnervated chronically distal nerve and denervated muscle. Fu & Gordon (1995a) found that prolonged axotomy reduced the number of regenerating motor axons making functional connections with muscle fibres: after immediate cross-suture, almost all motoneurons were able to reinnervate, but if axotomy had preceded cross-suture by longer than 6 months, only about 1/3 of the motoneurons were able to reinnervate. This was compensated by a three-fold increase in the size of the motor units, so that the muscles reinnervated by long-term axotomised motoneurons produced as much tension as those reinnervated by freshly-axotomised motoneurons. However, recovery of fine motor control would obviously be affected by the much smaller number of available motor units. The second paper (Fu & Gordon, 1995b) showed that a more important factor than the inability of long-term axotomised motoneurons to regenerate seemed to be the deterioration of the chronically denervated intramuscular nerve sheaths. They no longer served as a suitable substrate for axon growth. Further, the long-term denervated muscle fibres were unable to recover fully from denervation-induced atrophy after reinnervation.

The importance of Schwann cells to reinnervation of muscle fibres has been illustrated by several papers from the laboratory of W.J. Thompson. In beautiful immunohistochemical studies of whole-mounts of rat soleus muscle after a soleus nerve injury, double labelled with antibody to neurofilament (to show axons) and a monoclonal antibody to show Schwann cells, Son & Thompson (1995) showed that during the period of denervation, Schwann cells at the end-plate extended processes which often reached to adjacent end-plates. When the regenerating axons returned to the muscle they grew along the Schwann cell processes to re-innervate a second end-plate, resulting in polyneuronal innervation of muscle fibres.

It seems likely that the behaviour of terminal Schwann cells following denervation is dependent on neuregulins. In a later paper (Trachtenberg & Thompson, 1997) it was reported that application of GGF-2 to normal neonatal muscle fibres induced the loss of normal neuromuscular junctions and postsynaptic acetylcholine receptor clusters, simultaneously with the migration of terminal Schwann cells from the end-plates, and growth of axons along these newly-formed Schwann cell pathways. Interestingly, a similar re-arrangement of the neuromuscular junction was not seen in more adult (30 day) rats when GGF-2 was applied, and this may be correlated with the decreased expression, after the second postnatal week, of erbB2 receptors in Schwann cells. Denervation of adult Schwann cells causes them to re-express erbB2 receptors (Carroll *et al.*, 1997) so that once again they would respond to neuregulin released from muscle, regenerating axons or the Schwann cells themselves. As shown by the work of Fu & Gordon (1995b), long-term denervated nerve sheaths are a poor substrate for axon growth. Relatively few studies have explored the reasons for this. An exception is the work of Li *et al.* (1997) who focussed on the expression of GGF-2 and erbB2 receptors during long-term denervation. Consistent with previous studies, they found only small and transient changes in GGF-2 mRNA expression in dorsal root ganglion following nerve injury. While there was a robust increase in erbB2 expression in the distal nerve following sciatic nerve section and immediate suture of the proximal and distal stump, if the repair was delayed by 4 months, erbB2 expression, already low at the time the repair was performed, increased only slightly, and with a delay of six months between injury and repair there was barely any response to the ingrowth of axons. There was a good correlation between the degree of erbB receptor expression and the degree of reinnervation of the distal stump. Li *et al.* (1997) concluded that chronically denervated Schwann cells do not express erbB receptors. This makes them insensitive to neuregulin released from regenerating axons, and thus unable to become involved in the reciprocal signalling between axon and Schwann cell conducive to good regeneration (e.g., by production of neurotrophic factors, Mahanthappa *et al.*, 1996). More studies are needed on the comparison between acutely- and chronically-injured peripheral nerve in order to understand better the poor regeneration observed after peripheral nerve injury in humans.

Endnote

What's next in the study of peripheral nerve regeneration? Powerful molecular biology techniques will undoubtedly reveal more novel molecules involved in trophic, guidance, recognition and adhesion events. Transduction pathways through which many of these molecules act will gradually be unravelled. The signal for the cell body reaction will be determined. All these discoveries will provide therapeutic opportunities which may be applied to both peripheral and central nervous systems. What is badly needed, however, are more studies focussing on the longer-term events associated with prolonged axotomy and axonal deprivation in the distal nerve. These are needed to understand why human peripheral nerve regeneration is so poor. The greatest advance, however, would be to restore the specificity of axonal guidance which existed during development of the nervous system, so that injured axons are able to reinnervate their original targets and fully restore function.

Acknowledgement

I am grateful to Ms. Leslie Preston who allowed me to use her excellent term paper on the neuregulins as the basis for organizing my thoughts on the topic. However, any errors in the presentation of this information are entirely the fault of the author.

References

Agius Y E and Cochard P (1998) Comparison of neurite outgrowth induced by intact and injured sciatic nerves: a confocal and functional analysis. Journal of Neuroscience 18, 328–338

Araki T and Milbrandt J (1996) Ninjurin, a novel adhesion molecule, is induced by nerve injury and promotes axonal growth. Neuron 17, 353–361

Bisby M A (1995) Regeneration of peripheral nervous system axons. In The Axon: Structure, Function and Pathophysiology, edited by S G Waxman J D Kocsis P K Stys, pp. 553–578. New York: Oxford University Press

Bisby M A Tetzlaff W and Brown M C (1995) Cell body response to injury in motoneurons and primary sensory neurons of a mutant mouse, Ola (Wld), in which Wallerian degeneration is delayed. Journal of Comparative Neurology 359, 653–662

Brown M C Perry V H Hunt S P and Lapper S R (1994) Further studies on motor and sensory nerve regeneration in mice with delayed Wallerian degeneration. European Journal of Neuroscience 6, 420–428

Carroll S L Miller M L Frohnert P W Kim S S and Corbett J A (1997) Expression of neuregulins and their putative receptors, ErbB2 and ErbB3, is induced during Wallerian degeneration. Journal of Neuroscience 17, 1642–1659

Chadwick B P Heath S K Williamson J Obermayr F Patel L Sheer D and Frischauf A M (1998) The human homologue of the ninjurin gene maps to the candidate region of the hereditary sensory neuropathy type I Genomics 47, 58–63

Coleman M P Conforti L Buckmaster E A Tarlton A Ewing R M Brown M C Lyon M F and Perry V H (1998) An 85–kb tandem triplication in the slow Wallerian degeneration (Wlds) mouse. Proceedings of the National Academy of Sciences USA 95, 9985–9990

Crowder R J and Freeman R S (1998) Phosphatidylinositol 3–kinase and Akt protein kinase are necessary and sufficient for the survival of nerve growth factor-dependent sympathetic neurons. Journal of Neuroscience 18, 2933–2943

Doyle C A and Hunt S P (1997) Reduced nuclear factor kappa B (p65) expression in rat primary sensory neurons after peripheral nerve injury. Neuroreport 8, 2937–2942

Eilers A Whitfield J Babij C Rubin L L and Ham J (1998) Role of the Jun kinase pathway in the regulation of c-Jun expression and apoptosis in sympathetic neurons. Journal of Neuroscience 18, 1713–1724

Fu S Y and Gordon T (1995a) Contributing factors to poor functional recovery after delayed nerve repair - prolonged axotomy. Journal of Neuroscience 15, 3876–3885

Fu S Y and Gordon T (1995b) Contributing factors to poor functional recovery after delayed nerve repair — prolonged denervation. Journal of Neuroscience 15, 3886–3895

Fu S Y and Gordon T (1997) The cellular and molecular basis of peripheral nerve regeneration. Molecular Neurobiology 14, 67–116

Gitler D and Spira M E (1998) Real time imaging of calcium-induced localized proteolytic activity after axotomy and its relation to growth cone formation. Neuron 20, 1123–1135

Herdegen T Skene P and Bahr M (1997) The c-Jun transcription factor–bipotential mediator of neuronal death, survival and regeneration. Trends in Neurosciences 20, 227–231

Kenney A M and Kocsis J D (1998) Peripheral axotomy induces long-term c-jun amino-terminal kinase-1 activation and activator protein-1 binding activity by c-jun and JunD in adult rat dorsal root ganglia *in vivo*. Journal of Neuroscience 18, 1318–1328

Kobayashi N R Fan D-P Giehl K M Bedard A M Wiegand S J and Tetzlaff W (1997) BDNF and NT-4/5 prevent atrophy of rat rubrospinal neurons after cervical axotomy, stimulate GAP-43 and Tα1-tubulin mRNA expression, and promote axonal regeneration. Journal of Neuroscience 17, 9583–9595

Kwon Y K Bhattacharyya A Alberta J A Giannobile W V Cheon K Stiles C D and Pomeroy S L (1997) Activation of ErbB2 during Wallerian degeneration of sciatic nerve. Journal of Neuroscience 17, 8293–8299

Li H Terenghi G and Hall S M (1997) Effects of delayed re-innervation on the expression of c-erbB receptors by chronically denervated rat Schwann cells *in vivo*. Glia 20, 333–347
Livesey F J and Hunt S P (1996) Identifying changes in gene expression in the nervous system: mRNA differential display. Trends in Neurosciences 19, 84–88
Livesey F J O'Brien J A Li M Smith A G Murphy L J and Hunt S P (1997) A Schwann cell mitogen accompanying regeneration of motor neurons. Nature 390, 614–618
Ma W and Bisby M A (1998) Increased activation of nuclear factor kappa B in rat lumbar dorsal root ganglion neurons following partial sciatic nerve injuries. Brain Research 797, 243–254
Maggirwar S B Sarmiere P D Dewhurst S and Freeman R S (1998) Nerve growth factor-dependent activation of NF-κB contributes to survival of sympathetic neurons. Journal of Neuroscience 18, 10356–10365
Mahanthappa N K Anton E S and Matthew W D (1996) Glial growth factor 2, a soluble neuregulin directly increases Schwann cell motility and indirectly promotes neurite outgrowth. Journal of Neuroscience 16, 4673–4683
Minturn J E Fryer J L Geschwind D H and Hockfield S (1995) TOAD-64, a gene expressed early in neuronal differentiation in the rat, is related to unc-33 a C elegans gene involved in axon outgrowth. Journal of Neuroscience 15, 6757–6766
Mulderry P K and Dobson S P (1996) Regulation of VIP and other neuropeptides by c-jun in sensory neurons: implications for the neuropeptide response to axotomy. European Journal of Neuroscience 8, 2479–2491
Nakayama M Miyake T Gahara Y Ohara O and Kitamura T (1995) A novel RING-H2 motif protein downregulated by axotomy: its characteristic localization at the postsynaptic density of axosomatic synapse. Journal of Neuroscience 15, 5238–5248
Pasterkamp R J Giger R J and Verhaagen J (1998) Regulation of semaphorin III/collapsin-1 gene expression during peripheral nerve regeneration. Experimental Neurology 153, 313–327
Perez N L Sosa M A and Kuffler D P (1997) Growth cones turn up concentration gradients of diffusible peripheral target-derived factors. Experimental Neurology 145, 196–202
Povelones M Tran K Thanos D and Ambron R T (1997) An NF-κB-like transcription factor in axoplasm is rapidly inactivated after nerve injury in Aplysia. Journal of Neuroscience 17, 4915–4920
Riccio A Pierchala B A Ciarallo C L and Ginty D D (1997) An NGF-TrkA- mediated retrograde signal to transcription factor CREB in sympathetic neurons. Science 277, 1097–1100
Rohlmann A Laskawi R Hofer A Dermietzel R and Wolff J R (1994) Astrocytes as rapid sensors of peripheral axotomy in the facial nucleus of rats. Neuroreport 5, 409–412
Sendtner M Gotz R Holtmann B and Thoenen H (1997) Endogenous ciliary neurotrophic factor is a lesion factor for axotomized motoneurons in adult mice. Journal of Neuroscience 17, 6999–7006
Senger D L and Campenot R B (1997) Rapid retrograde tyrosine phosphorylation of trkA and other proteins in rat sympathetic neurons in compartmented cultures. Journal of Cell Biology 138, 411–421
Smith D S and Skene J H P (1997) A transcription dependent switch controls competence of adult neurons for distinct modes of axon growth. Journal of Neuroscience 17, 646–658
Son Y J and Thompson W J (1995) Nerve sprouting in muscle is induced and guided by processes extended by Schwann cells. Neuron 14, 133–141
Song H J Ming G L and Poo M M (1997) cAMP-induced switching in turning direction of nerve growth cones. Nature 388, 275–279
Strittmatter S M Fankhauser C Huang P L Mashimo H and Fishman M C (1995) Neuronal pathfinding is abnormal in mice lacking the neuronal growth cone protein GAP-43. Cell 80, 445–452

Su Q N Namikawa K Toki H and Kiyama H (1997) Differential display reveals transcriptional up-regulation of the motor molecules for both anterograde and retrograde axonal transport during nerve regeneration. European Journal of Neuroscience 9, 1542–1547

Takemura R Nakata T Okada Y Yamazaki H Zhang Z and Hirokawa N (1996) mRNA expression of KIF1A, KIF1B, KIF2, KIF3A, KIF3B, KIF4, KIF5, and cytoplasmic dynein during axonal regeneration. Journal of Neuroscience 16, 31–35

Tanelian D L Barry M A Johnston S A Le T and Smith G M (1997) Semaphorin III can repulse and inhibit adult sensory afferents in vivo. Nature Medicine 3, 1398–1401

Tetzlaff W Alexander S W Miller F D and Bisby M A (1991) Response of facial and rubrospinal neurons to axotomy: changes in mRNA expression for cytoskeletal proteins and GAP-43. Journal of Neuroscience 11, 2528–2544

Torigoe K Tanaka H F Takahashi A Awaya A and Hashimoto K (1996) Basic behavior of migratory Schwann cells in peripheral nerve regeneration. Experimental Neurology 137, 301–308

Trachtenberg J T and Thompson W J (1997) Nerve terminal withdrawal from rat neuromuscular junctions induced by neuregulin and Schwann cells. Journal of Neuroscience 17, 6243–6255

Vogelbaum M A Tong J X and Rich K M (1998) Developmental regulation of apoptosis in dorsal root ganglion neurons. Journal of Neuroscience 18, 8928–8935

Wang L H and Strittmatter S M (1996) A family of rat CRMP genes is differentially expressed in the nervous system. Journal of Neuroscience 16, 6197–6207

Yoon S O Casaccia-Bonnefil P Carter B and Chao M V (1998) Competitive signaling between TrkA and p75 nerve growth factor receptors determines cell survival. Journal of Neuroscience 18, 3273–3281

Zagrebelsky M Buffo A Skerra A Schwab M E Strata P and Rossi F (1998) Retrograde regulation of growth-associated gene expression in adult rat Purkinje cells by myelin-associated neurite growth inhibitory proteins. Journal of Neuroscience 18, 7912–7929

Ziv N E and Spira M E (1997) Localized and transient elevations of intracellular Ca^{2+} induce the dedifferentiation of axonal segments into growth cones. Journal of Neuroscience 17, 3568–3579

13

The role of macrophages in degeneration and regeneration in the peripheral nervous system

P M Hughes and V H Perry

Introduction

Degeneration and regeneration of the peripheral nervous system are frequently studied using a simple *in vivo* model. A peripheral nerve is cut or crushed so that the cell body is separated from a portion of the axon, this results in the degeneration of the distal isolated axon and its myelin sheath. Schwann cells respond rapidly to axonal degeneration and initiate myelin removal. The blood-nerve barrier, which consists of endoneurial blood vessels and the perineurial sheath, is disrupted for several weeks after nerve transection beginning at the second day after injury. Leukocyte recruitment to the degenerating axon begins 2–3 days after nerve injury and, unlike a typical acute inflammatory response, is macrophage specific. These events, which occur distal to the site of nerve transection, are referred to as Wallerian degeneration (Figure 1). Despite the superficial simplicity of this system the sequence of cellular and molecular events resulting in nerve degradation and subsequent regeneration are by no means completely understood.

Following nerve injury the cell body of the transected nerve reprograms its synthetic machinery so as to produce building material needed for axonal elongation. Successful regeneration, however, is also dependent on the local environment which must be permissive for regeneration and provide positive cues for regrowing axons. Wallerian degeneration prepares the environment for regeneration removing potentially inhibitory myelin debris, increasing the synthesis of growth factors and creating a matrix able to support axonal growth. Investigations into Wallerian degeneration may therefore lead to novel ways of modulating the nervous system response to injury, in the interest of promoting regeneration in the peripheral and central nervous system (PNS and CNS).

We will review here the sequence of cellular and molecular events which initiate axonal and myelin degradation during Wallerian degeneration and consider whether macrophages are essential contributors to the degradation of the myelin sheath, Schwann cell proliferation and subsequent regenerative events.

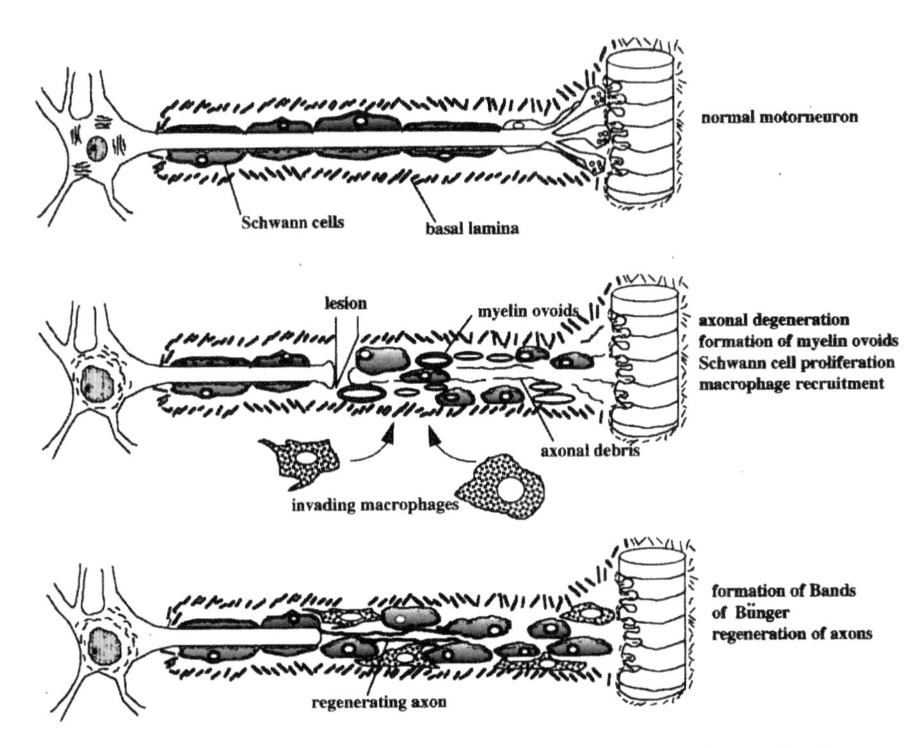

Figure 1 Sequence of cellular events during Wallerian degeneration in the peripheral nervous system. Within 24 hours after a lesion the axon degenerates with granular disintegration of the cytoskeleton. There is a retraction of the Schwann cell cytoplasm from the myelin sheath and myelin ovoids are formed. Between 1 and 3 days Schwann cells proliferate and blood-borne macrophages are recruited to the degenerating nerve. Axonal and myelin debris are removed by Schwann cells and macrophages. Ten-20 days later Schwann cells form the Bands of Bünger, along which regenerating axons grow. The basal lamina remains intact during Wallerian degeneration.

Axonal degeneration

Following axotomy, the first degenerative changes are seen in the axon itself. Within hours following injury, mitochondria and other organelles accumulate in the nodal and paranodal regions of the distal segment. At this initial stage of Wallerian degeneration fast axonal transport continues at normal rates and electrical conduction and synaptic transmission can be elicited normally with stimulation of the distal stump. The axonal cytoskeleton then breaks down abruptly, with simultaneous disintegration of axonal organelles, axolemma and cytoskeletal components (George & Griffin, 1994). This process, which is referred to as axonal granular disintegration, begins at the site of nerve section and progresses distally. Axonal degeneration progresses rapidly with complete axonal loss 3 days after nerve section (George & Griffin, 1994). Axonal degeneration is initiated by an active process, as opposed to resulting from lack of maintenance from the neuronal cell body. This was firmly established by studies on the murine mutant, C57BL/Wld, in which Wallerian degeneration is delayed after axotomy (Lunn *et al.*,

Key points

- Section of peripheral nerve so that cell body is separated from distal part of axon results in degeneration of the distal isolated axon and its myelin sheath.
- Schwann cells respond rapidly to axonal degeneration and initiate myelin removal.
- Blood-nerve barrier, is disrupted for several weeks after nerve transection.
- The injury induced acute inflammatory response is macrophage specific.
- Following nerve injury, cell body of transected nerve reprograms its synthetic machinery to produce building materials needed for axonal elongation.
- Wallerian degeneration prepares local environment distal to injury for regeneration, removing potentially inhibitory myelin debris, increasing synthesis of growth factors and creating a matrix to support axonal growth.

1989). Following transection the axons survived for weeks, rather than the usual 1–3 days (Lunn *et al.*, 1989).

An increase in intra-axonal calcium, leading to activation of calcium-activated proteinases (calpains) occurs rapidly following nerve injury. The calpains are capable of degrading components of the axonal cytoskeleton, including neurofilaments. Normally, following transection, neurites of cultured ganglia degenerate within 12–16 hours. Using cell permeable specific inhibitors of the calpains, severed axons in culture are preserved for up to 4 days. (George *et al.*, 1995). In comparison, cultured neurons from mutant C57BL/Wld mice remained intact for up to 7 days after transection (Buckmaster *et al.*, 1995). This suggests there are significant additional factor(s) to calcium and calpains which initiate axonal breakdown. The phenotype of the slow degeneration, observed in the C57BL/Wld mice, depends on properties which are intrinsic to the axon (Buckmaster *et al.*, 1995). The mutation in C57BL/Wld mice has recently been described and insights into the molecular events controlling Wallerian degeneration should follow (Coleman *et al.*, 1998).

Apoptosis is the term used to describe a process of self-inflicting rapid cell death. During apoptosis the cell shrinks, its chromosomal material is broken into short fragments and the debris is removed by tissue macrophages, without initiating an acute inflammatory response. Hallmarks of apoptosis include nuclear changes, such as chromatin condensation and DNA fragmentation (Wyllie, 1980). However, the nucleus is not essential for apoptosis. Cells that normally have a nucleus, such as anucleate cytoplasts, die with apoptotic characteristics. Axonal degeneration shares features with apoptosis, including blebbing of the axon plasma membrane (Buckmaster *et al.*, 1995) and cytoskeleton changes. Furthermore, manipulations that influence the onset of neuronal cell body apoptosis also influence the rate of degeneration of isolated neurites (Buckmaster *et al.*, 1995). These observations suggest common mechanisms in neuronal apoptosis and axonal degeneration but they are not identical as highlighted by the finding that over expression of B-cell lymphoma protein-2 (Bcl-2), an apoptosis regulatory gene,

Key points

- Axonal degeneration is initiated by an active process, as opposed to resulting from lack of maintenance from the neuronal cell body.
- Increase in intra-axonal calcium, leading to activation of calcium-activated proteinases (calpains) occurs rapidly following nerve injury.
- Calpains degrade components of axonal cytoskeleton, eg neurofilaments.
- Axonal degeneration shares features with apoptosis, including blebbing of axon plasma membrane and cytoskeleton changes.

prevented motoneuron cell body loss but did not affect axonal degeneration in a genetic mouse model of Motor Neuron Disease (Sagot *et al.*, 1995). The degeneration of the axon following transection appears to be dictated by a programmed cell death-like mechanism, intrinsic to the axon and at this stage macrophages are not involved.

Schwann cell response

Morphological changes in the Schwann cell and upregulation of the early response genes c-fos and c-jun occur within 24 hours after nerve injury and follow axonal granular disintegration. The work of Heath and colleagues suggests that Schwann cell changes are initiated by an active signal, rather than simply the loss of a viable axon (Kidd & Heath, 1991). They describe a very unusual situation in the mouse superior cervical ganglion, where double myelinated axons are located. Double myelination in the adult mouse is due to displacement of a myelinating Schwann cell from axonal contact, which results from the invasion of another Schwann cell at the node. Transection of these double myelinated axons resulted in breakdown of the inner myelin sheath but the outer sheath survives axonal degeneration (Kidd & Heath, 1991). After axonal damage, it is likely that a signal provided by the degenerating axon interacts with receptors on the Schwann cell surface. Subsequent changes in Schwann cell gene expression could regulate extrusion of the myelin sheath, entry into the cell cycle, alterations in phenotypic expression and the formation of longitudinal Schwann cell bands, the Bands of Büngner. Following axotomy the Schwann cell separates itself from the multiple layers of the myelin sheath. This complex event presumably occurs by a process of fusion of the cytoplasmic membranes in the region of the nucleus. Surprisingly, this important initiating event in myelin degeneration has not been explored.

During normal development Schwann cells proliferate to accommodate neurite elongation. This proliferation ceases during maturity and the Schwann cell population becomes quiescent. However, after nerve injury, the Schwann cells regain their proliferative phenotype and cell division starts at day 2 and peaks rapidly at day 3 post axotomy. At later time points examined (days 18 and 30) Schwann cell proliferation is markedly decreased (Carroll *et al.*, 1997). Numerous agents have been demonstrated to be mitogenic to Schwann cells *in vitro* but it is not clear that these play a functional role *in vivo*. Proposed mitogens include axonal derived signals, components of the extracellular matrix, cyclic AMP analogues, myelin debris and peptide growth factors (Kleitman & Bunge, 1995).

In vitro studies using neonatal rat Schwann cells led to the identification of a group of potent Schwann cell mitogens known as glial growth factors (GGFs) (Brockes *et al.*, 1980). Glial growth factors induce tyrosine phosphorylation of the neu proto-oncogene and therefore are referred to as neuregulins. During development axon associated neuregulins have been implicated in Schwann cell differentiation and proliferation (Dong *et al.*, 1995). Following axotomy of the adult rat sciatic nerve, the mRNA expression of neuregulins in the distal stump is induced at day 2, coinciding with the initiation of Schwann cell proliferation (Carroll *et al.*, 1997). Surprisingly, the Schwann cells themselves are the source of neuregulins rather than the neuron. Hence it appears neuregulins are not the axon derived signal which initiates Schwann cell proliferation; rather they mediate interactions between the Schwann cells themselves. Increases in neuregulin mRNA are detected for up to 30 days post axotomy, when Schwann cell proliferation has decreased and mRNA expression of the two neuregulin receptors increased concurrently 5 days post axotomy and remained elevated until day 18 (Carroll *et al.*, 1997). A factor in the medium, conditioned by cultured macrophages that have phagocytosed myelin membranes, is mitogenic for Schwann cells (Braichwal *et al.*, 1988) and transforming growth factor-β (TGF-β), a macrophage product, induces Schwann cell division *in vitro* (Ridley *et al.*, 1989). The presence of macrophages, however, is not essential for Schwann cell proliferation *in vitro* (Fernandez-Valle *et al.*, 1995). Although work studying the effect of macrophage depletion on Wallerian degeneration *in vivo* has not focused on the state of Schwann cell proliferation the timing of events indicates Schwann cell proliferation begins prior to significant recruitment of macrophages.

Denervated Schwann cells organise themselves into cords spanning the two injured ends and form the Bands of Büngner in the distal segment of the nerve, which consist of aligned Schwann cells surrounded by basement membrane tubes. In addition to the observed migration into the gap between injured stumps, Schwann cells also migrate within the distal stumps to facilitate axonal regrowth. Son and Thompson demonstrated that Schwann cells extend processes, which lead and guide peripheral nerve regeneration, across a gap in a nerve lesion and into the degenerating distal nerve segment (Son & Thompson, 1995).

Loss of physical contact with the axon induces a downregulation in Schwann cell myelin gene expression (Trapp *et al.*, 1988). However, changes in the Schwann cell phenotype may be initiated prior to axonal loss, due to alterations in axonal transport or biochemical signal(s) produced by the axon (Wu *et al.*, 1994). Axotomy of the sciatic nerve results in the levels of myelin protein mRNA falling by at least 95% (Trapp *et al.*, 1988). Following nerve injury, Schwann cells dedifferentiate and upregulate their expression of the neural and glial adhesion molecules, N-CAM (neural cell adhesion molecule), L1 and N-cadherin (Daniloff *et al.*, 1986; Martini & Schachner, 1988). In addition, Schwann cells upregulate the expression of the neurotrophins, including nerve growth factor, brain-derived neurotrophic factor (BDNF) (Meyer *et al.*, 1992) and insulin-like growth factor (IGF) (Pu *et al.*, 1995). Macrophages may be involved in Schwann cell dedifferentiation and the subsequent upregulation of neurotrophins and adhesion molecules which facilitate axon regrowth. This is discussed below, when considering the involvement of macrophages and Schwann cells in axonal regeneration.

Key points

- Morphological changes in Schwann cell and upregulation of early response genes c-fos and c-jun occur within 24 hours of nerve injury and follow axonal granular disintegration.
- Schwann cell changes are probably initiated by an active signal, rather than simply the loss of a viable axon.
- Glial growth factors (GGFs) are potent Schwann cell mitogens.
- GGFs induce tyrosine phosphorylation of the neu proto-oncogene and are referred to as neuregulins.
- Following axotomy of adult rat sciatic nerve, mRNA expression of neuregulins in Schwann cells of distal stump is induced at day 2, at time of onset of initiation of Schwann cell proliferation.
- Axotomy of the sciatic nerve results in the levels of myelin protein mRNA falling by at least 95%.
- Following nerve injury, Schwann cells dedifferentiate and upregulate expression of neural and glial adhesion molecules.

Macrophage response

The increase in total cell number observed in degenerating nerves is, in part, due to Schwann cell mitosis but there is also an increase in the number of other cells present. Immunohistochemical investigations confirmed that leukocytes invading the distal nerve segment are specifically macrophages (Perry *et al.*, 1987; Stoll *et al.*, 1989; Avellino *et al.*, 1995). There are no phenotypic markers to distinguish the infiltrating blood-borne macrophages and resident macrophages. In the rat sciatic nerve resident macrophages are known to account for up to 4% of the total number of endoneurial cells. Resident macrophages often lie near blood vessels but the majority are distributed throughout the endoneurial space. This random spacing makes it unlikely that each node is contacted by a resident macrophage in the normal PNS. Resident macrophages are ramified in their appearance and constitutively express a range of macrophage antigens including major histocompatability (MHC) class II and the complement type 3 receptor. As the non-resident macrophages greatly outnumber the resident cells during Wallerian degeneration, it is difficult to assess the changes in resident macrophages.

There is a delay in macrophage recruitment to the degenerating nerve, a significant increase in numbers not being detected until day 3 in the rat (Perry *et al.*, 1987; Stoll *et al.*, 1989). This was not found by Avellino and colleagues (1995) who detected a significant 1.4 fold increase above control numbers 1 day after nerve section (Avellino *et al.*, 1995). In this study a length (1.5 cm) of the sciatic nerve was freed from the surrounding tissue by blunt dissection (Avellino *et al.*, 1995). This procedure was not reported to have been carried out in the other studies (Perry *et al.*, 1987; Stoll *et al.*,

1989) and is likely to be responsible for the early recruitment of macrophages, probably due to disruption of surrounding tissues and blood supply.

A typical acute inflammatory response, in the skin for example, begins with an influx of neutrophils within 2 hours. Neutrophil numbers peak at 6 hours and decline thereafter. Macrophage recruitment shows similar kinetics, although fewer macrophages are recruited and they persist in the tissue longer. Two features of the acute inflammatory response in the distal segment of a damaged peripheral nerve make it distinct from that in non-neuronal tissue. Firstly, following nerve damage neutrophils are only recruited to the site of cut or crush and are rarely found in the degenerating distal nerve segment. Secondly, there is a delay of two to three days prior to a significant increase in macrophage numbers. Research has focused on determining the signals, or lack of them, responsible for this unusual leukocyte response to cell degeneration.

The 'signal' controlling the selective infiltration and activation of macrophages to degenerating axons during Wallerian degeneration has not been determined. Studies using the C57BL/Wld mouse demonstrate that the 'signal' depends on axonal degeneration. In this mutant mouse the macrophages only enter the nerve after axonal breakdown has begun at day 10, rather than day 3 as observed in normal mice (Lunn *et al.*, 1989). Macrophages are specifically attracted to the degenerating myelin-axon unit, ignoring intact nerves, during Wallerian degeneration. Acrylamide administration results in a small number of nerve fibres undergoing Wallerian degeneration. ED-1 positive macrophages are found both within the basal tubes and outside these degenerating fibres but are not associated with intact myelin-axon units (Stoll, 1989). Similarly, within a nerve partially damaged by crushing, recruited macrophages only accumulate around degenerating fibres. Axonal degeneration thus appears to trigger a secondary response in Schwann cells or possibly resident macrophages, which release molecules to attract macrophages. Obvious candidates to consider are members of the chemokine family, which can attract and stimulate specific subsets of leukocytes.

Chemokines are defined by a common structural motif, consisting of four conserved cysteine residues that form two characteristic intramolecular disulphide bridges. The chemokines are subdivided according to the position of the two most amino-proximal cysteine residues. In the C-X-C chemokines these residues are separated by an intervening amino-acid; in the C-C chemokines they are located next to one another. In general, the structural distinction between the C-X-C and C-C chemokines is mirrored by a functional difference. *In vitro* chemotaxis assays show that C-X-C chemokines tend to attract neutrophils, whereas C-C chemokines act preferentially on macrophages. C-C chemokines include macrophage chemotactic protein-1 (MCP-1), macrophage inhibitory protein-1α (MIP-1α) and macrophage inhibitory protein-1β (MIP-1β). Schwann cells may influence the macrophage response in ways other than chemoattraction. Following axotomy, prior to and during macrophage recruitment, there is an increase in the production of interleukin-10 (IL-10) by Schwann cells (Jander *et al.*, 1996). This cytokine is a potent immúnosupressive cytokine which downregulates major histocompatability (MHC) II expression and the production of pro-inflammatory cytokines and proteinases by macrophages. IL-10 present during Wallerian degeneration may create an immunosupressive environment in the nerve, regulating the activation state of recruited macrophages.

Another important stage in the recruitment of macrophages during inflammation is the upregulation of endothelial adhesion molecules, which enables leukocyte margination and diapedesis to occur. In other systems interactions between the adhesion molecules leukocyte functional antigen-1 (LFA-1) or the complement type-3 receptor (CR-3) and intercellular adhesion molecule-1 (ICAM-1) and between very late antigen-4 (VLA-4) and vascular cell adhesion molecule-1 (VCAM-1) have been shown to be important in creating a firm adhesion between leukocytes and endothelial cells. Adhesion is accompanied by a change in leukocyte shape which precedes them crossing the endothelium. Neutralising antibodies against the adhesion molecules ICAM-1, VLA-4 and CR-3 administered just prior to transection of the mouse sciatic nerve, does not prevent the entry of macrophages into degenerating nerves (Brown *et al.*, 1997). This surprising finding suggests that other, possibly novel, adhesion molecules are involved in the transendothelial migration of macrophages during Wallerian degeneration. Macrophage migration also requires adhesion to the endoneurial extracellular matrix surrounding myelin-axon units and using an *in vitro* assay system this adhesion was shown to involve a β1-integrin, but not $\alpha_4\beta_1$ or $\alpha_5\beta_1$ (Brown *et al.*, 1997).

Cellular mechanisms of myelin degradation

During Wallerian degeneration, myelin degradation appears to be initiated by proteinases and lipases which are intrinsic to the myelin sheath. Extracellular enzymes, synthesised by Schwann cells and macrophages, are then likely to continue the degradation prior to or in association with phagocytosis and removal of the debris.

Once isolated from the Schwann cell nucleus the myelin sheath segments into short portions and forms structures known as ovoids. Discrete ovoids are observed at day 3 post section, their lengths shortening rapidly with time (George *et al.*, 1995). Within these ovoids the precise spacing of the layers of myelin becomes disrupted. Ultrastructural studies during Wallerian degeneration reveal an early loosening of the myelin lamellae, suggesting that the structural alteration is due to the disruption of the proteins implicated in maintaining compact membrane adhesion. The loss of two

Key points

- Cell numbers increase in the degenerating distal stump of an injured nerve.
- Cell numbers increase because of mitogenesis of Schwann cells, infiltration of blood-borne macrophages and possibly an increase in resident macrophages.
- There are no specific markers that distinguish these 2 populations of macrophages, which makes it difficult to determine their differential contribution to axon degeneration.
- The 'signal' controlling the selective infiltration and activation of macrophages to degenerating axons during Wallerian degeneration has not been determined, but it depends on degeneration.
- Chemokines and interleukin-10 are likely candidates.

Table 1 Myelin proteins degraded by proteinases intrinsic to the myelin sheath

Proteinases associated with the myelin sheath	Myelin proteins degraded *in vitro*	Reference
lysosomal proteinases – cathepsins	MBP	(Hallpike & Adams, 1969)
Calpains (μ and m)	MBP	(Banik *et al.*, 1991)
MADM	MBP	(Howard, *et al.*, 1996)
uncharacterised metalloproteinase	P0	(Agrawal, *et al.*, 1990)

myelin-specific integral membrane glycoproteins, P_0 and myelin-associated glycoprotein, occurs 2 days after axotomy (George & Griffin, 1994). This initial disruption is likely to be attributable to intracellular proteolysis (Table 1). The first proteinases implicated in myelin breakdown during Wallerian degeneration were the acidic lysosomal proteinases (Hallpike & Adams, 1969), likely to be members of the cathepsin family. Intracellular proteolysis could also be carried out by non-lysosomal proteinases, such as the calpains (Banik *et al.*, 1991).

There has been controversy over the relative contribution made by macrophages and Schwann cells to myelin degradation and removal. Macrophages and Schwann cells have both been credited individually or jointly with carrying out myelin degradation. Beuche & Friede's (1984) experiments rekindled interest in the involvement of macrophages during Wallerian degeneration. They implanted chambers, containing nerve segments, into the peritoneal cavity of mice of the same strain. The pore size in the chambers was varied to either allow or exclude macrophage invasion. When the pore size was reduced, preventing entry of the macrophages, normal looking compact myelin was visible in the electron microscope for several weeks. At the same time the usual mitosis of Schwann cells was absent, suggesting that the Schwann cell response in this model was impaired because of the absence of macrophages (Beuche & Friede, 1984). In later experiments Beuche & Friede (1986) inhibited macrophage invasion into the nerve during Wallerian degeneration *in vivo* by treatment with silica dust. They reported that myelin degradation was delayed but the effect was slight compared with their previous study, using the chambers. These papers, in contrast to earlier reports, emphasised the essential nature of macrophages in Wallerian degeneration; without macrophages the degeneration process did not even begin. However, it should be noted that nerve segments within these chambers are likely to be hypoxic and indeed using a similar chamber system, with rat sciatic nerve, White and colleagues found that recruited macrophages did not significantly alter the early metabolic responses of the Schwann cells, which initiated the breakdown of the myelin sheath. Their results showed that in nerves undergoing Wallerian degeneration there is a decrease in overall lipid synthesis and in particular that of lipids targeted for myelin (White *et al.*, 1989). The authors suggest that the response of the Schwann cell is similar to a dedifferentiation brought about by axonal loss.

Whether macrophages are essential during Wallerian degeneration *in vivo* and the weight of their contribution to myelin degradation was addressed directly by Perry

Key points

- Myelin degradation initiated by proteinases and lipases intrinsic to myelin sheath.
- Schwann cells and macrophages continue degradation by production of extracelluar enzymes, prior to or in association with phagocytosis and removal of debris.
- Relative contribution of macrophages and Schwann cells to myelin degradation and removal has been controversial.
- Substantial amount of myelin removal occurs early in Wallerian degeneration, even in absence of invading macrophages, but are macrophages necessary for continuing rapid removal of all of debris.

and co-workers. Using whole body irradiation the production of short-lived circulating monocytes was destroyed, removing macrophages from the scene *in vivo* (Perry *et al.*, 1995). Electron microscopy and immunohistochemistry followed the rate of myelin degradation in the mouse sciatic nerve following axotomy. Wallerian degeneration proceeds at comparable rates in both a normal and irradiated mice for up to 5 days. At this stage there is considerable myelin degeneration, with the loss of over half of the myelin basic protein. At later time points examined (10 days after transection) removal of the remaining myelin is slower in the irradiated mice compared to normal controls. These findings are similar to those of Beuche & Friede and Brück and colleagues, who used silica dust or liposomes, which are toxic to macrophages, to deplete blood-borne monocytes (Beuche & Friede, 1986; Brück *et al.*, 1996). Thus, all studies confirm there is a substantial amount of myelin removal, early during Wallerian degeneration, even in the absence of invading macrophages, but macrophages are necessary for the continuing rapid removal of all of the debris. The macrophages appear to do this, in part, via a complement dependent pathway since complement depletion impairs myelin removal (Dailey *et al.*, 1998). In larger myelinated fibres macrophages may play a more prominent role in myelin degradation and removal.

Molecular mechanisms of myelin degradation

Macrophages can produce a battery of proteinases, capable of degrading components of the myelin sheath *in vitro*. Incubation of isolated myelin with conditioned medium from activated macrophages results in degradation of the myelin proteins, myelin basic protein and P_0. The activity responsible for this degradation was identified as plasminogen activator and an uncharacterised calcium-dependent neutral proteinase (Cammer *et al.*, 1981). More recently, other neutral proteinases called matrix metalloproteinases (MMPs) have been shown to degrade myelin basic protein *in vitro* (Chandler *et al.*, 1995) and macrophages have been reported to produce an array of proteinases, including numerous MMPs. The specific proteinases produced by macrophages during Wallerian degeneration and whether they participate in degradation of the myelin sheath and/or cellular invasion into the nerve has not been elucidated.

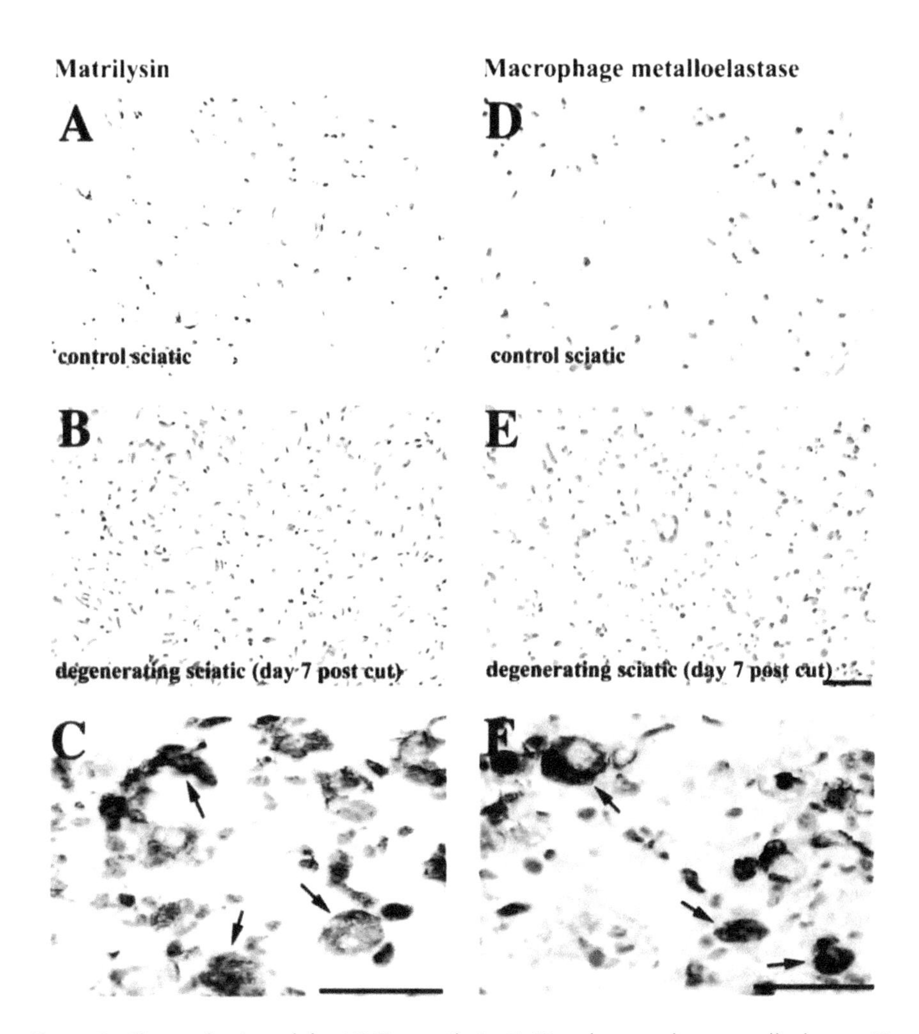

Figure 2 The production of the MMPs matrilysin (A-C) and macrophage metalloelastase (D-F) is increased during Wallerian degeneration. Matrilysin and macrophage metalloelastase immunoreactivity was scattered throughout control sciatic nerves and associated with Schwann cells (A & D). During Wallerian degeneration, coincident with macrophage invasion, the immunostaining for both matrilysin and macrophage metalloelastase was significantly increased (B & E). Double labelling revealed ED-1 positive macrophages (black 'foamy' cells) produced both matrilysin and macrophage metalloelastase (brown precipitate) during Wallerian degeneration (C & F). Scale bar = 50μm.

The MMP family consists of at least 17 members, which between them can degrade all the components of the extracellular matrix. All MMPs posses a catalytic domain with a zinc binding motif, in which three histidines bind to the catalytic zinc atom. They are extracellular, neutral proteinases which require calcium for activity and structural stability. Since MMPs can degrade all the protein constituents of the extra-cellular matrix, it is not surprising that disruption of the regulation mechanisms controlling MMP activity gives rise to a number of pathological situations (reviewed in Birkedal-Hansen, 1993). Uncontrolled proteolytic activity could contribute to the pathogenesis of nerve degeneration in several ways. Aberrant expression of MMPs by inflammatory and resident cells of the PNS could degrade the extracellular matrix macromolecules which constitute the basal lamina and the protective blood-nerve-barrier. Macrophages may use MMPs for transmigration through vascular and endoneurial endothelium. Distinct from their action on extracellular matrix components, MMPs could also directly degrade the myelin sheath. Myelin basic protein is degraded by 72kDa gelatinase, stromelysin-1, interstitial collagenase, matrilysin and 92kDa gelatinase *in vitro* (Chandler *et al.*, 1995).

During Wallerian degeneration we have shown that ED-1 positive macrophages are labelled for the several MMPs, including matrilysin and macrophage metalloelastase (Figure 2). Both myelinating and non-myelinating Schwann cells are capable of producing several MMPs (Hughes *et al.*, 1998). Zymography confirmed MMPs were elevated and present in their activated forms during Wallerian degeneration (Hughes, 1997). Changes in mRNA and protein expression and the cellular localisation of the MMPs during the first week of Wallerian degeneration are consistent with a role for these enzymes in myelin disruption. However, a broad spectrum matrix metalloproteinase inhibitor, given simultaneously with nerve section *in vivo*, did not affect the rate or extent of myelin loss during the first 10 days after injury (Hughes, 1997). Macrophage recruitment into the degenerating nerve was not affected. Several possible conclusions can be drawn from these findings. Firstly, macrophages are not dependent on matrix metalloproteinase activity to facilitate migration into the nerve during Wallerian degeneration. Secondly, matrix metalloproteinase activity does not appear to be involved in myelin degradation during Wallerian degeneration. It is possible that the synthetic matrix metalloproteinase inhibitor did not affect myelin degradation because the MMPs may be localised at the cell surface. This localisation is known to confine proteolytic activity and protect the enzymes from endogenous inhibitors (Owen *et al.*, 1995), which are also elevated during Wallerian degeneration (La Fleur *et al.*, 1996). Alternatively, access of the synthetic matrix metalloproteinase inhibitor into the nerve may have been restricted.

A transient increase in the activity of the serine proteinases plasminogen activator and thrombin is also observed following peripheral nerve injury in the distal nerve segment (Bignami *et al.*, 1982; Smirnova *et al.*, 1996). This elevation in serine proteinases suggests a potential involvement in the events occurring during Wallerian degeneration. Plasminogen activator activity was found to be increased in degenerating rat sciatic nerves. Its identification as plasminogen activator activity relied on its ability to activate plasminogen, which was then able to degrade the fibrin overlay on which the nerves were incubated (Bignami *et al.*, 1982). The increased activity of these serine proteinases occurred prior to the induction of protease nexin-1 (Smirnova *et al.*, 1996). Protease nexin-1 was found in a subpopulation of Schwann cells in the distal segment of the rat sciatic nerve following axotomy (Meier *et al.*, 1989) and may be part of a regulatory response to nerve injury, limiting proteolytic degradation of the extracellular matrix or adhesion molecules which are upregulated after axotomy. The cellular source and exact physiological functions of plasminogen activator and thrombin in Wallerian degeneration remains to be established.

Key points

- Macrophages can produce a battery of proteinases, capable of degrading components of the myelin sheath *in vitro*: plasminogen activator, an uncharacterised calcium-dependent neutral proteinase and MMPs (MMPs) produced by macrophages.

Phagocytosis of myelin

Phagocytosis of myelin debris can occur via two distinct mechanisms: (1) opsonin-dependent phagocytosis mediated via the Fc or complement receptor (CR3), (2) opsonin-independent phagocytosis via other receptors.

The Fc receptor does not appear to play a role in myelin removal, as Wallerian degeneration is not affected in the absence of immunoglobulins (Hann *et al.*, 1988). The application of the antibody Mac-1, which selectively recognises the a chain of the complement type 3 receptor, causes a detectable inhibition of myelin phagocytosis after axotomy (Brück & Friede, 1991) as does complement depletion using cobra venom factor (Dailey *et al.*, 1998).

The production of biologically active granulocyte macrophage-colony stimulating factor (GM-CSF) is induced in peripheral nerves within 5 hours after injury, too early to be the signal responsible for macrophage recruitment. GM-CSF has been shown to upregulate the expression of MAC-2/galectin-3, a galactose specific lectin cell surface receptor on macrophages and Schwann cells, *in vitro* (Saada *et al.*, 1996). Early in Wallerian degeneration the macrophage antigen MAC-2, but not other macrophage markers, is expressed by Schwann cells. Reichert *et al.* (1994) proposed that MAC-2 can mediate myelin phagocytosis through binding to sugar moieties on the myelin sheath, a process referred to as lectinophagoytosis. As MAC-2 is a galactose-specific lectin, they investigated this mechanism using galactose and other sugars to compete with the sugars in the myelin binding to MAC-2. High concentrations of galactose inhibited myelin removal by Schwann cells during Wallerian degeneration *in vitro*. The authors concluded that Schwann cells display a phagocytic phenotype during Wallerian degeneration, as a result of the upregulation of MAC-2 expression. During Wallerian degeneration MAC-2 could mediate myelin phagocytosis by Schwann cells and macrophages via lectinophagocytosis. Using a different *in vitro* system Brück and Friede reported that galactose did not inhibit macrophage phagocytosis of myelin in mouse nerve explants (Brück & Friede, 1990). Segments of desheathed peripheral nerves were placed in culture with macrophages. These macrophages rapidly invade the degenerating nerves. The addition of cytosine arabinoside to the culture medium, to inhibit the proliferation of cells intrinsic to the nerve, may also have interfered with the production

Key points

- Phagocytosis of myelin debris can occur via two mechanisms: (1) opsonin-dependent phagocytosis mediated via Fc or complement receptor (CR3), (2) opsonin-independent phagocytosis via other receptors.

of GM-CSF. Therefore, the macrophages entering the nerve may not have been MAC-2 positive and used alternative, or additional, routes to phagocytose myelin.

Participation of macrophages and Schwann cells in axonal regeneration

Studies on the mutant C57BL/Wld mouse confirm that Wallerian degeneration is pre-requisite for successful regeneration, but they do not tell us if macrophages are essential for regeneration to occur. Wallerian degeneration prepares the environment for regeneration partly through the disruption and removal of intact peripheral myelin and partly due to the production of factors which are upregulated as a consequence of the degeneration process. That intact peripheral nerve is inhibitory to axonal sprouting and regeneration was first noted in 1904 by Langley and Anderson, who observed that lesioned sensory axons did not grow into the intact proximal stump of a severed peripheral nerve. This has recently been re-emphasised by Brown and colleagues. Motor axons which usually grow vigorously into the degenerating distal stump failed to advance if they were confronted with a piece of nerve which had not degenerated (Brown *et al.*, 1991). When Wallerian degeneration failed to occur rapidly, as in the C57BL/Wld mouse, then axons did not regenerate despite vigorous sprouting at the lesion site and an apparently normal cell body response to injury (Brown *et al.*, 1994). When the distal stump started to degenerate (Brown *et al.*, 1994) or was induced to degenerate rapidly regeneration proceeded. In addition, neurites extending from the dorsal root ganglia of adult mice would not grow *in vitro* on thin sections of normal peripheral nerves but would grow on degenerated nerve (Bedi *et al.*, 1991). Purified peripheral myelin, however, has been reported to be a good substrate for axonal growth (Schnell *et al.*, 1994). This finding may have been due to laminin being co-purified with the myelin preparations, which was shown to even override the inhibitory effects of CNS myelin (David *et al.*, 1995). Work in this area implicates multiple inhibitory signals to be associated with intact peripheral myelin and/or the associated Schwann cells. The work of Alvarez and colleagues (Tapia *et al.*, 1995) implicates the Schwann cells in the repression not only of the sprouting response of axons but also the early tendency of lesioned axons to regenerate into an intact nerve prior to the onset of Wallerian degeneration. Killing the Schwann cells in the nerve segment immediately distal to the crush allows regeneration to proceed promptly after the injury. The macrophage does not appear to be involved in these early sprouting/regenerative interactions between the axon and Schwann cell.

During Wallerian degeneration Schwann cells and macrophages produce increased levels of enzymes, which may be necessary to clear the inhibitory debris, modify the extracellular matrix (see above) and facilitate regeneration. Two observations demonstrate that during Wallerian degeneration proteolytic activity is tightly regulated. Firstly, acrylamide administration results in only a few nerve fibres undergoing Wallerian degeneration. ED-1 positive macrophages are associated with these degenerating fibres, but not with intact fibres (Stoll *et al.*, 1989). If soluble proteinases, produced by macrophages or Schwann cells, are involved in myelin degradation one would expect these soluble products to diffuse through the endoneurium and affect neighbouring intact axons, which does not occur. Secondly, amongst the components of the Schwann cell basal lamina are collagen type IV, laminin, fibronectin and proteoglycan, all of which are readily degraded *in vitro* by members of the matrix metalloproteinase family.

During Wallerian degeneration the Schwann cell basal lamina remains relatively structurally intact (Thomas, 1964). Unregulated proteolytic activity would cause extensive degradation of the nerve extracellular matrix which could impede neuronal regeneration. Regrowing axons require specific matrix components to facilitate regrowth, such as laminin and collagen IV which have been shown to promote neurite outgrowth *in vitro* (Reichardt & Tomaselli, 1991).

How then is this proteolytic activity regulated? A recent study suggests that despite elevated levels of MMPs during Wallerian degeneration the basal lamina is not degraded because invading macrophages induce an increase in the production of the endogenous inhibitor of activated MMPs, tissue inhibitor of metalloproteinases-1 (TIMP-1) (La Fleur *et al.*, 1996). In culture degenerating sciatic nerves, removed prior to macrophage influx, produce excess MMP activity (Hughes *et al.*, 1995). Using a similar model La Fleur and colleagues placed nerves into culture 4 days after nerve crush, when macrophages were abundant, and now net MMP *inhibitory* activity is detected (La Fleur *et al.*, 1996). Using *in situ* hybridisation, TIMP-1 mRNA is localised to macrophages and Schwann cells, only after macrophage infiltration into the nerve. To confirm invading macrophages regulate TIMP-1 production, macrophage conditioned medium added to nerves degenerating in culture (which normally produce excess MMP activity) resulted in a subsequent increase in the levels of TIMP expression (La Fleur *et al.*, 1996). Macrophages may facilitate regeneration by limiting proteolytic degradation to the myelin-axon unit, leaving the basal lamina and endoneurium relatively intact. It is possible that proteinases, produced by macrophages, remodel the basal lamina in other subtle ways rather than physically destroying components of the extracellular matrix. For example, 72kDa gelatinase produced by breast epithelial tissue can specifically cleave laminin 5 to a fragment which induces cell migration (Giannelli *et al.*, 1997). MMPs, may therefore, cause structural rearrangements in the components of the basement membrane which are necessary for Schwann cell migration or axonal re-growth after nerve injury but the role of these molecules in regeneration has not been put directly to the test.

Macrophages may influence the production of neurotrophins during Wallerian degeneration. In culture, the non-neuronal cells of degenerating nerve segments upregulate the mRNA expression of nerve growth factor (NGF) (Heumann *et al.*, 1987). This is a biphasic response, the early phase being macrophage independent, the later response after 48 hours requiring the presence of macrophages (Heumann *et al.*, 1990). Heumann and colleagues demonstrated that the sustained rise in NGF mRNA is triggered *in vitro* by the macrophage derived cytokine IL-1β. IL-1β enhances the transcription and translation of NGF by endoneurial fibroblasts, however, Schwann cells are not sensitive to IL-1β (Matsuoka *et al.*, 1991). Although NGF may play a role in the regeneration of peripheral adrenergic axons (Bjerre & Rosengreen, 1974), its participation in regeneration of other types of axons is less clear. Sensory nerves regenerate normally in rats treated with antiserum to NGF, even when given in large doses that prevented collateral sprouting of the same axon (Diamond *et al.*, 1987). The potential role of IL-1 in regeneration was directly evaluated by assaying the regrowth of fibres along guide tubes filled with IL-1 receptor antagonist (Guenard *et al.*, 1991). When confronted with these high concentrations of the antagonist a modest effect was observed, with 10% fewer fibres regenerating than through the control tubes. IL-1β could influence the regerative events through the upregulation of other neurotrophins during Wallerian degeneration such as leukemia

Key points

- Wallerian degeneration failed to occur rapidly in the C57BL/Wld mouse and axons did not regenerate, despite vigorous sprouting at lesion site and apparently normal cell body response to injury.
- Onset or induction of degeneration lead to rapid regeneration.
- Multiple inhibitory signals thought to be associated with intact peripheral myelin and/or Schwann cells.
- Despite elevated levels of MMPs during Wallerian degeneration, basal lamina is not degraded because invading macrophages induce increase in production of the endogenous tissue inhibitor of metalloproteinases-1 (TIMP-1).
- Cytokines (eg IL-1β, IL-6) and secretory molecules of macrophages may promote regeneration directly or by influence on removal of debris or on Schwann cells.

inhibitory factor (LIF) (Carlson *et al.*, 1996). Although the elevation of these two neurotrophins is related to the invasion of macrophages other growth factors produced after axotomy, such as brain-derived neurotrophic factor (BDNF), show distinct temporo-spatial expression patterns (Meyer *et al.*, 1992).

In addition to IL-1 there are, of course, other cytokines and secretory molecules of macrophages that may promote regeneration directly or by their influence on the removal of the debris or their influence on Schwann cells. Studies on transgenic mice expressing human IL-6 and/or its receptor implicate this cytokine in regeneration by an unknown mechanism, although it unlikely that this is simply a macrophage mediated effect (Hirato *et al.*, 1986). It has been suggested that apolipoprotein-E might play a role in regeneration but in apolipoprotein-E deficient mice regeneration was normal (Popko *et al.*, 1993). Complement is involved in the removal of debris from the distal nerve and it has been found that regeneration in complement depleted rats was less efficient than normal (Dailey *et al.*, 1998). In the nerves of complement depleted animals macrophage recruitment was reduced and those that were recruited they did not appear to be activated as normal. It is thus unclear whether it is the persisting debris or the absence of a macrophage secretory product that impaired regeneration.

Overview and implications for CNS regeneration

The Schwann cell is a remarkable cell and it clearly the key to peripheral nerve regeneration with the macrophage acting as support player. Wallerian degeneration in the PNS is a necessary pre-requisite for successful nerve regeneration. Following nerve transection axonal breakdown occurs via programmed cell death-like mechanism along the whole of the distal nerve segment and is independent of macrophage recruitment. Schwann cells and macrophages participate in the subsequent degradation and removal of axonal and myelin debris but in this process macrophages play a secondary role to Schwann cells, for in their absence Schwann cells alone can degrade a substantial proportion of myelin. Macrophage derived factors have the potential to stimulate

Schwann cell proliferation following axotomy and although the exact mitogen's have yet to be identified it is clear that this step is intiated in the absence of macrophages. Cytokines produced by macrophages can influence the synthesis of growth factors, proteinases and enzymatic inhibitors by Schwann cells, endoneurial fibroblasts and/or other macrophages. The extent to which neurotrophins, other than NGF and leukaemia inhibitory factor, are under cytokine control remains to be defined. MMPs and their endogenous inhibitors, the TIMPs, are regulated by macrophage derived cytokines. It is important proteolytic activity is tightly regulated during Wallerian degeneration to prevent excess degradation of the extracellular matrix which could impede regeneration. The precise weighting of the macrophage contribution to PNS regeneration remains to be established but it is clear that Schwann cells can dedifferentiate to a state that supports regeneration without the contribution of macrophages. The Schwann cell is absolutely required for succesful PNS regeneration.

It is against this background that we can ask whether manipulating the macrophage response in the CNS is likely to facilitate regeneration. Monocytes are only slowly recruited to a degenerating CNS fibre tract and the resident macrophages, the microglia, are slow to participate in clearance of the debris. Despite the picture that has emerged from the studies described above, which indicate that PNS regeneration is orchestrated and dependent on the Schwann cell, it has recently been reported by Schwartz and colleagues that the transplantation of monocytes, into a damaged CNS fibre tract will dramatically promote functionally effective CNS regeneration (see Lazarov-Spiegler *et al.*, 1998 for references). The studies report that it is the prior activation of the monocytes by exposure to a piece of peripheral nerve that is the key component. Factors released by cells of the peripheral nerve segment are able to activate the monocytes so that they overcome the capacity of the CNS microenvironment to suppress a typical acute inflammatory response. Given the modest contribution of the macrophages in PNS regeneration it will be important to understand the cellular and molecular events underlying the paradigm described by Schwartz and colleagues in the CNS.

Key points

- Schwann cells the key to peripheral nerve regeneration; macrophages act as support player.
- Wallerian degeneration in PNS is necessary pre-requisite for successful nerve regeneration.
- Axonal breakdown occurs in nerve segment distal to injury via apoptotic-like mechanism independently of macrophage recruitment.
- Schwann cells (essential) and macrophages (secondary role) participate in subsequent degradation and removal of axonal and myelin debris.
- Schwann cell proliferation stimulated by unidentified factors from macrophages.
- Cytokines produced by macrophages can influence synthesis of growth factors, proteinases and enzymatic inhibitors by Schwann cells, endoneurial fibroblasts and/or other macrophages.

Acknowledgement

We thank the BBSRC (UK) and British Biotech Pharmaceuticals Ltd for support for PMH.

References

Agrawal H C Agrawal D and Strauss A W (1990) Cleavage of the PO glycoprotein of the rat peripheral nerve myelin: tentative identification of cleavage site and evidence for the percusor-product relationship. Neurochemistry Research 15, 993–1001

Avellino A D Hart D Dailey A T Mackinnon M Ellegala D and Kliot M (1995) Differential macrophage response in the peripheral and central nervous system during Wallerian degeneration of axons. Experimental Neurology 136, 183–198

Banik N L DeVries G H Neuberger T Russell T Chakrabarti A K and Hogan E L (1991) Calcium-activated neutral proteinase (CANP:calpain) activity in Schwann cells: immunofluorescence localisation and compartmentation of μ and mCANP. Journal of Neuroscience Research 26, 346–354

Bedi K Winter J Berry M and Cohen J (1991) Adult rat dorsal root ganglion neurons extend neurites on predegenerated but not on normal peripheral nerves *in vitro*. European Journal of Neuroscience 4, 193–200

Beuche W and Friede R L (1984) The role of non-resident cells in Wallerian degeneration. Journal of Neurocytology 13, 767–796

Beuche W and Friede R L (1986) Myelin phagocytosis in Wallerian degeneration of peripheral nerves depends on silica-sensitive, Bg/Bg- negative and Fc-positive monocytes. Brain Research 378, 97–106

Bignami A Cella G and Chi N H (1982) Plasminogen activators in rat neural tissues during development and in Wallerian degeneration. Acta Neuropathologica (Berlin) 58, 224–228

Birkedal-Hansen H Moore W G I Bodden M K Windsor L J Birkedal-Hansen B DeCarlo A and Engler J A (1993) MMPs: A review. Critical Reviews in Oral Biology and Medicine 42, 197–250

Bjerre B and Rosengreen E (1974) Effects of nerve growth factor and its antiserum on axonal regeneration of short adrenergic neurons in the male mouse. Cell and Tissue Research 150, 299–322

Braichwal R R Bigbee J W and DeVries G H (1988) Macrophage-mediated myelin-related mitogenic factor for cultured Schwann cells. Proceedings of the National Academy of Sciences USA 85, 1701–1705

Brockes J P Lemke G E and Balzer D R (1980) Purification and preliminary characterisation of a glial growth factor from the bovine pituitary. Journal of Biological Chemistry 255, 8374–8377

Brown M C Lunn E R and Perry V H (1991) Poor growth of mammalian motor and sensory axons into intact proximal nerve stumps. European Journal of Neuroscience 3, 1366–1369

Brown M C Perry V H Hunt S P and Lapper S (1994) Further studies on motor and sensory nerve regeneration in mice with delayed Wallerian degeneration. European Journal of Neuroscience 6, 420–428

Brown H C Castano A Fearn S Townsend M Edeards G Streuli C and Perry V H (1997) Adhesion molecules involved in macrophage responses to Wallerian degeneration in the murine peripheral nervous system. European Journal of Neuroscience 9, 2057–2063

Brück W and Friede R L (1990) L-Fucosidase treatment blocks myelin phagocytosis by macrophages *in vitro*. Journal of Neuroimmunology 27, 217–227

Brück W and Freide R L (1991) The role of complement in myelin phagocytosis during PNS Wallerian degeneration. Journal of Neurological Sciences 103, 182–187

Brück W Huitinga I and Dijkstra C D (1996) Liposome-mediated monocyte depletion during Wallerian degeneration defines the role of hematogenous phagocytes in myelin removal. Journal of Neuroscience Research 46, 477–484

Buckmaster E A Perry V H and Brown M C (1995) The rate of Wallerian degeneration in cultured neurons from wild type and C57BL/Wlds mice depends on time in culture and may be extended in the presence of elevated K+ levels. European Journal of Neuroscience 7, 1596–1602

Carlson C D Bai Y Jonakait G M and Hart R P (1996) Interleukin-1β increases inhibitory leukemia inhibitory factor mRNA levels through transient stimulation of transcription rate. Glia 18, 141–151

Cammer W Brosnan C F Bloom B R and Norton W T (1981) Degradation of the P0, P1, and Pr proteins in peripheral nervous system myelin by plasmin: Implications regarding the role of macrophages in demyelinating diseases. Journal of Neurochemistry 36, 1506–1514

Carroll S L Miller M L Frohnert P W Kim S S and Corbet J A (1997) Expression of neuregulins and their putative receptors, ErbB2 and ErbB3, is induced during Wallerian degeneration. Journal of Neuroscience 15, 1642–1659

Chandler S Coates R Gearing A Lury J Wells G and Bone E (1995) MMPs degrade myelin basic protein. Neuroscience Letters 201, 223–226

Coleman M P Conforti L Buckmaster E A Tarlton A Ewing R M Brown M C Lyon M F and Perry V H (1998) An 86–kb tandem triplication in the slow Wallerian degeneration (Wlds) mouse. Proceedings of the National Academy of Sciences USA 95, 9985–9990

Dailey A T Avellino A M Bentham L Silver J and Kliot M (1998) Complement depletion reduces macrophage infiltration and activation during Wallerian degeneration and regeneration. Journal of Neuroscience 18, 6713–6722

Daniloff J K Levi G Grumet M Rieger F and Edelman G M (1986) Altered expression of neuronal cell adhesion molecule induced by nerve injury and repair. Journal of Cell Biology 103, 929–945

David S Braun P E Jackson D L Kottis V and McKerracher L (1995) Laminin overrides the inhibitory effects of the PNS and CNS myelin- derived inhibitors of neurite growth. Journal of Neuroscience Research 42, 594–602

Diamond J Coughlin M Macintyre L Holmes M and Visheau B (1987) Evidence that endogenous beta nerve growth factor is responsible for collateral sprouting, but not the regeneration, of nociceptive axons in the adult rat. Proceedings of the National Academy of Sciences USA 84, 6596–6600

Dong Z Brennan A Liu N Yarden Y Lefkowitz G Mirsky R and Jessen K R (1995) Neu differentiation factor is a neuron-glia signal and regulates survival, proliferation and maturation of rat Schwann cell precursors. Neuron 15, 585–596

Fernandez-Valle C Bunge R P and Bunge M B (1995) Schwann cells degrade myelin and proliferate in the absence of macrophages: evidence from in vitro studies of Wallerian degeneration. Journal of Neurocytology 24, 667–679

George E B Glass J D and Griffin J W (1995) Axotomy-induced axonal degeneration is mediated by calcium influx through ion specific channels. Journal of Neuroscience 15, 6445–6452

George R and Griffin J W (1994) The proximo-distal spread of axonal degeneration in the dorsal columns of the rat. Journal of Neurocytology 23, 657–667

Giannelli G Falk-Marzillier J Schiraldi O Stetler-Stevenson W and Quaranta V (1997) Induction of cell migration by matrix metalloproteinase-2 cleavage of laminin-5. Science 277, 225–228

Guenard V Dinarello C A Weston P J and Aebischer P (1991) Peripheral nerve regeneration is impeded by interleukin-1 receptor antagonist released from a polymeric guidance chamber. Journal of Neuroscience Research 29, 396–400

Hallpike J F and Adams C W M (1969) Proteolysis and myelin breakdown: a review of recent histochemical and biochemical studies. Histochemical Journal 1, 559–578

Hann P G Beuche W Neumann U and Friede R L (1988) The rate of Wallerian degeneration in the absence of immunoglobulins. A study in chick and mouse peripheral nerves. Brain Research 451, 126–132

Heumann R L Korsching S Bandtlow C and Thoenen H (1987) Changes of nerve-growth factor synthesis in non-neural cells in response to sciatic nerve transection. Journal of Cell Biology 104, 1623–1631

Heumann R L Hengerer B Lindholm D Brown M C and Perry V H (1990) Mechanisms leading to increases in nerve growth factor synthesis after peripheral nerve lesion. In: Advances in Neural Regeneration Research, Edited by F J Seil, pp. 125–145, Wiley-Liss, New York

Hirota H Kiyama H Kishimoto T and Taga T (1996) Accelerated nerve regeneration in mice by upregulated expression of interleukin (IL) 6 and IL-6 receptor after trauma. Journal of Experimental Medicine 183, 3227–2634

Howard L Lu X Mitchell S Griffith S and Glynn P (1996) Molecular cloning of MADM: a catalytically active mammalian disintegrin-metalloprotease expresses in various cell types: Biochemical Journal 317, 45–50

Hughes P M Ward G A Tsao J W Miller K M and Brown M C (1995) A role for MMPs in peripheral myelin breakdown. Journal of Physiology 487, 68P

Hughes P M (1997) The role of MMPs in inflammatory demyelination of the peripheral nervous system. DPhil. Thesis, University of Oxford, Oxford U K

Hughes P M Wells G M A Clements J M Gearing A J H Redford E J Davies M Smith K J Hughes R A C Brown M C and Miller K M (1998) Matrix metalloproteinase expression during experimental autoimmune neuritis. Brain 121, 481–494

Jander S Pohl C and Stoll G (1996) Differential expression of interleukin- 10 mRNA in Wallerian degeneration and immune-mediated inflammation of the rat peripheral nervous system. Journal of Neuroscience Research 43, 254–259

Kidd G J and Heath J W (1991) Myelin sheath survives following axonal degeneration in doubly myelinated nerve fibers. Journal of Neuroscience 11, 4003–4014

Kleitman N and Bunge R P (1995) The Schwann cell: Morphology and development. In The Axon: Structure, Function and Pathophysiology edited by S G Waxman J D Kocsis, and P K Stys, pp. 97–115. Oxford University Press, Oxford

La Fleur M Underwood J Rappolee D A and Werb Z (1996) Basement membrane and repair of injury to peripheral nerve: Defining a potential role for macrophages, MMPs, and tissue inhibitor of metalloproteinases-1. Journal of Experimental Medicine 184, 2311–2326

Langley J N and Anderson H K (1904) The union of different kinds of nerve fibres. Journal of Physiology 31, 365–391

Lazarov-Spiegler O Rapalion O Agranov G and Schwartz M (1998) Restricted inflammatory reaction in the CNS: a key impediment to axonal regeneration? Molecular Medicine Today 4, 337–342

Lunn E R Perry V H Brown M C Rosen H and Gordon S (1989) Absence of Wallerian degeneration does not hinder regeneration in the peripheral nerve. European Journal of Neuroscience 1, 27–33

Martini R and Schachner M (1988) Immunoelectron microscopic localization of neural cell adhesion molecules (L1 and NCAM and myelin-associated glycoprotein) in regenerating adult mouse sciatic nerve. Journal of Cell Biology 106, 1735–1746

Matsuoka I Meyer M and Thoenen H (1991) Cell-type-specific regulation of nerve growth factor (NGF) synthesis in non-neuronal cells: comparison of Schwann cells with other cell types. Journal of Neuroscience 11, 3165–3177

Meier R Spreyer P Ortmann R Harel A and Monard D (1989) Induction of glial-derived nexin after lesion of a peripheral nerve. Nature 342, 548–550

Meyer M Matsuoka I Wetmore C Olson L and Thoenen H (1992) Enhanced synthesis of brain-derived neurotropic factors in the lesioned peripheral nerve: Different mechanisms are responsible for the regulation of BDNF and NGF. Journal of Cell Biology 119, 45–54

Owen C A Campbell M A Sannes P L Boukedes S S and Campbell E J (1995) Cell surface-bound elastase and cathepsin G on human neutrophils: A novel, non-oxidative mechanism by which neutrophils focus and preserve catalytic activity of serine proteinases. Journal of Cell Biology 131, 775–789

Perry V H Brown M C and Gordon S (1987) The macrophage response to central and peripheral nerve injury. Journal of Experimental Medicine 165, 1218–1223

Perry V H Tsao J W Fearn S and Brown M C (1995) Radiation-induced reductions in macrophage recruitment have only slight effects on myelin degradation in sectioned peripheral nerves of mice. European Journal of Neuroscience 7, 271–280

Popko B Goodrum J F Bouldin T W Zhang S H and Maeda N (1993) Nerve regeneration occurs in the absence of apolipoprotein E in mice. Journal of Neurochemistry 60, 1155–1158

Pu S F Zhuang H X and Ishhii D N (1995) Differential spatio-temporal expression of the insulin-like growth factor genes in regenerating sciatic nerve. Molecular Brain Research 34, 18–28

Reichardt L F and Tomaselli K J (1991) Extracellular matrix molecules and their receptors: functions in neural development. Annual Review of Neuroscience 14, 531–570

Reichert F Saada A & Rotshenker S (1994) Peripheral nerve injury induces Schwann cells to express two macrophage phenotypes: phagocytosis and the galactose-specific lectin MAC-2. Journal of Neuroscience 14, 3231–3245

Ridley A J Davis J B Stroobant P and Land H (1989) Transforming growth factors-beta1 and beta 2 are mitogens for rat Schwann cells. Journal of Cell Biology 109, 3419–3424

Saada A Reichert F and Rotshenker S (1996) Granulocyte macrophage colony stimulating factor produced in lesioned peripheral nerves induces the upregulation of cell surface expression of MAC-2 by macrophages and Schwann cells. Journal of Cell Biology 133, 159–167

Sagot Y Dubois-Dauphon M Tan S A de Bilbao F Aebischer P Martinou J-C and Kato A C (1995) Bcl-2 overexpression prevents motoneuron cell body loss but not axonal degeneration in a mouse model of neurodegenerative disease. Journal of Neuroscience 15, 7727–7733

Schnell L Schneider R Kolbeck R Barde Y A and Schwab M E (1994) Neurotrophin-3 enhances sprouting of corticospinal tract during development and after spinal cord injury. Nature 367, 170–173

Smirnova I V Ma J Y Citron B A Ratzlaff K T Gregory E J Akaaboune M and Festoff B W (1996) Neural thrombin and protease nexin 1 kinetics after murine peripheral nerve injury. Journal of Neurochemistry 67, 2188–2199

Son Y J and Thompson W J (1995) Schwann cell processes guide regeneration of peripheral axons. Neuron 14, 125–132

Stoll G Griffin J W Li C Y and Trapp B D (1989) Wallerian degeneration in the peripheral nervous system: participation of both Schwann cells and macrophages in myelin degradation. Journal of Neurocytology 18, 671–683

Tapia M Inestrosa N C and Alvarez J (1995) Early axonal regeneration: repression by Schwann cells and a protease? Experimental Neurology 131, 124–132

Thomas P K (1964) The deposition of collagen in relation to Schwann cell basement membrane during peripheral nerve regeneration. Journal of Cell Biology 23, 375–382

Trapp B D Hauer P and Lemke G (1988) Axonal regulation of myelin protein mRNA levels in actively myelinating Schwann cells. Journal of Neuroscience 8, 3515–3521

White F V Toews A D Goodrum J F Novicki G L and Bouldin T W (1989) Lipid metabolism during the early stages of Wallerian degeneration in the rat sciatic nerve. Journal of Neurochemistry 52, 1085–1092

Wu W D Toma J G Chan H Smith R and Miller F D (1994) Disruption of fast axonal transport in vivo leads to alterations in Schwann cell gene expression. Development 163, 423–439

Wyllie A H (1980) Glucocorticoid-induced thymocyte apoptosis is associated with endogenous endonuclease activation. Nature 284, 555–556

14

The response of the somatosensory system to peripheral nerve injury

Peter D Kitchener and Peter Wilson

Introduction

In the nervous system, the response to injury involves not only the neurons that are lesioned, but also the response of the entire system to which they belong. Lesions in one part of a neural system might produce degenerative changes, or compensatory plasticity, in neurons at other locations in the system; such changes may be beneficial or detrimental to the systems overall function. Full characterisation the response to injury requires information on many different levels of analysis; ultimately, the relationship between the type, location and extent of lesions, the cellular processes that underlie plastic or degenerative changes in damaged or disconnected neurons, and the extent and adaptive value of these changes in the connections between neurons need to be understood. While far from complete, considerable progress in characterising the responses of nervous systems to injury has been obtained from studying sensory pathways. The relatively tractable interconnections of the somatosensory system have, for many decades, provided a means of studying plastic reorganisation of neurons in response to injury. More recently, studies of somatosensory lesions have explored the relationship between the resultant functional changes in somatosensory representations and their perceptual consequences.

As with all sensory neurons, the function of a sensory neuron in the somatosensory system can be defined by the type and intensity of stimulus, and the region of the peripheral receptor array, that activates the neuron. These characteristics are known, respectively, as the neuron's modality and its receptive field. Because neighbouring somatosensory neurons in the central nervous system (CNS) tend to have receptive fields on adjacent (or overlapping) regions of the body, there exists a topographic mapping of body to the population of responsive neurons. Topographic relations between inputs appear to be preserved throughout the somatosensory projection. However, these somatosensory representations are not static but can undergo dynamic changes in response to alterations in sensory input. Although the importance of topographic maps *per se* in the elaboration of somaesthesia is largely unknown, there is increasing evidence that changes in representational maps are directly related to altered somaesthesia.

Changes in representational maps are, by definition, due to neurons changing their receptive fields — in other words, neurons become responsive to inputs from regions of the periphery to which they were not previously responsive. This implies underlying changes in neuronal connectivity, which are referred to as neuronal plasticity. Such changes have been investigated in the context of the response to injury (particularly peripheral nerve injury), and more recently, as a consequence of changes in physiological activation of particular somatosensory pathways. There is evidence, not always incontrovertible, for plasticity at many levels of somatosensory processing including the spinal dorsal horn, the dorsal column nuclei (DCN), the ventroposterior (VP) thalamus and somatosensory areas of the cerebral cortex.

This review is concerned with the use of representational maps to detect the changes in somatosensory organisation in adult mammals that occur in response to physiological alterations in somatosensory stimulation and as a response to nervous system injury. In focusing on these key issues and approaches to studying somatosensory plasticity in adult mammals, we have neglected consideration of phenomena and mechanisms that might pertain only to the developing somatosensory system. It is, of course, possible that adult and developmental plasticity share common mechanisms. Also, the plasticity of high threshold cutaneous modalities will only be mentioned in as much as they impact on hypothesised mechanisms of low threshold cutaneous reorganisation. For a comprehensive review of somatosensory plasticity in both adult and developing mammals, see Snow & Wilson (1991).

There is ample evidence that changes in somatosensory input, due to lesions or altered afferent activation, can lead to changes in the receptive fields of somatosensory neurons (see Wall, 1988; Snow & Wilson, 1991; O'Leary *et al.*, 1994 for reviews). However, amongst the numerous reports of plasticity there are almost equally numerous differences and even inconsistencies in the observed plasticity. Changes in cutaneous representations might occur rapidly (within minutes or hours) after lesions, or may evolve within weeks or months; the new representations that they contribute to may be complete and contiguous maps, or the new maps may be fragmented or incomplete. In some cases, the extent of reorganisation is vast, in other cases re-organisation is small or negligible. In addition to characterising the extent and variability of plastic change in the somatosensory system in response to various manipulations, studies of representational plasticity have raised, and in some cases pursued, the underlying questions of 1) where in the somatosensory pathway (i.e. at which synapses) the changes in the observed receptive field properties occur, and 2) what are the mechanisms by which neurons change their receptive fields.

Just as the method of detecting changes in representations relies on knowledge of the normal representation pattern, so too are the answers to the underlying questions of location and mechanism informed by (and raise question about) the normal structure of the somatosensory system. The following section considers how the organisation of cutaneous representations can be viewed.

Low-threshold cutaneous representations

Topographic representations of low threshold cutaneous inputs from the periphery have been found at the spinal, brainstem, thalamic, and cortical levels of the somatosensory system (see Box 1). The majority of studies on the somatosensory representations have described the organisation and reorganisation of low threshold cutaneous inputs from the body (especially the limbs and digits). The cutaneous representations of the whiskers of the mystacial pad of rodents has also been used to examine issues of organisation and reorganisation of cutaneous representations due to the regularity of the representation and correspondence of the peripheral whisker array with cytological parcellation of neuronal groupings in the somatosensory pathway (see Box 1).

Box 1. Central Pathways of Low-threshold Cutaneous Somatosensory Information

Cutaneous primary sensory afferents have their cell bodies in the dorsal root ganglia (DRG) and make central projections to second order neurons in two locations: there is a spinal projection to neurons in laminae III and IV of the dorsal horn, and an ascending projection which travels in the dorsal columns terminating in the dorsal column nuclei (DCN) of the medulla. Studies of the organisation and plasticity of cutaneous sensory afferents of the trigeminal system have been facilitated by the orderly arrangement of the mystacial vibrissae and the discrete organisation of vibrissa-related neural centres in the somatosensory system. Vibrissal and vibrissa-related afferents project to four regions of the medulla: the interpolaris, caudalis and oralis divisions of the spinal trigeminal complex, and to the principle trigeminal nucleus (PrV). Cutaneous somatosensory information projects from the trigeminal nuclear complex and the DCN across the midline (in the medial lemniscus) to the ventral posterior thalamus where the face is represented in VPm and the body in VPl. Some cutaneous information reaches VPm via projections from spinal trigeminal nuclei and cutaneous information to VPl arrives via the spinothalamic and spinocervical tracts. VPm and VPl project to ipsilateral somatosensory cortex.

The usual method of obtaining a representational map is to make microelectrode recordings from a large number of individual neurons, or clusters of neurons, in the somatosensory regions of the CNS. In this way the distribution (and by inference the proportion) of the neuronal population that represent a particular region of the periphery can be determined. In principle, the number and density of recording sites determines the resolution of the resulting representational map. Studies of gross changes in representational maps are complemented by those in which individual neurons and their receptive fields are studied before and after experimental manipulations of somatosensory input. A variation that combines these approaches is to use electrode arrays where a number of electrodes (typically between 10 and 20) are positioned in the relevant region and the activity of individual neurons is identified by passing digitised recordings through analysis software which is designed to identify the characteristic wave forms contributed by single neurons. The electrode array can also have the advantage of being implanted permanently, thus allowing chronic recording from awake animals.

Rather than estimate representations by sampling, or examining individual receptive fields, there are techniques that allow the activity of the entire population of neurons that contribute to representation in regions of the CNS to be visualised. Godde *et al.* (1995) have examined the activation of somatosensory cortex by looking at changes in reflectance associated with increased metabolic activity. Because it is known from numerous microelectrode studies that receptive fields of somatosensory neurons overlap considerably with the receptive fields of other (i.e. neighbouring) neurons, is not

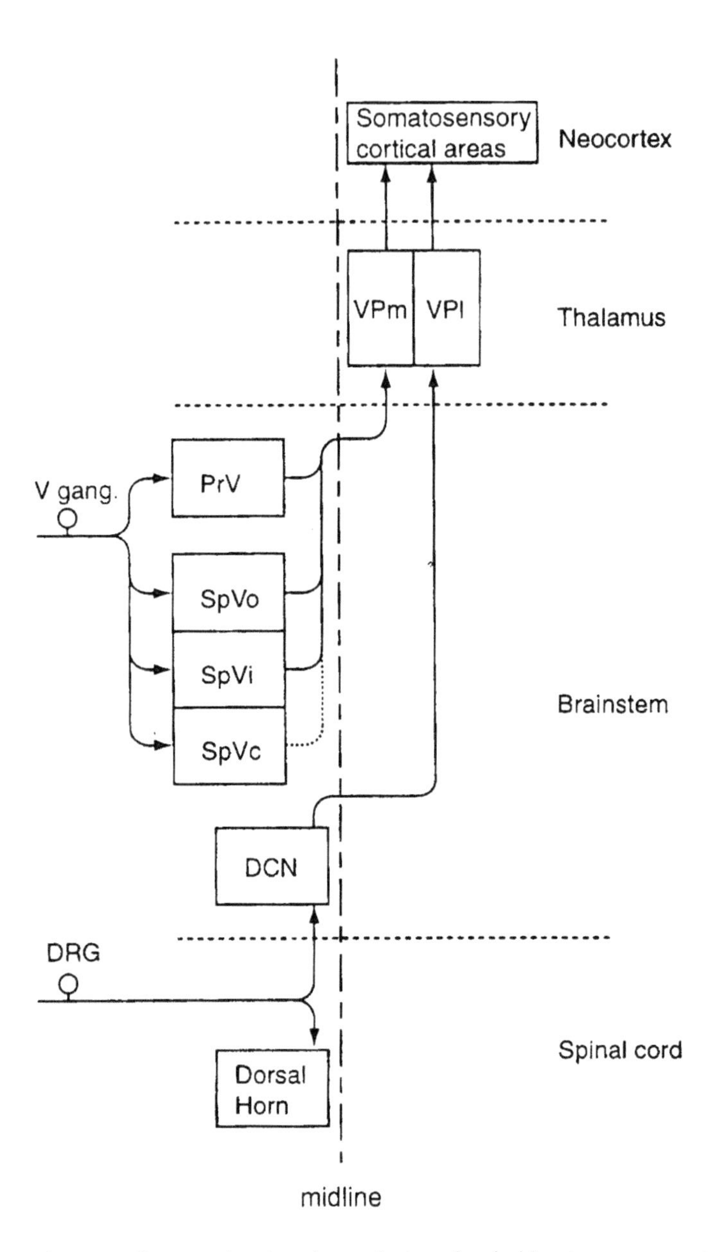

Figure 1 Schematic diagram showing the main low-threshold cutaneous somatosensory pathways. Abbreviations: DCN, dorsal column nuclei; DRG, dorsal root ganglion; PrV, principal trigeminal nucleus; SpVc, spinal trigeminal nucleus caudalis; SpVi, spinal trigeminal nucleus interpolaris; SpVo, spinal trigeminal nucleus oralis; VPm and VPl, medial and lateral ventroposterior nucleus of the thalamus; V gang, trigeminal ganglion.

surprising that there is considerable overlap of the populations of neurons activated by different regions of the body. Like optical recording, the method of examining the pattern of 2-deoxyglucose uptake is also a population-based assessment of somatosensory representations (see Welker *et al.*, 1992 for example). Given the uncertainties about the resolution of these techniques, there is good agreement between population based estimates (optical functional imaging and 2-deoxyglucose) and sample based (neuronal recording) estimates of the extent of cutaneous representations in somatosensory cortex.

The representational maps of low threshold cutaneous input are most easily appreciated when mapped to sheet or layered neural structures such as the laminae of the spinal dorsal horn and the somatosensory cortex. Representational maps are also present in the dorsal column nuclei and in the VP thalamus where the neurons are organised into volumes rather than sheets or laminae. Not surprisingly perhaps, less is known about the orientation and grouping of neurons that represent cutaneous input in the DCN and VP. In the case of the vibrissal representation of rodents and marsupials, there are discrete cytologically identifiable sub-volumes for each whisker of the mystacial pad. Similarly, there are regions of the rat and mouse somatosensory cortex that correspond to representations of the hindlimb, forelimb and trunk, which may be identified by cytochrome oxidase staining, but more investigative work will have to be undertaken to determine the degree of association between cytological parcellation and the cutaneous regions they represent in subcortical regions and how this might vary between species.

That differences might exist in the detailed somatotopic organisation in different species is illustrated by considering the VP thalamus: in macaque monkeys the face is represented (in VPm) in a collection of rostro-caudally oriented rods; neurons within a single rod all tend to have receptive fields on the same region of the body (Rausell & Jones, 1991). However, not all rods run the entire length of VP, and rods may not be separate throughout their entire course. This is especially true in VPl (Rausell *et al.*, 1992). The "isorepresentational rods" of the macaque may be a secondary level of organisation overlaid on an overall somatotopic representation, and the grouping into such strict rod subregions may not occur in other animals. Mapping the squirrel monkey VP thalamus also reveals regions of isorepresentation but also abrupt discontinuities in receptive field location when making electrode penetrations (Kaas *et al.*, 1984). As we have recently described in the marmoset monkey VP (Kitchener *et al.*, 1997; Wilson *et al.*, unpublished observations), the somatotopic representation in the squirrel monkey VP is best described as a single continuous representation (Kaas *et al.*, 1984); the abrupt discontinuities in receptive field location appear to be the result of folds in the body representation in VPl. This folding of an essentially 2-dimensional sheet into the 3-dimensional space of VP may also best describe the limited data from the rat. Angel & Clarke (1975) described the representation of the forepaw in rat VPl as being folded in the form of a closed fist. The paucity of knowledge of VP (and DCN) representations, and the extent to which organisational details that are known can be generalised across species, have tended to confound interpretation of plasticity studies of these regions.

The uncertainty regarding the orientation details of representational maps is an important consideration in detecting changes in the size and proportions of representational maps. Additional considerations, regarding the components of the receptive fields that make up maps, are especially important when addressing the questions of location and mechanisms of plastic changes. It has to be appreciated that somatosensory representations are more complicated than a simple mapping of excitatory inputs to arrays of neurons. The definition of the receptive field, as the region of the periphery that activates the cell, suggests that activation means to make the neuron fire action potentials (or to increase the basal rate of action potentials), but neurons may also receive sub-threshold excitatory inputs from peripheral regions adjacent to or outside the receptive field. In addition, a neuron may receive inhibitory inputs from within or beyond its receptive field; possibly from afferents of different modality. The variation in the type and strength of synaptic inputs also raises the issue that anaesthetics are known to alter receptive field properties (how they effect the expression of different types of inputs is largely unknown). A final caveat is that inputs that effect the receptive field of neurons could include the possibility that plastic changes at one level of the pathway might, through descending projections, or even commissural connections influence the neurons in other regions of the somatosensory system.

Key Points

- The receptive fields of cutaneous sensory inputs are organised into continuous topographic maps of the body surface: these maps are relatively easy to demonstrate experimentally in the sheet-like structures of the somatosensory cortex and in the dorsal laminae of the spinal cord. Topographic maps also exist in somatosensory nuclei of the brainstem and thalamus but are more difficult to map due to the compact nature of these structures.
- The representation maps of cutaneous input are not static organisations but can exhibit plastic changes in response to damage of peripheral or central neurons, and to functional changes in sensory input.
- Changes in somatosensory input (such as caused by peripheral nerve or CNS injury) may have repercussions elsewhere in the somatosensory system: injury to part of the pathway can result in plastic reorganisation of connection between neurons throughout the system.

The evidence for plasticity of somatosensory representations

Spinal Dorsal Horn

The first synapse in the somatosensory projection from the body surface is made by the primary sensory afferents to neurons in the dorsal horn of the spinal cord. As well as projecting their main axon (in the dorsal columns) to the DCN of the medulla, low threshold cutaneous afferents also emit collaterals to the deeper laminae of the spinal dorsal horn. An advantage of examining the representations of cutaneous input in the post-synaptic neurons in the dorsal horn is that the neurons are arranged in sheet-like structures, corresponding to the anatomical laminae IV-V, and thus are more amenable to mapping than the post-synaptic neurons of the DCN, which are contained in a comparatively compact three-dimensional volume.

There are two fundamentally different approaches to the permanent removal of peripheral input to the central somatosensory centres. The central projections of primary afferents can be destroyed by cutting the dorsal roots (dorsal rhizotomy) with the resulting complete degeneration of all central projections of the affected dorsal root ganglion and loss of input from the dermatome which it innervated. Peripheral input can also be eliminated by lesioning peripheral nerves, which, for nerves innervating the limbs, include afferents from more than one dermatome. The important difference with peripheral nerve lesions is that while a sub-population of afferents are destroyed by peripheral axotomy, the majority of afferents survive and their central projections are maintained (in fact, important changes occur in the central projections of different classes of peripherally axotomized afferent, see Box 2). A variation on the rhizotomy approach is a technique that allows the destruction of all of the central projections from a peripheral nerve: micro-injection of neurotoxins or proteolytic enzymes into the peripheral nerves results in these substances being transported back to the cell bodies in the dorsal root ganglion or trigeminal ganglion where they kill the cell (Yamamoto *et al.*, 1983; LaMotte & Kapadia, 1987).

In common with many early claims of extensive plasticity in various regions of the somatosensory system, the results of studies in the dorsal horn of the spinal cord proved to be controversial. Using suppressed silver stains, Liu & Chambers (1958) observed a degeneration pattern of a spared dorsal root which suggested that afferent in the spared root had significantly extended their termination zone within the spinal grey matter. In these studies, several dorsal roots rostral and caudal to a single spared root were transected, after several months the spared root itself was sectioned and the resultant degeneration of primary afferent terminals in the dorsal horn was detected with the silver stain. The apparent extensive expansion of spared root afferents was not supported in a number of subsequent studies using silver staining of degenerating axons or transganglionic tracers, and such extensive reorganisation of connections of afferents in spared roots is also at odds with the results of many (but not all) functional studies that have examined low threshold cutaneous input in the dorsal horn (see Snow & Wilson 1991 for review). The reasons for the conflicting observations, especially the unsuitability of degeneration silver staining for determining the extent of primary afferent arborisations, have been discussed in a number of reviews (Micevych *et al*.. 1986; Brown, 1987; Snow & Wilson, 1991).

As with the initial reports of anatomical sprouting of spared root afferents, those claiming substantial *functional* reorganisation of primary afferent input to dorsal horn neurons following peripheral nerve lesions have also required modification in the light of subsequent studies. Devor & Wall (1978, 1981a,b) reported that following denervation of the lower hindlimb in the cat and rat, neurons in the medial dorsal horn, which normally receive only lower hindlimb inputs, acquired (over the next 4 weeks) inputs from the upper hindlimb. A number of subsequent studies of failed to find any significant mediolateral shifts of somatotopy of low threshold cutaneous receptive fields following peripheral nerve injury. The issue of whether mediolateral reorganisation in the dorsal horn of the spinal cord occurs after peripheral nerve transection has been considered in previous reviews (see Snow & Wilson, 1991) and would not warrant reiteration here were it not for the fact that many current authors continue to pay more attention to the initial reports of such plasticity rather than to the bulk of evidence which suggests that such extensive reorganisation does not occur.

In contrast, there is considerable evidence that reorganisation, in the form of an expansion of primary afferent central projections, does occur in the rostro-caudal direction of the dorsal horn after peripheral nerve lesions. Transection of the sciatic nerve in rats denervates all but the saphenous nerve territory in the lower hindlimb. This produces a zone (about 2mm rostro-caudally) in the dorsal horn where neurons, which normally receive sciatic inputs, can no longer be activated by peripheral stimuli (Markus *et al.*, 1984). After three weeks however, the "deafferented" neurons in the sciatic region had acquired inputs from saphenous nerve afferents, which normally only activate dorsal horn neurons in a region 3 mm further rostral. In the adult cat, sciatic nerve transection was not followed by re-organisation of the inputs to the deprived region of the dorsal horn Wilson (1987), but when smaller deafferentations were made – by cutting the nerves innervating single digits, spinocervical tract neurons deprived of input by the lesion acquired (after several weeks) receptive fields on adjacent digits that were represented rostral and caudal of the denervated digit (Wilson & Snow, 1987).

The relatively slow evolution of the new receptive fields following peripheral nerve lesion in rats and cats seems to imply that the new inputs derive from growth of collaterals from intact adjacent (rostral and caudal) afferents. However, previous analysis of the relationship between primary afferent terminal fields and the somatotopic representation of low threshold cutaneous afferents suggested another explanation. Meyers & Snow (1984) had shown that single afferents labelled by intra-axonal injection of horseradish peroxidase (HRP) showed collaterals that projected (rostrally and caudally) beyond the regions of the dorsal horn that contained neurons that could be activated by stimulation of the afferent's receptive field. The afferent's collaterals that projected beyond the region where the afferent's receptive field is representational in the somatotopic map in the dorsal horn were termed "somatotopically inappropriate" projections. The intra-axonal horse-radish peroxidase labelled collaterals that projected to somatotopically inappropriate regions of the dorsal horn appeared to be simple in structure and possessed few boutons. Thus, a possible mechanism for the acquisition of saphenous inputs by neurons in the sciatic fields was that these simple somatotopically inappropriate collaterals, which already existed in a non-functional form, made functional synapses with neurons in the sciatic regions of the dorsal horn. Soon after the demonstration of long range collateral projections of primary afferents, a new method of intra-axonal labelling (utilising the small molecule Neurobiotin™ instead of the enzyme HRP) was employed and a very much expanded picture of the extent of collateral projection has emerged. While Neurobiotin™ labelling confirmed that afferents project a number of complex collaterals with numerous varicosities to a central zone of their projection, all afferents examined gave rise to projections that extended surprisingly long distances rostral and caudal to the central region of terminals (Wilson *et al.*, 1996, 1997). The superior labelling obtained with Neurobiotin™ also showed that many of the distant collaterals possessed varicosities characteristic of synaptic boutons (see Figure 2). Interestingly, such long-range projections of central collaterals of primary afferents are not evident in the great majority of transganglionic tracing studies (see Wilson & Kitchener, 1996 for review). Typically, transganglionic labelling produces an estimate of the terminal field of peripheral nerves that is approximately the same as the electrophysiologically derived representational maps

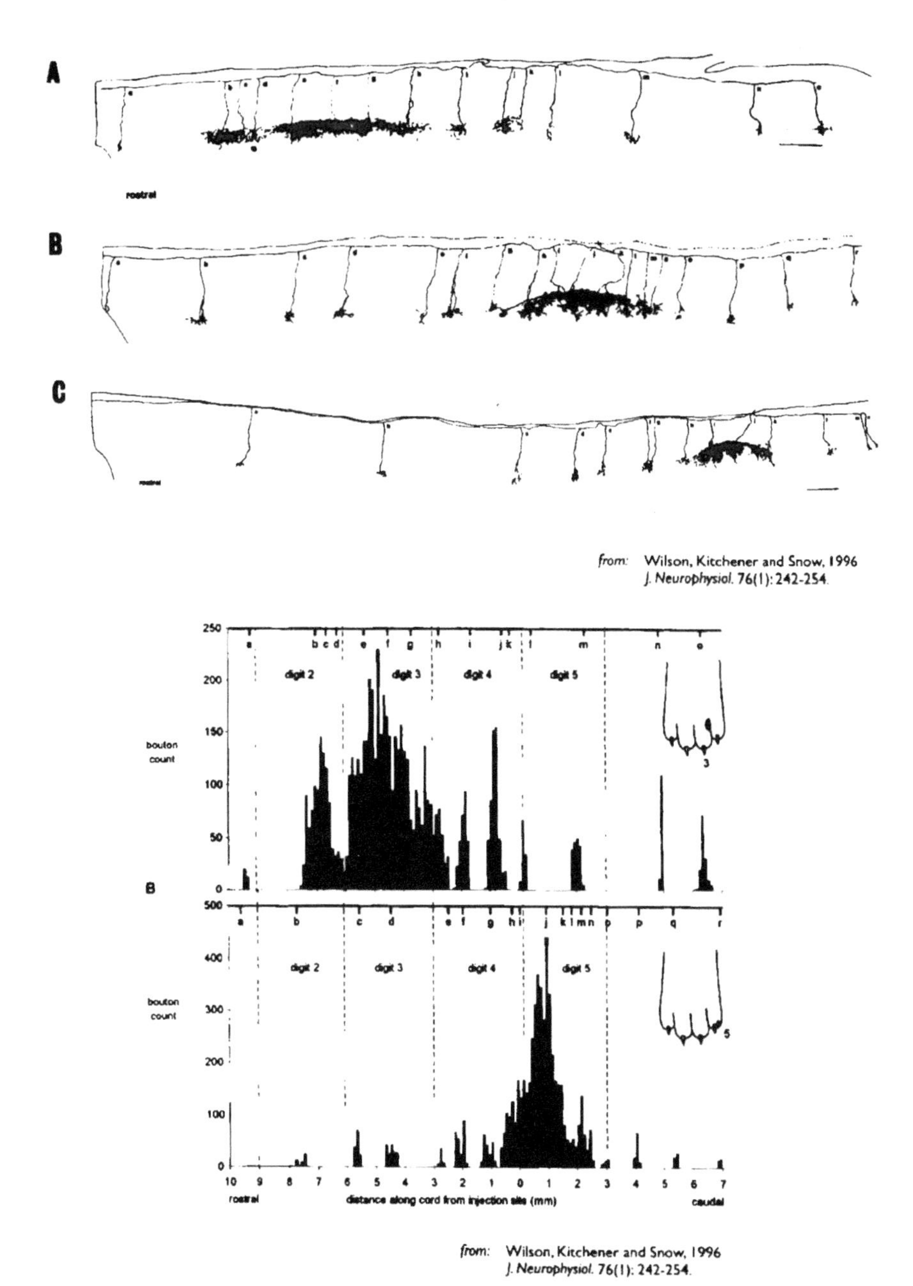

Figure 2 The upper figures (A,B,C) show parasagital views of single hair follicle afferents in the cat spinal cord intra-axonally labelled with Neurobiotin. Note that these afferents give rise to terminals on collaterals spread over a rostro-caudal range of more than 15 mm (the scale bars are 1 mm; scale bar in A applies to B). The relationship between these collaterals and the somatotopic map in the dorsal horn is shown by the 2 examples in the lower figure. The inset drawing of the paw shows the location of the afferents receptive field, note that while most of the afferents terminals are in the dorsal horn region representing the digit on which the receptive field is located, there are also numerous terminals in areas of the dorsal horn representing other digits (the letters and tick marks on the upper borders show the location of the descending collateral branches).

(see for example, Woolf & Fitzgerald, 1986). Perhaps long-range projections have different properties in regard to the axonal transport of tracers. Alternatively, the density of such projections might be very low compared to the dense central terminals of primary afferents.

The revelation, by Neurobiotin™ intra-axonal labelling, of the existence of projections well beyond the region the dorsal horn where the afferent activates dorsal horn neurons raises the question of what the function of numerous, long-range, sub-threshold inputs to the dorsal horn could be in the normal operation of the dorsal horn. A second question is what might the role of these projections be in the re-organisation of inputs following peripheral nerve injury.

A number of possible explanations of the role of collateral projections of afferents to regions of the dorsal horn where the afferent does not activate neurons (at least under the conditions of the mapping experiments) have been considered (see Wilson & Kitchener, 1996 for review). The action potential in the primary afferent may not propagate into the distant regions of the arborisation, perhaps due to intrinsic features of the axons or to synaptic mechanisms that can block propagation at certain locations along the axon. Alternatively, the synapses of the distant collaterals may be silent or weak (i.e. capable only of sub-threshold activation of the post-synaptic neuron), or they may be functional but have their influence suppressed by inhibitory mechanisms. The existence of long-range functionally ineffective (or at least sub-threshold) projections outside the region were they activate dorsal horn neurons could be said to fulfil the prediction of studies showing that primary afferents in a dorsal root could be antidromically activated by micro-stimulation in the dorsal horn many segments rostral and caudal to the recorded root (see Wall, 1995). However, the long-range collaterals (as currently visualised by Neurobiotin labelling) are not extensive enough to explain the activation of dorsal horn neurons by antidromic stimulation from far caudal segments (Wall, 1995). This apparent inconsistency may reflect under or over estimates of the true extent of primary afferent collaterals in intra-axonal labelling and micro-stimulation studies. Wall & Bennett's (1995) studies of activation of afferents from remote locations in the dorsal horn indicated that tonic modulation of impulse propagation by GABAergic mechanisms could limit the transmission of action potentials into the distant collateral branches of primary afferent fibres. This could explain why primary afferents appear only to activate dorsal horn neurons in the central region of their projection.

The existence of inhibitory mechanisms in the dorsal horn that are associated with primary afferent activity has been extensively characterised (see Burke & Rudomin, 1977 for review). More recently, direct evidence that tonic GABA-mediated mechanisms shape the spatial extent of primary afferent input has been provided by Biella & Sotgiu (1995) who recorded from dorsal horn neurons while removing their afferent input and also blocking GABA receptors. These experiments demonstrated that saphenous afferents, which do not normally contribute to the activation of neurons in the sciatic nerve terminal field, could be shown to activate sciatic field neurons when sciatic afferents are temporarily inactivated with lignocaine. Biella & Sotgiu (1995) concluded that activity of sciatic afferents suppresses, through inhibitory interneurons, the excitatory input from saphenous nerve primary afferents that enter the spinal cord several segments rostrally. Because some sciatic field neurons showed changes in activity when neurons in the saphenous field of the dorsal horn were activated or inhibited by micro-injection of *N*-methyl D aspartate (NMDA) or glycine (respectively), it appears that at least some inputs from saphenous afferents to sciatic field neurons are poly-synaptically mediated (Biella & Sotgiu, 1995).

The idea that the topographic organisation of the receptive fields of dorsal horn neurons is the result of inhibitory pruning of a more extensive and overlapping rostro-caudally oriented terminal projection of primary afferents could provide an mechanism (or at least a substrate) for the reorganisation of inputs following peripheral nerve lesions. Following the loss of activation by peripherally axotomized afferents, the synapses on the distant projections of intact afferents might become more efficacious and thus able to activate the denervated neurons. A Hebbian modulation of synaptic transmission (i.e. changes in synaptic efficacy due to temporal association of pre and postsynaptic neuronal activity) which is dependent on the operation of NMDA subtype of glutamate receptors, would be consistent with the mechanisms shown to underlie the receptive field organisation that follows peripheral nerve regeneration. A role for the NMDA receptor in this plasticity was demonstrated by Lewin *et al.* (1994), who found that the normal sequence of forming focused, contiguous receptive fields from the more diffuse receptive fields that are present immediately after peripheral nerve lesion, was prevented by administration of the NMDA receptor antagonist MK801 during nerve regeneration.

The role, if any, of long-range projections in somatotopic reorganisation remains to be determined (see Wilson & Kitchener, 1996 for discussion). One difficulty in speculating on how long-range projections of primary afferents might be involved in plastic changes in dorsal horn somatotopy is that the full range of primary afferent central projections is much greater than the relatively modest amount of rostro-caudal reorganisation that occurs following denervation (Markus *et al.*, 1984; Wilson & Snow, 1987). Not only in the dorsal horn, but also in many studies of plastic reorganisation of inputs at supra-spinal levels, there seems to be evidence of limits to the extent to which inputs can extend to make connections with denervated neurons (as discussed below). The low threshold cutaneous primary afferents are particularly interesting in regards to what might limit the range over which new inputs may form because they have two central targets: the dorsal horn and the DCN. While it seems that interneuronal circuits might act to differentially alter the functional activity of the dorsal horn and the DCN projections of a single afferent (Lomeli *et al.*, 1998), it is not clear whether peripheral lesion-induced sprouting of afferents (if this does occur) could occur at one projection and not the other. In contrast, a mechanism of plastic change in primary afferent inputs to post-synaptic neurons that depended on the unmasking of existing synapses, under influence of input related inhibitory mechanisms, would presumably result in autonomous reorganisation at the spinal and DCN projections of a primary afferent fibre. Why comparatively little reorganisation occurs after peripheral nerve lesions (and perhaps no reorganisation following dorsal rhizotomy) is unknown. The dorsal horn representations can be viewed, in broad terms, as a collateral of the main somatosensory projection; with the main pathway running from primary afferents, through DCN and VP thalamus to the cortex (see Figure 1). Reasoning teleologically, perhaps the role of spinal cord sensory processing in spinal reflex modulation (from simple segmental reflexes to complex changes such as nocifensive behaviours) allows less scope for plastic changes in connectivity than do the relatively simple relay connections elsewhere in the pathway.

Box 2. Robust Central Sprouting of Peripherally axotomized Primary Afferents

One of the most striking examples of somatosensory plasticity in the CNS is one which has little effect on somatotopic relationships but profoundly changes the sources of input to dorsal horn neurons that convey nociceptive information. Woolf and colleagues (reviewed by Woolf & Doubell, 1994) have shown that within several weeks of peripheral nerve cut or crush in the adult rat, the central projections of the large myelinated fibres that terminate in laminae III and IV of the dorsal horn sprout dorsally into laminae II. Lamina II does not normally receive any terminals of myelinated afferents. Because the dorso-ventral axis of the dorsal horn is somatotopically neutral (i.e. the receptive fields of neurons in the superficial and deeper dorsal horn are, approximately at least, in register) invasion of myelinated afferents into lamina II changes the modality rather than somatotopic location of inputs to lamina II neurons. The invasion by myelinated afferents may be facilitated by the fact that the projections to lamina II (which are predominantly nociceptive) do not sprout after peripheral axotomy, but their central terminals exhibit signs of atrophy. The implication of these findings is that the chronic neuropathic pain that can follow peripheral nerve lesions in humans could be due to the abnormal convergence of low threshold afferents to neurons conveying nociceptive information.

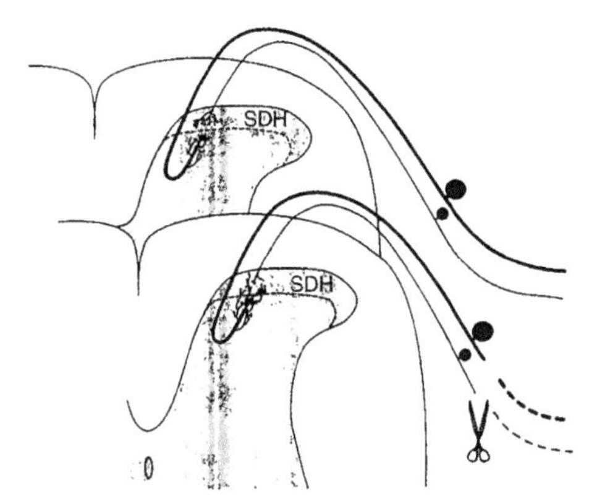

Figure 3 Low-threshold cutaneous afferents (thick lines) terminate in layers ventral to the superficial dorsal horn (SDH) in the spinal grey matter, whereas high-threshold afferents (thin lines) terminate in the SDH. After peripheral nerve lesion, high-threshold terminals atrophy and low threshold afferents sprout into the SDH.

This form of plastic reorganisation of primary afferent input is clearly mediated by the growth of new collaterals and synapses. Morphological details of myelinated afferent terminals that had sprouted into lamina II have been demonstrated with transganglionic tracing and intra-axonal labelling of individual primary afferent fibres (see Doubell & Woolf, 1994 for references). In the examples of functional reorganisation discussed elsewhere in this review, the basis for the plasticity has not yet been demonstrated directly.

Until fairly recently, only in the dorsal horn had the relationship between the extent of afferent projections, somatotopic maps, and inhibitory interneuronal activities been directly examined. Recent studies of peripheral nerve lesions and plasticity in supra-spinal somatosensory regions have begun to furnish direct and indirect evidence that similar properties of the afferent inputs and inhibitory interactions exist, and these properties might underlie the observed plastic changes.

Dorsal column nuclei

While there exist several descriptions of the terminal arbours of single primary afferents and the extent of dendritic arborisation of DCN neurons (Fyffe *et al.*, 1986a,b), there are no studies relating the extent of single arbours to the somatotopic representation of cutaneous inputs in DCN. However, it is clear from the descriptions of the representational maps in the gracile and cuneate nuclei that the dimensions of afferent terminal projections (as visualised by intra-axonal injection of horseradish peroxidase) might well encompass considerably more of the DCN than the region where they activate DCN neurons; inhibitory processes related to the activity of primary afferents are also evident.

The inhibitory contributions to the receptive fields of DCN neurons, originally characterised by Janig *et al.* (1977), have been studied in the context of the acute changes to DCN neuron receptive fields that follow elimination of activity in peripheral nerves innervating the skin. Single unit recordings of cuneate neurons during anaesthetic block of the peripheral afferents that innervate a region of skin that included the receptive field of the neuron, have revealed rapid, dramatic and reversible changes in these receptive fields (Pettit & Schwark, 1993). The expansion of the receptive field following temporary denervation of the local primary afferents is interpreted as being due to the unmasking of previously suppressed (inhibited) inputs from afferents innervating nearby regions of skin, i.e. innervating the new, expanded receptive field (Pettit & Schwark, 1993). This suggests that the afferents innervating the original receptive field were inhibiting the inputs of afferents innervating nearby (or overlapping) skin regions, thus sharpening the receptive field of cuneate neurons.

Similarly dramatic results were obtained by applying capsaicin to peripheral nerves. Capsaicin is the pungent vanilloid found in chilli peppers and has long been known to powerfully activate polymodal nociceptors (see Holzer, 1991 for review). Application of capsaicin to peripheral nerves of mature animals selectively blocks transmission of high-threshold afferents. While activity of high threshold afferents has effects on dorsal horn receptive fields that persist for minutes to hours after the stimulation (Cook *et al.*, 1987), the role of C-fibres in the plasticity of low threshold tactile representations is unclear. Several studies have employed the application of capsaicin to eliminate input from capsaicin sensitive afferents (see Snow & Wilson, 1991 for review), but the balance of evidence suggests that capsaicin sensitive afferents do not play a major role in the long-term (weeks) changes in dorsal horn somatotopy that follow peripheral nerve lesions. In the DCN, however, the acute changes in cuneate neuron receptive fields produced by peripheral afferent anaesthesia can be mimicked by application of capsaicin to the skin area that includes the cuneate neuron's receptive field (Pettit & Schwark, 1996). That topographical reorganisation can be induced by capsaicin suggests that it is the reduced activity of the thin myelinated and unmyelinated afferents (the capsaicin sensitive populations) that unmasks the inputs that contribute to the expanded receptive fields in lignocaine and capsaicin treated nerves. In many neurons the expanded receptive fields included response properties (modalities) different from those of the original receptive field, suggesting that afferents of different modalities as well as different receptive field location are unmasked (Pettit & Schwark, 1996). This unmasking of somatosensory inputs by decreased activity of capsaicin sensitive afferents has previously been described in VP thalamus (Rasmusson *et al.*, 1993) and primary somatosensory cortex (SI) cortex (Calford & Tweedale, 1991).

The demonstration of acute plasticity following local blockade of primary afferents in the periphery (Pettit & Schwark, 1993, 1996) reveals the probable mechanism of reorganisation (unmasking of normally inhibited inputs) but does not examine the limits of DCN plasticity. In the studies of acute changes to cuneate neuron receptive fields (Pettit *et al.*, 1993, 1996) the new receptive fields bordered the original receptive field. Thus the "shift" required would be in the order of 100s of micrometers, well within what is considered to be the extent of terminal arbours and dendritic trees in the cuneate nucleus. The difficulty of interpretation of the apparent rapid reorganisation obtained in such studies is highlighted by recent conflicting findings of Zhang & Rowe (1997). These authors used cold block of the median nerve to produce rapid, reversible, deafferentation of cuneate neurons in the cat. In neurons whose control receptive fields were within the cutaneous territory of the median nerve, this procedure produced neither a discernible expansion of the control receptive field nor an emergence of novel receptive fields. Although they could not completely rule out continuing impulse conduction by some unmyelinated afferents, the possibility that cooling of the nerve failed to block the discharge of capsaicin sensitive nociceptive afferents was discounted by Zhang & Rowe (1997) as an explanation of their negative findings.

The majority of information relating to of the extent of reorganisation of the representation of cutaneous inputs within the DCN comes from studies of long-term deafferentations by peripheral nerve lesions or by dorsal rhizotomies. One exception to this is a study by McMahon & Wall (1983) who compared the activity of locations in DCN to peripheral stimulation in acute and chronic deafferentation of the lower hindlimb of rats: no differences where found in the number of sites (in a 50 μm spaced grid of recording sites) that were silenced by the acute or chronic lesions. However, in this study there was no evidence of significant acute reorganisation by unmasking of inputs. Dostrovsky *et al.* (1976) had previously used the approach of recording from an array of positions regularly spaced throughout the DCN in order to show acute and long-term changes in the entire cutaneous representation following removal of hindlimb afferents by rhizotomy. Eight months after the rhizotomies, the DCN was mapped by noting the general receptive field location of multiunit activity recorded at defined positions within a grid. The authors concluded that many more grid locations on the operated side than on the intact side responded to abdominal stimuli, suggesting abdominal inputs were now activating gracile neurons in what was previously the hindlimb region of the nucleus; there was, however, considerable variation in the number of abdomen sites in the two animals from which the data was obtained. Dostrovsky *et al.* (1976) also examined acute changes by the same grid recording method and found immediate changes (compared to the contralateral gracile nucleus) in the proportion of recording sites responding to abdominal inputs, which they interpreted as unmasking of pre-existing latent inputs. While the region of the gracile nucleus responding to abdominal inputs appeared to have increased markedly, the actual distance by which inputs had shifted is difficult to estimate from multi-unit receptive fields made at a set grid of recording loci and comparing relative sizes, but not absolute locations, of lesioned and intact representations.

Because the DCN relay primary afferent input through to the VP nucleus of the thalamus (which provides the major somatosensory input to the cortex), changes in the representational maps in DCN would presumably have a considerable effect on the representations further along the somatosensory pathway. As discussed below, there is considerable interest in the capacity and the consequences of changes in representational maps in somatosensory cortex. Significant reorganisation of representational maps at the cortical level could be due to changes in the interconnections of cortical neurons, or could be entirely a reflection of reorganisation of VP thalamus or DCN. Due to the compact nature of the VP and DCN representations, comparatively small increases in the extent of functional innervation by an afferent could (potentially) greatly increase the range of the somatotopic representation the afferent could influence.

Deafferentation of DCN neurons by removal of afferents (by dorsal rhizotomy) rather than disconnection of afferent inputs by peripheral nerve lesions, has been shown to alter the gross and fine structure of the DCN and its afferent input. Recently, Sengelaub *et al.* (1997) used transganglionic tracing to examine the extent of projections of afferents in peripheral nerves to the rat DCN and found expanded representations several months after dorsal rhizotomies. These authors suggested that the extent of these changes could, through divergence to higher levels, lead to significant reallocation of cortical territory. An important difference between deafferenting DCN by removing afferents (by rhizotomy) or disconnecting afferents from peripheral receptors (by nerve lesions) has been shown by Jones & Pons (1998): long-tern dorsal rhizotomy (C2 to T4) that deafferented the forelimb in monkeys resulted in large changes in cortical representations (as discussed in the following section) but also resulted in considerable shrinkage of the cuneate nucleus as a result of the degeneration of the central processes of forelimb primary afferents. Interestingly, the medial region of VPl thalamus had also shrunk, due to the trans-neuronal degeneration of the region that normally represents the forelimb (Jones & Pons, 1998). Although similarly extensive cortical reorganisation has been reported to follow peripheral nerve lesions in monkeys (Florence *et al.*, 1998), there is no evidence of trans-neuronal degeneration in brainstem or thalamic nuclei following this type of lesion. Degenerative changes (loss of neurons and down-regulation of neurochemical markers) have, however, been quantified in cat DCN following peripheral nerve lesions in the cat (Avendano & Dykes, 1996).

Certainly, the acute changes in DCN after peripheral nerve lesions, and the indications of long-term changes in afferent arborisation and DCN gross structure, suggest that DCN changes may potentially underlie the plasticity observed at subsequent loci of the somatosensory projection. Recent studies of DCN somatotopy after peripheral nerve lesions support this proposal. Rasmusson *et al.* (1997) have shown that the forelimb digits of the raccoon are represented in separate, discrete regions of the cuneate nucleus. Utilising this knowledge of the normal organisation, the digit representation was studied by single unit recordings 4 months after amputation of a single digit. Neurons judged to be in the region where the amputated digit was previously represented had large receptive fields on one or more adjacent digits. In the normal raccoon DCN, neurons representing the digit that was to be amputated in the experimental animals did not show responses to stimulation of the adjacent digits, thus new inputs to deafferented neurons had been made. The basis of the new inputs (whether they were previously existing silent synapses, or newly sprouted synapses, or perhaps the synapses under tonic inhibition that are unmasked by denervation) is not known, but the involvement of pre-existing connections is supported by the studies on the hand representation in DCN of monkeys where reorganisation may occur rapidly after peripheral nerve lesions (Xu & Jain, 1997).

VB Thalamus

Like the studies on acute changes in cuneate receptive fields (but nearly 30 years earlier) Nakahama *et al.* (1966) demonstrated rapid reversible increases in VP neuron receptive fields when the original receptive field was anaesthetised by local subcutaneous injection of procaine. Nakahama *et al.* (1966) also observed that the modality of new receptive fields was not necessarily the same as that of the original receptive field. As in the DCN, Janig *et al.* (1979) have characterised the receptive fields of VP neurons as comprising excitatory and inhibitory regions — with the inhibitory region sometimes extending beyond the excitatory field. The acute effects of local deafferentation on the receptive fields of single VP neurons has been examined recently by Shin *et al.* (1995) who compared deafferentation by local anaesthesia with deafferentation by digit amputation. Both manipulations resulted in the rapid appearance of new receptive fields on adjacent digits, with the new receptive fields emerging immediately after amputation and about 30 minutes after anaesthesia of the digit containing the original receptive field.

The effect of removing DCN neurons that convey somatosensory inputs to VP neurons has been examined by Alloway & Aaron (1996); this approach requires that lesions in the DCN destroy neurons that have the same peripheral inputs as the recorded neuron in VP (i.e. coextensive or at least overlapping receptive fields). The majority of VP neurons showed alterations in their evoked activity after cuneate lesions indicating a change in inputs; some neurons exhibited expansion of their receptive fields, while others underwent shrinkage of their receptive field. Clearly, these plastic responses to acute loss of a major source of input from the DCN suggests a dynamic unmasking or recruitment of inputs from nearby intact DCN neurons that were not previously contributing to the receptive field of the VP neuron.

Studies of the effect of removal of inputs, by lesioning primary afferents or DCN neurons, on the organisation of representational maps in VP (as opposed to the receptive fields of single neurons) have often been motivated by attempts to characterise subcortical changes that might underlie plasticity of cortical maps. Although the cutaneous representations of the body surface in VP have been described in only a few of the species commonly used to study plasticity, it is evident that changes in thalamic representations could produce significant changes in cortical maps without necessarily involving changes in thalamo-cortical connections. Many years ago Wall & Egger (1971) found that neurons with hindlimb receptive fields in both VP thalamus and primary somatosensory cortex (SI) cortex could be activated by forelimb afferents several days after dorsal column or gracile nucleus lesions. It should be noted however, that this study, like others that assume no change in recording conditions within a deafferented region of the somatosensory system, has been criticised on technical grounds (see Rhoades *et al.*, 1987; Snow & Wilson, 1991) and is at odds with more recent studies which have examined the effects of dorsal column lesions. Like Wall & Egger (1971), Jain *et al.* (1995) studied urethane anaesthetised rats that had complete lesions of one dorsal column at thoracic level, and failed to find any evidence that forelimb inputs activate cortical neurons representing a denervated hindlimb, even as long as 3 months after the deafferentation. Jain *et al.* (1995) suggest that the results reported by Wall & Egger (1971) may apply only to lesions which destroy DCN neurons, not to dorsal column lesions. One of the few attempts to document morphological changes in connectivity of VP inputs after large partial lesions of the DCN has provided clear evidence of reorganisation. The number of synapses (of the type determined to be derived from DCN neurons) in VPl fell to about 50% of its initial value, but increased again after 30 days, rising to normal (pre lesion) values by 46 days after DCN lesions (Wells & Tripp, 1987).

While it is clear that lesions of the hindlimb component of the dorsal columns in rats, cats and monkeys do not result in significant substitution of hindlimb afferents by afferents in the intact forelimb pathway (see Jain *et al.*, 1995 for references), changes in VP somatotopy have been reportedly found in several studies where portions of the forelimb input were removed by peripheral nerve lesions or by dorsal rhizotomy. Garraghty & Kaas (1991) denervated the dorsal and lateral aspect of the hand in three squirrel monkeys and after 2–5 months found changes in VPl and primary somatosensory cortex (SI) such that territory previously devoted to the denervated regions had assumed inputs from regions supplied by intact nerves. The large and discrete forepaw digit representation in the raccoon VPl has been utilised by Rasmusson (1996) to show that 2–5 months after amputation of a single digit, neurons in the region of primary somatosensory cortex (SI) cortex representing the missing digit had acquired receptive fields on adjacent digits and palm. Similar reallocation of territory was also recorded in the VPl nucleus of the thalamus, suggesting that much of the change in primary somatosensory cortex (SI) may have been due to reorganisation at the thalamic level. Long-term deafferentation of the forelimb in macaque monkeys by dorsal rhizotomy leads to extensive reorganisation of somatosensory cortex which is accompanied by marked shrinkage of the DCN and VP thalamus (Jones & Pons 1998). The thalamic change is such that the shrunken VP thalamus contains apparently normal sized face and hindlimb representations that abut each other, rather than being separated by a forelimb region as in the normal VP thalamus.

A study that combined the analysis of individual receptive fields of VP thalamic neurons with acute and chronic peripheral lesions and lesions, that also allowed the overall representation to be examined, was implemented by Nicolelis *et al.* (1993), using chronically implanted multi-electrode arrays in the whisker representation in rat VPm. The electrode array was positioned in VPm and recordings were made before, during and after the temporary deafferentation of whiskers by local anaesthesia. The majority of single units showed immediate unmasking of previously unexpressed inputs from adjacent whiskers (often from those located more caudally in the same row). The overall effect of changes in the inputs to VPm neurons was a reduction of short latency responses from the central anaesthetised region but a relative increase in the longer latency responses from this region, which can be viewed as compensatory and, as suggested by Nicolelis *et al.* (1993), might be the seed for long-term compensatory reorganisation.

An interesting finding in the study by Nicolelis *et al.* (1993) was that the normal VPm neurons show preferential activation from one whisker, but also show strong activation from up to several nearby whiskers (particularly from those in the same row). This differs from the established view of VPm receptive fields in which inputs arise exclusively from one or sometimes 2 whiskers (Waite *et al.*, 1984). The basis of this difference is not immediately obvious, but may be due in part to the lighter level of anaesthesia used. It must be stressed that, no matter how reliable such pattern recognition techniques might appear to be, they are required to discriminate the response of a single unit from complex multiunit signals recorded by a large number of microelectrodes, and would thus seem to be inherently less reliable than the single unit recording technique.

The neuronal basis for the changes in the receptive fields of VP neuron would appear to be related to the role of tonic inhibitory mechanisms of the expression of inputs. Rhoades *et al.* (1987) examined lesions of brainstem nuclei that relay trigeminal inputs to VPm. New receptive fields in VPm neurons emerged 6 days after excitotoxic lesion of spinal trigeminal nucleus caudalis (SpVc) and increased in number over the next 3 months. The new receptive fields had wide dynamic range properties (rather than purely low threshold), suggesting that previously unexpressed inputs to VPm from SpVc had taken over (refer to Figure 1). This suggestion was supported by the finding that lesion of SpVc (subsequent to the initial reorganisation that followed SpVc lesions) abolished the receptive fields in VPm (Rhoades *et al.*, 1987). Additional characterisation of the expanded receptive fields in VPm has indicated that they were due to the expression of SpVc inputs that are normally inhibited by neurons of the thalamic reticular nucleus (Lee *et al.*, 1994). The increase in size of VPm neuron receptive fields obtained by lesioning SpVc was similar to the expansion of VP receptive fields both immediately and several weeks after lesioning the ipsilateral thalamic reticular nuclei. The receptive fields of VPm neurons returned to their original size if the SpVc was subsequently lesioned. It therefore appears that the principle trigeminal nucleus (PrV) inputs inhibit, via the reticular nucleus, the expression of SpVc inputs to VP neurons (Lee *et al.*, 1994).

Somatosensory Cortex

Plasticity of somatosensory representations in the cerebral cortex has been the subject of a great many studies — especially in comparison to studies of plasticity in the DCN and in the VP nuclei of the thalamus. A number of recent reviews (Kaas & Garraghty, 1991; Snow & Wilson, 1991; Garraghty & Kaas, 1992; Buonomano & Merzenich, 1998; Calford *et al.*, 1998) provide a more comprehensive coverage of cortical plasticity than the sample of reports that are reviewed here, which have been selected to identify the main findings and the main issues.

The most dramatic evidence of expanded cutaneous representations following peripheral manipulations has been obtained in the somatic sensory cortex. A landmark report in this literature is the study by Pons *et al.* (1991) showing that monkeys which had survived a year or more after forelimb deafferentation (by dorsal rhizotomies) had undergone massive cortical reorganisation comprising an invasion of the forelimb region by face inputs. Although the possibility that long-term large deafferentations resulted in extensive reorganisation had been indicated by the experiments of Rasmusson *et al.* (1985) in the raccoon, prior to the study by Pons *et al.* (1991) it was considered that the maximum extent of cortical reorganisation (i.e. shift of somatotopic boundaries tangential to the cortical surface) was in the range of 1 or 2 mm, not the 10 mm or more that Pons *et al.* (1991) reported. Massive reorganisation has been observed in many subsequent studies that have addressed the questions: (a) what mechanisms might determine the limits of plastic reorganisation, (b) at what level of the somatosensory pathways are changes occurring, and (c) what mechanisms might underlie such reorganisation?

It is apparent that some lesions result in considerable reorganisation of cortical representations while others are followed by little or none. Jain *et al.* (1997) have demonstrated that neurons in area 3b of the somatosensory cortex of adult monkeys become, and remain, unresponsive after removal of all contralateral dorsal column afferent inputs (thus other ascending pathways appear not to contribution to primary somatosensory cortex (SI) neuronal activation) but cortical neurons in the deafferented region were found to be activated by afferents innervating the face 6 months after the dorsal column lesions. This is similar to the finding by Pons *et al.* (1991) where dorsal rhizotomy of forelimb afferents was shown to allow face afferents to activate cortical neurons that were in the region of cortex previously representing the forelimb; the results of Jain *et al.* (1995) show that removal of afferents that are conveyed by the dorsal column is alone sufficient to allow face afferents to activate neurons in what was originally the forelimb representation. While forelimb and hindlimb afferents do not normally appear to influence neurons outside their own representations in primary SI cortex, afferents that remain after partial dorsal column lesions can expand their influence over a greater area of the cortex, but only within that limb's representation.

With peripheral nerve lesions, the extent of the resulting cortical reorganisation depends more on the pattern than the extent of peripheral denervation. As discussed by Garraghty *et al.* (1994), when regions of the hand are denervated by peripheral nerve lesions in monkeys, little cortical reorganisation is observed. If however, the entire glabrous surface of the hand is denervated, extensive reorganisation follows such that neurons in the glabrous region acquire inputs from the hairy skin of the dorsum. By way of explanation, Florence *et al.* (1994) have suggested that glabrous inputs are segregated into domains in the DCN, and that this limits the ability of afferents from glabrous skin to activate neurons deafferented by denervation of adjacent glabrous skin regions. In contrast, inputs from dorsal (volar) skin do not show such segregation, allowing them more easily to activate deafferented neurons.

Recent studies by Florence *et al.* (1998) and Jones and Pons (1998) have suggested that the nature of the deafferentation profoundly affects the way that cortical reorganisation is achieved. Chronic removal of forelimb inputs by dorsal rhizotomy leads to significant cortical reorganisation and equally dramatic changes in thalamus and DCN (Jones & Pons, 1998). In contrast, when Florence *et al.* (1998) deafferented the forelimb by peripheral nerve lesions they found that while there was significant cortical plasticity, it was not accompanied by changes in the thalamic representation. The difference between these two types of deafferentations is presumably that primary afferent terminals in the DCN survive after peripheral nerve lesions but degenerate after dorsal rhizotomy (as discussed in the section on dorsal horn plasticity), resulting in a physical loss of synaptic inputs to DCN neurons. This is consistent with the long-term representational plasticity of the thalamic VP nucleus, which is seen after partial lesions of the DCN but not after partial removal of inputs by dorsal column lesions. Rather than being simply the physical loss of synapses, it might be the pattern of the resulting deafferentation or inactivation of inputs that is significant. Any single region of the representational map receives input from afferents running in several dorsal roots, but generally only a single peripheral nerve trunk. Consequently, dorsal root lesions do not remove as many inputs to a particular region of the spinal dorsal horn or the DCN as do peripheral nerve lesions; perhaps the patterns of lost and spared synaptic inputs after peripheral nerve lesions are more conducive to reorganisation than the patterns of synaptic loss that follow dermatomal lesions. This might be why DCN lesions (which are somatotopic), but not dorsal column lesions (which are dermatomal), are followed by plastic changes in the thalamic representation. It is perhaps worth noting, however, that the conditions that produce reorganisation of DCN and VP are different from those that lead to plasticity in the dorsal horn: as discussed in section on dorsal horn plasticity, relatively small deafferentations by peripheral nerve lesions result in rostro-caudal increases in the functional projection of intact afferents whereas the bulk of evidence suggests that dorsal rhizotomy does not appear to be followed by any dorsal horn reorganisation. An additional consideration as to why some lesions produce plastic reorganisation and others do not could be that reorganisation of inputs occurs in the direction of the long range projections of afferent arbours; this possibility can not be assessed until a great deal more is known about the fine structure of afferent organisation in central somatosensory representations.

Changes in cortical somatotopy could result from changing connections at any level of the somatosensory pathway up to and including the cortex itself. The role of cortical versus subcortical changes has been most thoroughly examined in the vibrissal representation of the rat trigeminal system. Support for a cortical based reorganisation has been found following neonatal denervations and differential activity regimes in adult rats. Again, the vibrissal representation within the rat trigeminal system has facilitated such investigations, and there is experimental evidence that points towards the involvement of horizontal corticocortical connections in use-dependent plasticity in barrel cortex. In both supragranular (layers I-III) and infragranular (layers V-VI) layers of neocortex, there are dense horizontal plexuses formed by axon collaterals of layer II/III pyramidal cells and of deep layer V pyramidal neurons, respectively. In the rat, neurons within barrel columns may be activated by several vibrissae in addition to their principal vibrissa. However, the response to all but the principal vibrissa depends on convergence of activity from surrounding barrel columns. This in turn depends on the plexus of horizontal circuits that pass through the inter-barrel septa and the supragranular and infragranular layers (reviewed by Armstrong-James, 1995). Changes in the receptive fields of barrel cortex neurons due to increased activity of subsets of whisker afferents in adult rats has been shown to occur earlier in supragranular and infragranular layers than in layer IV (Diamond *et al.*, 1994). As thalamic input projects to layer IV, the plasticity observed in the barrel cortex must be due to altered connectivity between cortical rather than subcortical neurons. By placing micro-lesions around the cortical barrel representing a spared whisker, Fox (1994) showed that the denervation-induced enhanced responses of neurons in neighbouring barrels to stimulation of the spared whisker are abolished. These results suggest that the development of increased responsiveness of neighbouring barrels to inputs from the barrel representing the spared vibrissa is due to changes corticocortical rather than subcortical connections (reviewed by Armstrong-James, 1995). Significantly, it is the supragranular layers of adult neocortex which contain the greatest concentration of NMDA receptors (reviewed by Fox & Daw, 1993), suggesting that synapses in layers II/III might be more plastic than those elsewhere. That long term potentiation can be generated at these synapses has been amply demonstrated *in vitro* by Hess *et al.* (1996). Clearly, an understanding of cortical integration and plasticity in barrel cortex requires a more detailed knowledge of the morphology and synaptic function of the horizontal projections between barrel columns.

Studies in the adult raccoon have suggested that the increased input from adjacent digits to cortical neurons located within the representation of a previously amputated digit is accompanied by increased numbers of short latency inputs from adjacent intact digits to neurons in the region where the amputated digit was represented, but not by increased numbers of corticocortical inputs to these neurons. This suggests that, in this case, the plasticity is mediated by changes in subcortical connections. Reasoning that the new receptive fields of cortical neurons might be due to the increased efficacy of previously ineffective synapses from inputs supplying adjacent digits, Smits *et al.* (1991) made intracellular recordings from primary SI neurons in normal raccoons while activating adjacent digit afferents with electrical and natural stimuli; these stimuli produced excitatory post-synaptic potentials in all cortical neurons examined. Subthreshold inputs were also elicited by microstimulation of nearby cortex that represented adjacent digits. Clearly, there is evidence for both cortical and subcortical sources of the new inputs to cortical neurons undergoing plastic changes (see Donoghue, (1995) for review of corticocortical plasticity). The relative contribution of these sources to plastic changes in cortical representations is not known. The expansion of thalamocortical projections could utilise pre-existing but unexpressed long-range collaterals of thalamo-cortical neurons. The range of the terminals made by thalamocortical neurons has been estimated by antidromic microstimulation in cats (Snow *et al.*, 1988), and by retrograde tracing in monkeys (Rausell *et al.*, 1995), and found to be considerably more extensive than the region of the somatotopic map to which they contribute.

A somewhat different experimental approach to cortical plasticity has been to examine the consequences of changes in the level or temporal patterning of sensory inputs (rather than reducing activity by lesions). In summary, temporal pairing of stimulus inputs in raccoon and monkey (to adjacent digits for example) results in a confluence of their cortical representations, while asynchronous activation causes disjunction (Zarzecki *et al.*, 1993; Wang *et al.*, 1995). The study of Xerri *et al.* (1994) is a particularly good example as it examined the consequences of an ethological manipulation of somatosensory input. The ventral surface of the trunk of adult female rats undergoes a great increase in stimulation during suckling their young. Xerri *et al.* (1994) examined the cortical representation of the ventrum in nursing rats, and in age matched virgin and non-lactating post partum controls. A significantly larger cortical area was dedicated to the ventrum (and especially the nipples) of nursing rats compared to non-nursing controls, and individual receptive fields of neurons representing the nipples of nursing rats were significantly smaller than in control rats. As discussed by

to the details of the representation to which they contribute in the DCN, thalamic VP nucleus or somatosensory cortex. There is, however, indirect evidence that, as in the dorsal horn, primary afferent projections to the DCN and thalamocortical projections to somatosensory cortex might extend beyond the regions where they contribute to the receptive fields of neurons (for review, see Snow & Wilson, 1991). The existence of long-ranging but functionally silent (or silenced) projections at all levels of the somatosensory pathways implies that the cutaneous representations that are routinely determined in physiological studies are in fact the result of a sharpening process whereby overlapping inputs are disguised by inhibitory interconnections, producing greater apparent resolution of the representational map. Of course, it is not clear to what extent the relative activity of excitatory and inhibitory inputs are affected by the anaesthetic conditions of the experiments from which these finding have come. Because receptive fields are known to be larger in the absence of anaesthetic (for review, see Snow & Wilson, 1991), the discrete, contiguous grouping of body part representations is somewhat artifactual, in as much as more overlap would be expected in the absence of the anaesthetic agents. Moreover, as shown by Dubuisson *et al.* (1979) in lamina II of the rat dorsal horn and by Nicolelis & Chapin (1994) for neurons of the vibrissa representation of the thalamic VB complex in the rat, in unanaesthetised animals receptive fields may undergo dynamic changes in size in the absence of any peripheral manipulation.

In the spinal dorsal horn, there is direct anatomical evidence for the existence of synaptic varicosities on collaterals which extend beyond the region where their receptive fields are represented. There is also electrophysiological evidence that these long-range projections possess functional synapses (reviewed by Wilson & Kitchener, 1996) but the possibility remains that rather than being tonically suppressed by inhibitory inputs, these synapses may be relatively or completely ineffective and not involved in the unmasking of receptive fields described above. The existence of silent or weak synapses at the distal regions of afferent arbours might be said to provide an explanation for longer term changes in representations: hypothetically, normally ineffective or silent connections might become functional as a consequence of the loss of nearby synapses after peripheral lesions. Alternatively, the sparse distal arbours might sprout new collaterals and thus increase the synaptic input to postsynaptic neurons.

The emergence of new receptive fields over a period of weeks or months is usually taken as possible evidence of new connections forming, either the increased efficacy of previously ineffective synapses, or the formation of new synaptic connection due to collateral sprouting. Garraghty *et al.* (1994) suggest that immediate reorganisation is due to modulation of GABA mediated inhibition whereas longer term reorganisation is due to increasing efficacy of previously ineffective inputs via an NMDA receptor mediated Hebbian process. It seems likely that unmasking and Hebbian changes due to altered input patterns must be related: the basis for NMDA mediated changes in synaptic efficacy would be an interrelated result of reduced excitatory input from lesioned afferents and "release from inhibition" of the weak inputs from nearby intact afferents. Thus, the acute changes following peripheral nerve lesions could be the basis for the emergence of longer term changes in inputs.

Given the extent of both the long range projections of primary sensory afferents into "inappropriate" representational regions of the dorsal horn and the dendritic arbours of the neurons with which they synapse (Brown, 1981), it would seem that a large degree of rostrocaudal reorganisation is theoretically possible. However, the extent of changes to representational maps in the dorsal horn is minor compared to the massive reorganisations observed in somatosensory cortex. If pre-existing long range projections are the substrate of long term reorganisation, the question arises as to why the extent of reorganisation in the spinal dorsal horn is small compared to the extent of the long range collateral projections.

An alternative, and not necessarily exclusive possibility, is that collateral sprouting and synaptoneogenesis underlie long term changes in somatosensory neuron receptive fields. The slow time-course (months to years) of some examples of reorganisation to reach its full extent makes it seems not unreasonable to suggest that these changes are mediated by degenerative and regenerative process that might involve neuronal death, atrophy of terminals and the growth of new collateral sprouts with the formation of new synapses (rather than purely functional changes in pre-existing connections). It is difficult to envisage how direct evidence for such a mechanism might be obtained, given that the spatial extent of afferent terminals does not necessarily indicate the functional extent of inputs, and also the emergence of new inputs via Hebbian strengthening of ineffective pre-existing synapses, and the growth of new synapses via collateral sprouting, might well appear similar when determined by the changes in neuron receptive fields and changes in representational maps. Complementing microelectrode recording and neuronal tracing, newer methods, such as the use of activity-related neurochemical markers and targeted alterations to neuronal genes will no doubt be increasingly employed to provide evidence for the underlying basis of somatosensory plasticity.

Perceptual consequences of changes in representational maps

Functional brain imaging studies on humans have demonstrated that the representations of body surface regions in the somatosensory cortex are related to the elaboration of somaesthesia for the particular region of the periphery that is represented. But does changing the inputs to regions of cortex change the apparent localisation of the sensations these new inputs evoke, or does a permanent dysaesthesia result when the topography of inputs to adult somatosensory cortex changes?

As recently discussed by Doestch (1998), available evidence suggests that plastic changes in cortical representations do not result in a change of cortical identity: changing the inputs to cortical neurons does not change the experiential consequences evoked by their activation. This is a most important issue for cases of nerve injury as new inputs to cortical neurons will generate (via largely unknown connections with other regions of the CNS) sensations that will appear to originate from regions of the periphery that previously innervated the now remapped region of cortex.

The continued activity of cortical somatosensory cortex that has been deprived of its peripheral input seems to provide a basis for the continued perception of a body part that has been amputated. Reorganisation of cortex such that afferents from one body region come to activate neurons representing other body regions seems also to explain the findings that phantoms can be evoked by stimulating other regions of the periphery. Such phantom sensations can be activated by stimulation of skin in regions of the body remote from the amputation site. Halligan *et al.* (1992) report a clinical case where a patient experienced the perception of an amputated hand when the ipsilateral face was stimulated. The existence of the phantom without activation would seem to be explained as a product of the continued activity of the denervated regions of the CNS that produces the sensation of the missing body part. The cortical re-mapping may be the cause of the changes in the perceived form of the missing body part over time.

The role of plastic changes in representations in the experiencing of phantoms is not certain but it appears that changes in cortical representational maps degrade rather than produce the sensation of the phantom. Plastic changes in inputs to cortical regions may explain why phantoms can be activated from other regions of the body. Given that injury-related changes in cortical inputs are apparently not associated with respecification of cortical identity, it will be interesting to determine whether the identity of cortex with new inputs is changed under conditions where the cortical changes are a result of manipulating the amount and pattern of physiological peripheral input.

Key Points

- **There is evidence for both a cortical and a sub-cortical basis of cortical reorganisation. A clear picture of what type of manipulations produce what combination of cortical and sub-cortical contributions to reorganisation has yet to emerge.**
- **Changing inputs to somatosensory cortex by peripheral lesions probably does not change cortical identity: the perception of the peripheral stimulus depends on what region of cortex is activated rather than which afferents activate the cortical neurons. It is not known if this applies to cortical changes induced by physiological alterations of afferent activity.**

Conclusions

Representational maps of cutaneous inputs have been used to demonstrate plasticity of neuronal connectivity at all levels of the somatosensory pathways, from the spinal cord to the primary somatosensory cortex. But the idea that once a brain region is inactivated, adjacent representations expand into the inactivated zone, is overly simplistic. In addition to the difficulty in determining the location and mechanism of changes, the temporal and spatial limits, and the conditions under which deafferentation does and does not result in the expansion of adjacent representation, have only begun to be explored. The nature of the plastic changes that have been documented has led to an increasing appreciation that these representations are not static organisations, but rather, dynamic instantiations of divergent, convergent, parallel, excitatory, inhibitory, ascending and descending, spontaneous and evoked activity. The capacity for plastic changes should be considered when devising therapeutic strategies aimed at maintaining and restoring the connections of damaged or deafferented neurons. Factors such as the location and the extent of the injury, and the delay in initiating a restoration of lost connections, may well have important consequences for the connectivity of damaged neurons and also for the connectivity undamaged neurons in the same pathway.

The demonstrations of cortical plasticity that follow changes in the amount or organisation of physiological stimulation indicates that plasticity of connections is a feature of normal somatosensory system operation. Whether the plasticity (or the lack of plasticity) that follows lesions of the peripheral nervous system have any adaptive value is less clear. In addition to determining where in the system changes in connectivity are occurring and what mechanisms underlie the changes following different lesions, future work should also address the relation between the type of peripheral injury and the extent to which the resultant reorganisation has a beneficial, neutral, or detrimental effect on somatosensory perception.

References

Alloway K D and Aaron G B (1996) Adaptive changes in the somatotopic properties of individual thalamic neurons immediately following microlesions in connected regions of the nucleus cuneatus. Synapse 22, 1–14

Angel A and Clarke K A (1975) An analysis of the representation of the forelimb in the ventrobasal thalamic complex of the albino rat. Journal of Physiology 249, 399–423

Armstrong-James M (1995) The nature and plasticity of sensory processing within adult rat barrel cortex. In: Jones E G and Diamond I T (eds) Cerebral Cortex Vol 11, The Barrel Cortex of Rodents, New York, Plenum, pp. 333–373

Avendano C and Dykes R W (1996) Quantitative analysis of anatomical changes in the cuneate nucleus following forelimb denervation: a stereological morphometric study in adult cats. Journal of Comparative Neurology 370, 491–500

Biella G and Sotgiu M L (1995) Evidence that inhibitory mechanisms mask inappropriate somatotopic connections in the spinal cord of normal rat. Journal of Neurophysiology 74, 495–505

Buonomano D V and Merzenich M M (1998) Cortical Plasticity: from synapses to maps. Annual Review of Neuroscience 21, 149–186

Burke R E and Rudomin P (1977) Spinal Neurons and Synapses. In Handbook of Physiology, Section 1: The Nervous System, vol. I, part 2, edited by E Kandel, pp877–944. Baltimore: Waverly Press Inc

Brown A G (1981) Organisation in the Spinal Cord. The Anatomy and Physiology of Identified Neurons. Berlin: Springer-Verlag

Brown P B (1987) A reassessment of evidence for primary afferent sprouting in the dorsal horn. In: Effects of injury on trigeminal and spinal somatosensory systems, edited by Pubols L M and Sessle B J pp273–280. New York: Liss

Calford M B Clarey J C and Tweedale R (1998) Short-term plasticity in adult somatosensory cortex. In: Morley, J W (ed) Neural Aspects of Tactile Sensation, Amsterdam, Elsevier, pp. 299–350

Calford M B and Tweedale R (1991) C-fibres provide a source of masking inhibition to primary somatosensory cortex. Proceedings of the Royal Society of London B 243, 269–275

Calford M B and Tweedale R (1988) Immediate and chronic changes in responses of somatosensory cortex in adult flying-fox after digit amputation. Nature 332, 446–448

Calford M B and Tweedale R (1990) Interhemispheric transfer of plasticity in the cerebral cortex. Science 249, 805–807

Cook A J Woolf C J Wall P D and McMahon S B (1987) Dynamic receptive field plasticity in rat spinal cord dorsal horn following C-primary afferent input. Nature 325, 151–153

Devor M and Wall P D (1978) Reorganisation of spinal cord sensory map after peripheral nerve injury. Nature 276, 75–76

Devor M and Wall P D (1981a) Effect of peripheral nerve injury on receptive fields of cells in the cat spinal cord. Journal of Comparative Neurology 199, 277–291

Devor M and Wall P D (1981b) Plasticity in the spinal cord sensory map following peripheral nerve injury in rats. Journal of Neuroscience 1, 679–684

Diamond M E Huang W and Ebner F F (1994) Laminar comparison of somatosensory cortical plasticity. Science 265, 1885–1888

Doestch G S (1998) Perceptual significance of somatosensory cortical reorganization following peripheral nerve denervation. Neuroreport 9, 29–35

Donoghue J P (1995) Plasticity of adult sensorimotor representations. Current Opinion in Neurobiology 5, 749–754

Dostrovsky J O Millar J and Wall P D (1976) The immediate shift of afferent drive to dorsal column nucleus cells following deafferentation: a comparison of acute and chronic deafferentation in gracile nucleus and spinal cord. Experimental Neurology 52, 480–495

Dubuisson D Fitzgerald M and Wall P D (1979) Ameboid receptive fields of cells in laminae 1, 2 and 3. Brain Research 177, 376–378

Florence S L Garraghty P E Wall J T and Kaas J H (1994) Sensory afferent projections and area 3b somatotopy following median nerve cut and repair in macaque monkeys. Cerebral Cortex 4, 391–407

Florence S L and Kaas J H (1995) Large-scale reorganization at multiple levels of the somatosensory pathway follows therapeutic amputation of the hand in monkeys. Journal of Neuroscience 15, 8083–8095

Florence S L Taub H B and Kaas J H (1998) Large-scale sprouting of cortical connections after peripheral injury in adult macaque monkeys. Science 282, 1117–1121

Fox K (1994) The cortical component of experience-dependent synaptic plasticity in the rat barrel cortex. Journal of Neuroscience 14, 7665–7679

Fox K and Daw N W (1993) Do NMDA receptors have a critical function in visual cortical plasticity? Trends in Neuroscience 16, 116–122

Fyffe R E Cheema S S Light A R and Rustioni A (1986) Intracellular staining study of the feline cuneate nucleus. II. Thalamic projecting neurons. Journal of Neurophysiology 56, 1284–1296

Fyffe R E Cheema S S and Rustioni A (1986) Intracellular staining study of the feline cuneate nucleus. I Terminal patterns of primary afferent fibers. Journal of Neurophysiology 56, 1268–1283

Garraghty P E Hanes D P Florence S L and Kaas J H (1994) Pattern of peripheral deafferentation predicts reorganizational limits in adult primate somatosensory cortex. Somatosensory and Motor Research 11, 109–117

Garraghty P E and Kaas J H (1992) Dynamic features of sensory and motor maps. Current Opinion in Neurobiology 2, 522–527

Garraghty P E and Kaas J H (1991) Functional reorganization in adult monkey thalamus after peripheral nerve injury. Neuroreport 2, 747–750

Godde B Hilger T von Seelen W Berkefeld T and Dinse H R (1995) Optical imaging of rat somatosensory cortex reveals representational overlap as topographic principle. Neuroreport 7, 24–28

Halligan P W Marshall J C Wade D T Davey J and Morrison D (1993) Thumb in cheek? Sensory reorganization and perceptual plasticity after limb amputation. Neuroreport 4, 233–236

Hess G Aizenman C D and Donoghue J P (1996) Conditions for the induction of long-term potentiation in layer II/III horizontal connections of the rat motor cortex. Journal of Neurophysiology 75, 1765–1778

Holzer P (1991) Capsaicin: cellular targets, mechanisms of action, and selectivity for thin sensory neurons. Pharmacological Reviews 43, 143–201

Jain N Florence S L and Kaas J H (1995) Limits on plasticity in somatosensory cortex of adult rats: hindlimb cortex is not reactivated after dorsal column section. Journal of Neurophysiology 73, 1537–1546

Jain N Catania K C and Kaas J H (1997) Deactivation and reactivation of somatosensory cortex after dorsal spinal cord injury. Nature 386, 495–498

Jänig W Schoultz T and Spencer W A (1977) Spatial and temporal parameters of excitation and inhibition in the cuneothalamic relay neurons. Journal of Neurophysiology 40, 882–835

Jänig W Spencer W A and Younkin S G (1979) Spatial and temporal features of afferent inhibition of thalamocortical relay cells. Journal of Neurophysiology 42, 1450–1560

Jones E G and Pons T P (1998) Thalamic and brainstem contributions to large-scale plasticity of primate somatosensory cortex. Science 282, 1121–1124

Kaas J H and Garraghty P E (1991) Hierarchical, parallel, and serial arrangements of sensory cortical areas: connection patterns and functional aspects. Current Opinion in Neurobiology 1, 248–51

Kaas J H Nelson R J Sur M Dykes R W and Merzenich M M (1984) The somatotopic organization of the ventroposterior thalamus of the squirrel monkey, Saimiri sciureus. Journal of Comparative Neurology 226, 111–140

Kitchener P D Hutton E J Wilson P and Snow P J (1997) Cutaneous receptive field organisation in the ventral posterior nucleus of the thalamus in the marmoset. Abstracts, 31st meeting of the International Congress of Physiological Sciences. P083.03

LaMotte C C and Kapadia S E (1987) Deafferentation-induced alterations in the rat dorsal horn: II. Effects of selective poisoning by pronase of the central processes of a peripheral nerve. Journal of Comparative Neurology 266, 198–208

Lee S M Friedberg M H and Ebner F F (1994) The role of GABA-mediated inhibition in the rat ventral posterior medial thalamus. I Assessment of receptive field changes following thalamic reticular nucleus lesions. Journal of Neurophysiology 71, 1702–1715

Lewin G R Mckintosh E and McMahon S B (1994) NMDA receptors and activity- dependent tuning of the receptive fields of spinal cord neurons. Nature 369, 482–485

Lomeli J Quevedo J Linares P and Rudomin P (1998) Local control of information flow in segmental and ascending collaterals of single afferents. Nature 395, 600–604

Liu C N and Chambers W W (1958) Intraspinal sprouting of dorsal root axons. Archives of Neurology and Psychiatry 79, 46–61

Markus H Pomeranz B and Krushelnycky D (1984) Spread of saphenous somatotopic projection map in spinal cord and hypersensitivity of the foot after chronic sciatic denervation in adult rat. Brain Research 296, 27–39

McMahon S B and Wall P D (1983) Plasticity in the nucleus gracilis of the rat. Experimental Neurology 80, 195–207

Micevych P E Rodin B E and Kruger L (1986) The controversial nature of the evidence for neuroplasticity of afferent axons in the spinal cord. In: Spinal Afferent Processing, pp. 417–443. Ed. T K Yaksh, Plenum, New York

Meyers D E and Snow P J (1984) Somatotopically inappropriate projections of single hair follicle afferent fibres to the cat spinal cord. Journal of Physiology 347, 59–73

Nakahama H Nishioka S and Otsuka T (1966) Excitation and inhibition in ventrobasal thalamic neurons before and after cutaneous input deprivation. Progress in Brain Research 12, 180–196

Nicolelis M A and Chapin J K (1994) Spatiotemporal structure of somatosensory responses of many-neuron ensembles in the rat ventral posterior medial nucleus of the thalamus. Journal of Neuroscience 14, 3511–3532

Nicolelis M A Lin R C Woodward D J and Chapin J K (1993) Induction of immediate spatiotemporal changes in thalamic networks by peripheral block of ascending cutaneous information. Nature 361, 533–536

Northgrave S A and Rasmusson D D (1996) The immediate effects of peripheral deafferentation on neurons of the cuneate nucleus in raccoons. Somatosensory and Motor Research 13, 103–113

Pettit M J and Schwark H D (1993) Receptive field reorganization in dorsal column nuclei during temporary denervation. Science 262, 2054–2056

Pettit M J and Schwark H D (1996) Capsaicin-induced rapid receptive field reorganization in cuneate neurons. Journal of Neurophysiology 75, 1117–1125

Pons T P Garraghty P E Ommaya A K Kaas J H Taub E and Mishkin M (1991) Massive cortical reorganization after sensory deafferentation in adult macaques. Science 252, 1857–1860

Rasmusson D D Turnbull B G and Leech C K (1985) Unexpected reorganization of somatosensory cortex in a raccoon with extensive forelimb loss. Neuroscience Letters 55, 167–172

Rasmusson D D Louw D F and Northgrave S A (1993) The immediate effects of peripheral denervation on inhibitory mechanisms in the somatosensory thalamus. Somatosensory and Motor Research 10, 69–80

Rasmussen D D (1996) Changes in the organisation of the ventroposterior lateral thalamic nucleus after digit removal in the adult raccoon. Journal of Comparative Neurology 364, 92–103

Rasmusson D D and Northgrave S A (1997) Reorganization of the raccoon cuneate nucleus after peripheral denervation. Journal of Neurophysiology 78, 2924–2936

Rausell E Bae C S Vinuela A Huntley G W and Jones E G (1992) Calbindin and parvalbumin cells in monkey VPL thalamic nucleus: distribution, laminar cortical projections, and relations to spinothalamic terminations. Journal of Neuroscience 12, 4088–4111

Rausell E and Jones E G (1991) Histochemical and immunocytochemical compartments of the thalamic VPM nucleus in monkeys and their relationship to the representational map. Journal of Neuroscience 11, 210–225

Rausell E and Jones E G (1995) Extent of intracortical arborization of thalamocortical axons as a determinant of representational plasticity in monkey somatic sensory cortex. Journal of Neuroscience 15, 4270–4288

Sengelaub D R Muja N Mills A C Myers W A Churchill J D and Garraghty P E (1997) Denervation-induced sprouting of intact peripheral afferents into the cuneate nucleus of adult rats. Brain Research 769, 256–262

Shin H C Park S Son J and Sohn J H (1995) Responses from new receptive fields of VPL neurones following deafferentation. Neuroreport 7, 33–66

Smits E Gordon D C Witte S Rasmusson D D and Zarzecki P (1991) Synaptic potentials evoked by convergent somatosensory and corticocortical inputs in raccoon somatosensory cortex: substrates for plasticity. Journal of Neurophysiology 66, 688–695

Snow P J and Wilson P (1991) Plasticity in the Somatosensory System of Developing and Mature Mammals – The Effects of Injury to the Central and Peripheral Nervous System. Progress in Sensory Physiology, vol. 11. Springer Verlag: Berlin

Waite P M (1984) Rearrangement of neuronal responses in the trigeminal system of the rat following peripheral nerve section. Journal of Physiology 352, 425–445

Wall J T (1988) Variable organization in cortical maps of the skin as an indication of the lifelong adaptive capacities of circuits in the mammalian brain. Trends in Neuroscience 12, 549–557

Wall P D (1995) Do nerve impulses penetrate terminal arborizations? A pre- presynaptic control mechanism. Trends in Neuroscience 18, 99–103

Wall P D and Bennett D L (1994) Postsynaptic effects of long-range afferents in distant segments caudal to their entry point in rat spinal cord under the influence of picrotoxin or strychnine. Journal of Neurophysiology 72, 2703–2713

Wall P D and Egger M D (1971) Formation of new connections in adult rat brains after partial deafferentation. Nature 232, 542–545

Wang X Merzenich M M Sameshima K and Jenkins W M (1995) Remodelling of hand representation in adult cortex determined by timing of tactile stimulation. Nature 378, 71–75

Welker E Rao S.B Dörfl J Melzer P and van der Loos H (1992) Plasticity in the barrel cortex of the adult mouse: effects of chronic stimulation upon deoxyglucose uptake in the behaving animal. Journal of Neuroscience 12, 153–170

Wells J and Tripp L N (1987) Time course of reactive synaptogenesis in the subcortical somatosensory system. Journal of Comparative Neurology 255, 466–475

Wilson P (1987) Absence of mediolateral reorganisation of dorsal horn somatotopy after peripheral deafferentation in the cat. Experimental Neurology 95, 432–447

Wilson P and Snow P J (1987) Reorganization of the receptive fields of spinocervical tract neurons following denervation of a single digit in the cat. Journal of Neurophysiology 57, 803–818

Wilson P Kitchener P D and Snow P J (1996) Intraaxonal injection of neurobiotin reveals the long-ranging projections of A beta-hair follicle afferent fibers to the cat dorsal horn. Journal of Neurophysiology 76, 242–254

Wilson P and Kitchener P D (1996) Plasticity of cutaneous primary afferent projections to the spinal dorsal horn. Progress in Neurobiology 48, 105–129

Woolf C J and Doubell T P (1994) The pathophysiology of chronic pain – increased sensitivity to low threshold Ab-fibre inputs. Current Opinion in Neurobiology 4, 525–534

Woolf C J and Fitzgerald M (1986) Somatotopic organization of cutaneous afferent terminals and dorsal horn neuronal receptive fields in the superficial and deep laminae of the rat lumbar spinal cord. Journal of Comparative Neurology 251, 517–531

Xerri C Stern J M and Merzenich M M (1994) Alterations of the cortical representation of the rat ventrum induced by nursing behavior. Journal of Neuroscience 14, 1710–1721

Xu J and Wall J T (1997) Rapid changes in brainstem maps of adult primates after peripheral injury. Brain Research 774, 211–215

Yamamoto T Iwasaki Y and Konno H (1983) Retrograde axoplasmic transport of toxic lectins is useful for transganglionic tracings of the peripheral nerve. Brain Research 274, 325–328

Zarzecki P Witte S Smits E Gordon D C Kirchberger P and Rasmusson D D (1993) Synaptic mechanisms of cortical representational plasticity: somatosensory and corticocortical EPSPs in reorganized raccoon SI cortex. Journal of Neurophysiology 69, 1422–1432

Zhang S P and Rowe M J (1997) Quantitative analysis of cuneate neurone responsiveness in the cat in association with reversible, partial deafferentation. Journal of Physiology 505, 769–783

Index

K

L